Tyres, Suspension and Handling

John C. Dixon

Senior Lecturer in Engineering Mechanics
The Open University

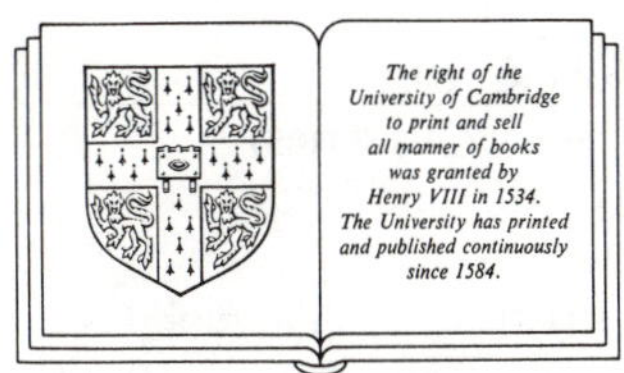

CAMBRIDGE UNIVERSITY PRESS

Cambridge
New York Port Chester
Melbourne Sydney

Disclaimer: This book is not intended as a guide for vehicle modification, and anyone who uses it as such does so entirely at their own risk. Testing vehicle performance may be dangerous. The author and publisher are not liable for consequential damage arising from application of any information in this book.

Published by the Press Syndicate of the University of Cambridge
The Pitt Building, Trumpington Street, Cambridge CB2 1RP
40 West 20th Street, New York, NY 10011, USA
10 Stamford Road, Oakleigh, Melbourne 3166, Australia

First published 1991

Printed in Great Britain at the University Press, Cambridge

British Library cataloguing in publication data
Dixon, John C. (John Charles) *1948 –*
Tyres, suspension and handling.
1. Motor vehicles. Design
I. Title
629.283

Library of Congress cataloguing in publication data
Dixon, John C., Ph. D.
Tyres, suspension and handling / John C. Dixon.
p. 432 cm. - - (Cambridge engineering series)
Includes bibliographic references (p. 412) and index
ISBN 0-521-40194-1 (hardcover)
1. Automobiles--Handling characteristics. 2. Automobiles--Springs and suspension. 3. Automobiles--Tires. I. Title. II. Title: Tires, suspension, and handling. III. Series.
TL 154.D55 1991 91-8861
623.23'1--dc20 CIP

ISBN 0 521 40194 1 hardback

GH

To Anne

Contents

Preface

Some years ago I wanted to read a book explaining the principles of vehicle handling, with supporting information on tyres and suspensions. Although there were two books available, they did not entirely meet my needs. This is my attempt to fill the gap.

I have emphasised physical understanding rather than mathematics, although I have been surprised by the number of equations that have been required in the chapters on handling.

I have included a fairly large number of questions, with answers for most of the quantitative ones. These questions should help self-organised study or act as stimulants for those using the book as a teaching aid.

I have been fortunate to have received constructive criticism of early drafts from a number of friends and colleagues. Thank you all. Especially I must mention John Dominy, Keith Martin, Rod Mansfield and John Whitehead. I have not always taken their advice. Of course, in a work of this kind one owes a great deal to the enormous number of authors of the vast research literature. No doubt there are still some technical faults in the material, for which I must remain responsible, and I would be delighted to be advised of corrections or to receive suggestions for possible improvements.

Certainly much more remains to be said on the subject area of this book, and I hope only that it will be seen as a reasonably thorough introduction.

Finally, thanks to Carla Walton, Mavis Beard and Rebecca McCormack for help with word processing, to David Greenway for the diagrams, and to Garry Hammond who performed the final text preparation and editing on behalf of Cambridge University Press.

John Dixon, Faculty of Technology, The Open University,
Milton Keynes, Buckinghamshire, England, UK

Acknowledgements

Figures 1.1.1, 1.1.2, and 2.4.1 are reproduced by permission of the Society of Automotive Engineers, Inc. Figure 1.1.1 is from Olley, M. (1934) 'Independent wheel suspensions', *Trans. S.A.E.*, Vol. 34, No. 2. Figure 1.2.2 is from Evans, R.D. (1935) 'Properties of tyres affecting riding, steering and handling', *Trans. S.A.E.*, Vol. 36, No. 2. Figure 2.4.1 is from *S.A.E. Recommended Practice J670e – Vehicle Dynamics Terminology*.

Figures 1.12.1, 3.7.1, 3.7.2, 6.22.1, 7.12.1, 7.12.2 and 7.12.3 are reproduced by permission of the Council of the Institution of Mechanical Engineers. Figure 1.12.1 is from Smith, J.G. and Smith, J.E. (1967) 'Lateral forces on vehicles during driving', *Automobile Engineer*, December. Figure 3.7.1 is from *Automobile Engineer*, letter from R. Prevost, September 1928. Figure 3.7.2 is from Irving, J.S. (1930) 'The Golden Arrow and the world's land speed record', *Automobile Engineer*, May. Figure 6.22.1 is from Milliken, W.F. and Rice, R.S. (1983) 'Moment method', *I.Mech.E. Conference C113/83 'Road Vehicle Handling'*. Figures 7.12.1, 7.12.2 and 7.12.3 are from Segel, L. (1957) 'Theoretical prediction and experimental substantiation of the response of the automobile to steering control', *Proc. I.Mech.E. (A.D.)*.

The computer graphic on page 388 appears by courtesy of The Peugeot Talbot Motor Company Ltd.

1

Introduction

1.1 Introduction and history

The automobile evolved from the horse-drawn carriage towards the end of the 19th century. At that time there was already a body of knowledge regarding ride quality and suspension systems. There was, however, virtually no published material concerning investigation of handling qualities.

It is interesting to speculate upon what sort of concept of handling the early chariot engineers may have had. Perhaps the most obvious requirement of a high-performance chariot is a small mass. Certainly, the importance of this was appreciated at least 2000 years ago. The wheels were so light that they were removed if not in use (e.g. overnight) to obviate eccentricity and lack of roundness due to creep.

Early carts had steering in which the entire front axle pivoted about a vertical central pivot. Lightweight chariots were two-wheeled, so yawing could be achieved by relative rotation of the wheels, and the need for steering as such did not arise. The traditional horse-drawn carriage of the 18th and 19th centuries had rigid axles front and rear; steering of the complete front axle about its central pivot naturally left the wheels perpendicular to radii from a notional centre of the path arc. However, steering by movement of the complete axle was inconvenient, requiring large clearance around the wheels, so steering by pivoting the wheels separately on stub axles was introduced. In 1816 Georges Langensperger laid down the geometric condition to maintain the wheels perpendicular to their arc of motion for stub-axle steering. Such an arrangement was highly desirable to minimise friction at the wheels during low-speed manoeuvring on small radii. This is still the case, although conditions are different at higher speeds. Ackermann recognised the importance of this invention; by agreement with Langensperger, acting as his agent in London, he took out British patents, and hence this arrangement is widely known as Ackermann steering. Some sixty years later, Amadée Bollée, in designs of 1873 and 1878, achieved a similar effect, possibly independently. In 1881 Jeantaud arrived at the same arrangement, and performed some more scientific assessments. As a result, Langensperger's principle is known in France as the Jeantaud diagram. In 1893 Benz was granted a German patent for another arrangement, similar to Bollée's later type.

The above steering designs were all kinematic in concept, in that there was no recognition of a need for sideforces or of the means by which such forces might arise. The dynamic concept of cornering only began at the start of the 20th century. In 1907 Lanchester used the term 'oversteer' in a paper to the Institution of Automobile Engineers (I.A.E.):

> "In practice even the inertia of the hand and arm of the driver of a car with bath-chair steering tells its tale in the slightly zigzag course to which such cars are liable, each small steering effort becomes overdone, the car 'oversteers', and its track is reminiscent of the motion of a water-fly."

The first significant steps towards a modern dynamic, rather than kinematic, concept of cornering had to be the recognition of need for lateral forces and the presence of the slip angle that produces them. The term 'slip' was used in an I.A.E. paper by Kersey (1921), in relation to toe-in and camber, with emphasis on the effective lateral velocity component. (References are generally given at the end of the book; early historical references are not all given here, but may be found in Milliken & Whitcomb (1956).) Healey (1924) included several pages on handling effects, but no mention of the terms understeer or oversteer. Credit for recognition of the concept and significance of slip angle is usually given to Georges Broulhiet, who called it *envirage* in his 1925 paper to the French Institution of Civil Engineers.

Bradley and Allen (I.A.E. 1930), investigated the friction properties of road surfaces, and published what is probably the first graph of tyre side force against angle, although the angle shown is not precisely that of the wheel relative to its ground motion, but the rigging angle of a motorcycle sidecar wheel. In fact, they were not interested in cornering *per se*, but in braking, and simply considered a rotating wheel at a large slip angle to be a good way to investigate the effects of road surface properties on maximum grip, the rotation preventing local overheating and excessive wear. This paper was conspicuously more scientific than

This book is a detailed introduction to control, stability, handling and cornering behaviour of four-wheeled vehicles. Control is the vehicle response to driver steering inputs. Stability is ability to maintain course despite external influences. Cornering and handling are the ability of a vehicle to achieve high lateral accelerations, and the ease with which a driver can attain the potential performance. The book begins with a historical introduction and overview of the subject covering the basic details of the vehicle body, differentials, wheels, roads and drivers. Following chapters discuss the tyre cornering characteristics, aerodynamics and atmospheric influences, suspension (including a full discussion of roll centres), steady-state handling (with an extensive discussion of oversteer and understeer) and unsteady-state (transient) handling. There is an extensive bibliography and over 400 problems are included – many with solutions.

Tyres, Suspension and Handling will be of value to all those in industry or universities with an interest in vehicle cornering behaviour.

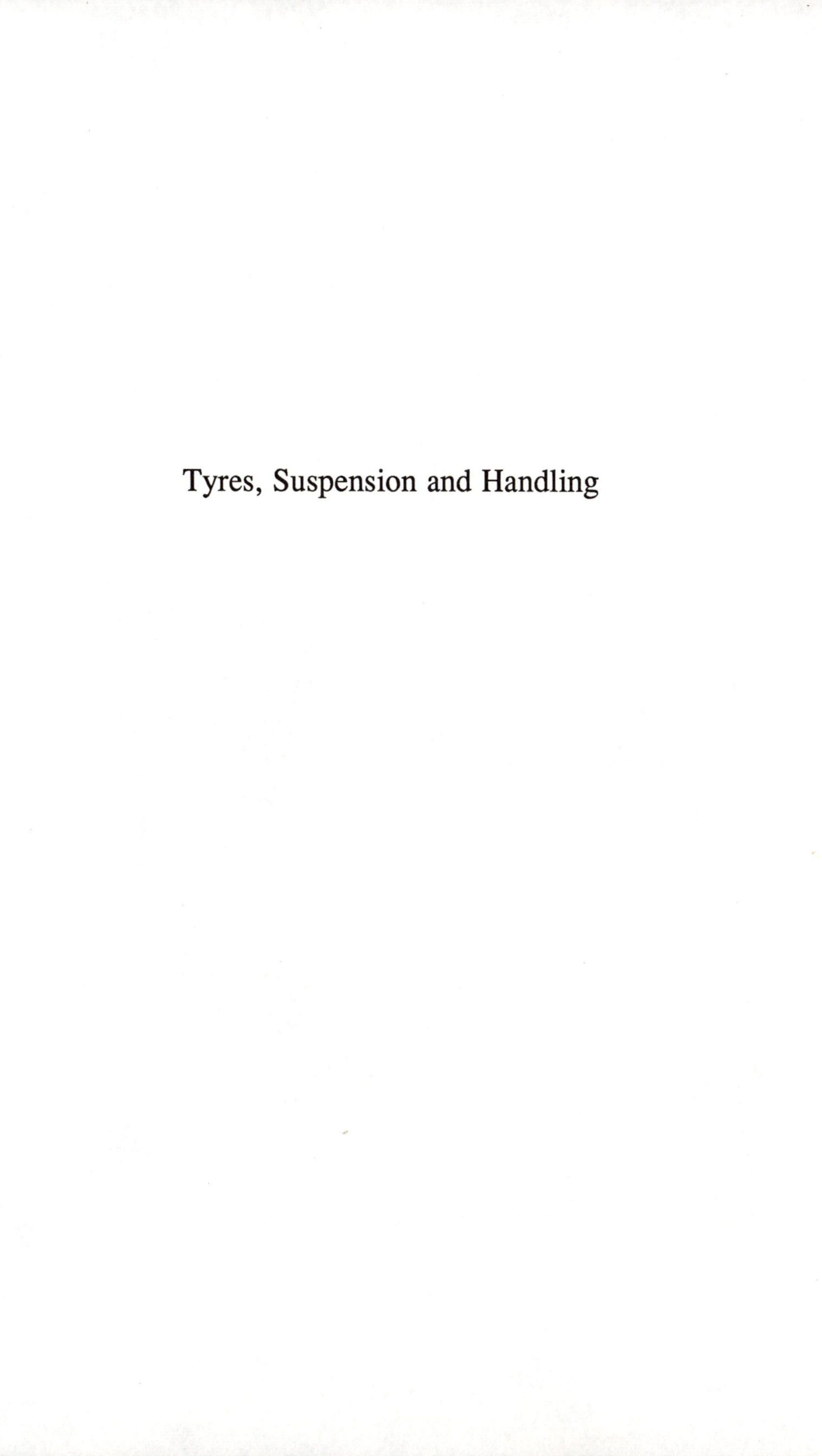

Tyres, Suspension and Handling

much that went before regarding tyre forces, and includes investigation of the effect of tread form and speed on tyre friction with various surfaces.

In 1931, a study of steering vibrations by Becker, Fromm and Maruhn drew further attention to the mechanical properties of the tyre. This included results of tests of tyres on a rotating steel drum, and was the first paper to record some of the characteristics of tyre cornering stiffness. Broulhiet presented a paper to the Society of Automotive Engineers (S.A.E.) in 1933 on European ideas about independent suspension, including description of his *envirage*. Olley (1934), describing work at Cadillac, included a diagram, Figure 1.1.1, and some discussion that indicates that the significance of the relationship between front and rear slip angles was understood, at least in principle.

Figure 1.1.1. Early handling analysis (Olley, 1934).

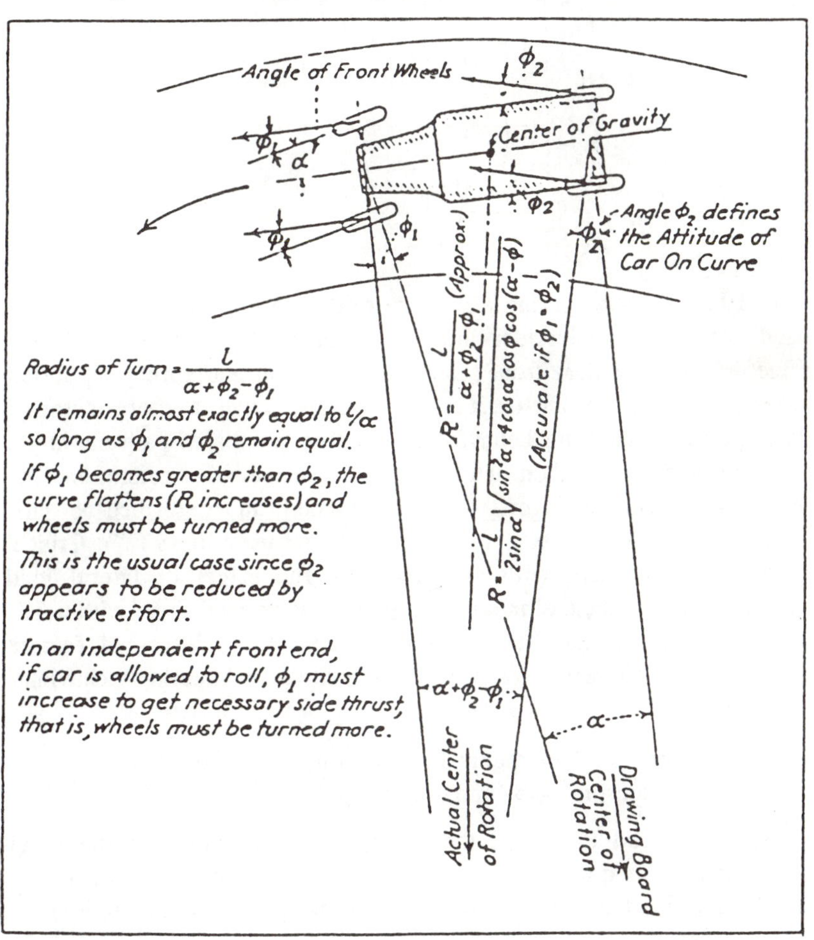

It is interesting that the terms understeer and oversteer were not used, although they were used in an unpublished General Motors report by Olley in 1937, which indicated that in 1931 roll-steer and tyre pressures were known to be very important in controlling stability. Olley's 1934 paper shows that the basic relationships governing tyre forces and slip angles were known, but it remained for Evans of Goodyear to make comprehensive investigations, which were published in 1935; Figure 1.1.2 shows his principal result.

Figure 1.1.2. First tyre cornering–force curve (Evans, 1935).

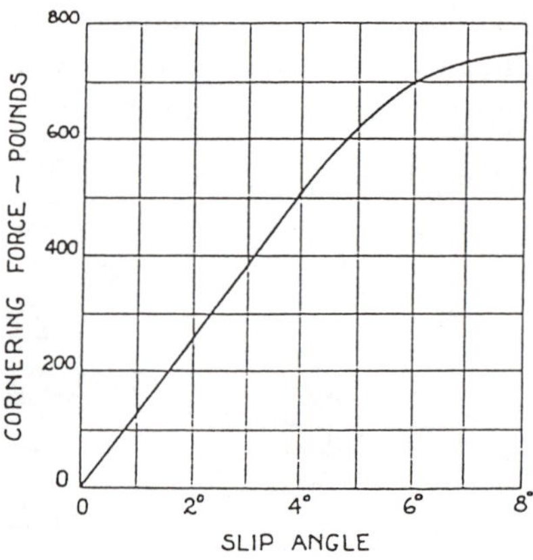

In 1937, Bastow, giving a forward reference to Olley's 1938 paper, used the terms understeer and oversteer, and discussed roll-steer effects on independent front suspensions, and on leaf-spring supported axles. In 1938, in an I.A.E. paper, Olley gave a more detailed description of handling behaviour, including graphs of path curvature against speed, and of front wheel angle against lateral acceleration. The terms understeer and oversteer were included, and described as "just coming into general use" ("oversteer was first used in 1932"). Critical speed was included, with comments on the effect of lateral load transfer. Mathematical simulation results were claimed to be within 5%, presumably for steady state. Later in 1938, Law described factors affecting tyre cornering stiffness, commented on handling behaviour, and stated:

> "Automobile engineers have become familiar with the terms 'over-steering' and 'under-steering'..."

Thus, by the mid-1930s, all the important ingredients of a theory of handling were available, and by the late 1930s reasonable mathematical steady-state models were in use, although not published in detail (Fox,

1937, and Olley, 1938). Attention soon turned to transient behaviour (Stonex, 1941), although it was to be three decades before a comprehensive transient-response analysis was achieved, even for low lateral accelerations.

German activity of the 1930s had a strong aerodynamic element. In 1938 Forster described a truck-towed tyre tester and gave results, and in 1940 Rieckert & Schunk, Dietz & Harling, and Huber & Sawatzki all published details, mostly working under the direction of Kamm who reviewed this work in 1953. In 1941, von Schlippe published his model of the tyre based on a stretched string analogy, successfully explaining the relaxation distance, relevant to the handling dynamics of the tyres.

There was also activity in France during this period, with papers emerging from deSeze in 1937 and Gratzmüller in 1942, and then with substantial progress after the war, from Rocard in 1946, Julien in 1947 and 1948, and Lozano in 1951.

International political difficulties and language barriers obviously created problems, and the significance of the tyre in vehicle behaviour was not appreciated in Russia for a decade. By 1946, however, the European and American work was known to them, and, as examples, Pevsner in 1946 described testing techniques and Chudakov in 1947 included tyre properties in his theoretical analysis. In the U.K., a number of papers appeared after the war (Olley 1947, Hardman 1949, Gough 1949). Progress became more rapid, not least in the area of tyres (for example Joy *et al.*, 1956). The classic set of papers by Milliken, Segel and Whitcomb (I.Mech.E) in 1956 and 1957, thirty years after Broulhiet's introduction of *envirage*, applied linear control theory to give a frequency response analysis. This gave a description of the linear response regime corresponding to normal driving, with lateral accelerations of up to 3 m/s^2 (0.3g), complete apart from aerodynamics.

The following decades saw attention turning to the high lateral acceleration regime, requiring non-linear analysis, and increasingly of late to the role of the driver, with closed-loop control analysis including the driver supplementing the driverless open-loop analysis. However, despite all this effort, there are still many interesting and unsolved problems in vehicle handling.

In this book we shall be concerned principally with the four-wheeled vehicle – the car or light truck – operating on metalled roads. Vehicles with other than four wheels, and road surfaces of a loose or viscous kind such as gravel, mud or snow, will be mentioned only briefly. In this first chapter we shall gather together some of the ideas necessary for use in the detailed handling analyses to follow. The vehicle body will be examined for both structural stiffness and its contribution to inertia. The engine and brakes will be considered because acceleration and deceleration forces must be produced by the tyre, and these affect its ability to produce lateral forces for cornering. The various types of differential are considered, and road characteristics will be studied because this is one half of the tyre–road interaction, and on a large scale it is the principal cause of manoeuvring requirements. The atmosphere will be considered because this affects aerodynamics, through density variations and wind and wind gusts, and affects the tyre–road contact patch through the presence of water, ice and snow. This chapter then considers the driver, and finally introduces testing methods.

1.2 Control, stability and handling

Control is action by the driver intended to influence the car's speed or path. Stability refers to the unwillingness of a car to be deflected from its existing path – usually a desirable trait, in moderation. Handling is the ability of a car to round corners successfully, the study of how this occurs, and the study of the driver's perception of the vehicle's cornering behaviour. The vehicle as a whole can only be influenced by forces exerted by the road, by the atmosphere and by gravity, and as a consequence we shall be interested in the aerodynamic properties of the body and the cornering force characteristics of the tyres. For most practical cases the tyre forces are the dominant ones, with aerodynamics playing a secondary role in handling behaviour.

The road vehicle, contrasted with the air-borne or water-borne vehicle, is characterised by the all-important presence of the tyre; it is essentially for this reason that road vehicle dynamics can be regarded as a separate subject in its own right. The forces exerted on the tyre by the road depend upon many factors including the dimensions, structure and materials of the tyre, and the angle of presentation of the tyre to the road. Hence it is necessary to consider in detail the tyre and the suspension system characteristics. Vehicle dynamics involves the study of both lateral and longitudinal motions. It is the former which pose most of the more interesting problems, so most of the standard results of longitudinal motion will be taken for granted here. 'Cornering', then, is the main subject. The term 'roadholding' is used to mean the ultimate ability of a car to execute a given manoeuvre given a perfect driver. 'Handling' is used in a narrow sense to mean the ease with which a real driver can realise desired manoeuvres, including straight running when subject to disturbances, and is also used in a wider sense to mean the same as cornering. Handling, then, is a qualitative assessment of the 'manners' of a car, and will depend on both the driver and the application. We shall deal here with roadholding and handling rather than with ride, which is the ability of a car to provide comfort despite road roughness.

Figure 1.2.1. Driver–vehicle system block diagram.

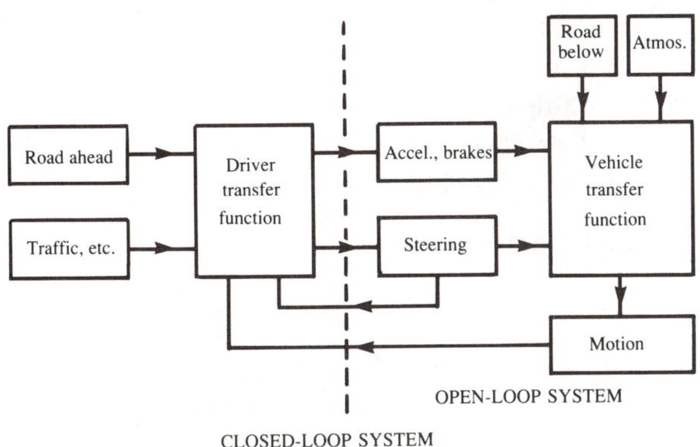

Figure 1.2.1 shows the driver–vehicle system as a block diagram. The section to the right of the dotted line is the open-loop system; this represents the vehicle and how it responds to control inputs or disturbances from the road or atmosphere. The complete system includes the driver. Because the driver takes feedback information from the vehicle motion and the steering, this is called the closed-loop system. Handling analysis may be based on the open loop alone, where tests are intended to reveal the vehicle characteristics, or on the closed loop when the performance of the driver is also involved.

1.3 Axis systems and notation

To study the response of a vehicle to control inputs or to disturbances, it is necessary to specify one or more coordinate systems to measure the position of the vehicle. The method recommended by the S.A.E. will be described here (see S.A.E. J670e). It was originally taken from aeronautical practice. Broadly similar methods are used throughout dynamics.

First there is an Earth-fixed axis system XYZ, Figure 1.3.1. (Note the use of upper-case letters to denote Earth-fixed coordinates.). For all ground vehicles the Earth may be considered to be stationary. This is not true in an absolute sense – the Earth rotates, and moves around the Sun – but for our purposes the *XYZ* system is an inertial coordinate system, i.e. it has negligible acceleration. The *X*-axis is chosen longitudinally forward in the horizontal plane; *Y* is then 90° clockwise from *X* as viewed from above, and also in the horizontal plane. To form a right-hand coordinate system, the *Z*-axis is vertically downwards, because this is the direction of motion of a right-hand screw turning *X* to *Y*. The origin of the *XYZ* system may be at any convenient point, typically in the ground plane. This is the S.A.E. system. In the International Standards Organisation system, the *Y*-axis is to the left and the *Z*-axis is upwards. There is also an axis system *xyz* (lower-case letters) fixed to the vehicle, Figure 1.3.2.

Figure 1.3.1. Axis systems.

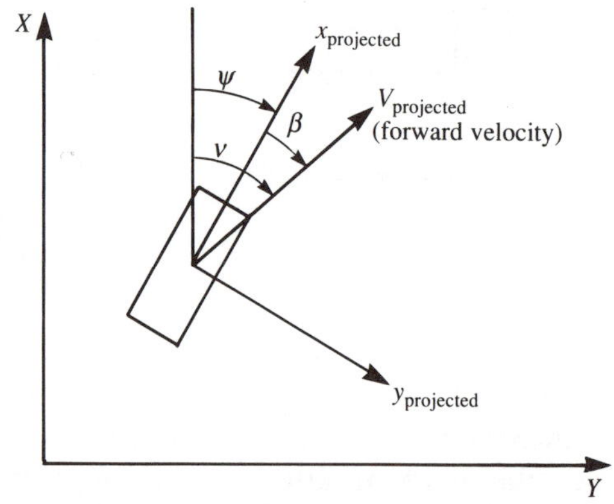

Figure 1.3.2. Vehicle-fixed axes.

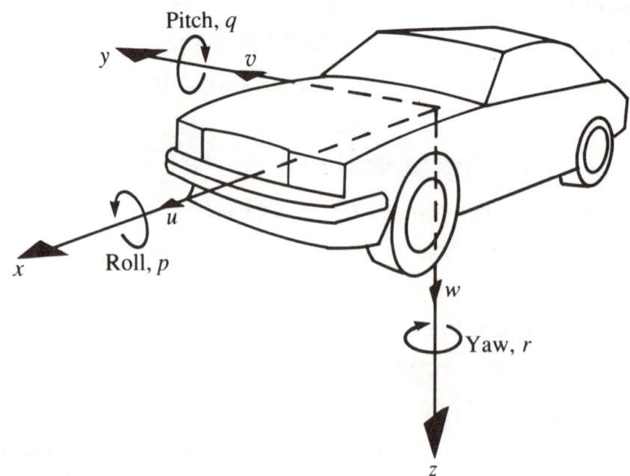

The use of upper-case XYZ for the Earth-fixed inertial system and lower-case xyz for the vehicle-fixed system follows widely accepted conventions. The origin of the xyz system is usually placed at the vehicle centre of mass. The x-axis is approximately in the central plane, pointing forward, and is horizontal when the vehicle is in its usual pitch attitude; thus the pitch angle is the angle of x to the horizontal plane. The y-axis points to the driver's right, and is horizontal when the vehicle has zero roll angle; thus the roll angle is the angle between the y-axis and the horizontal plane. The z-axis is downwards, again to give a right-hand system of mutually perpendicular axes. In general the vehicle has some acceleration, so the xyz system is a non-inertial system.

The position of the vehicle, i.e. of the moving xyz axes, relative to the Earth-fixed XYZ axes, is measured by three coordinates giving the position of the origin of xyz in XYZ, and three rotations of xyz in XYZ. Angular rotations are taken as right-handed by convention, so a positive rotation corresponds to the rotation of a right-hand screw when advancing in the positive direction of the corresponding axis. The three rotations defined as standard by the S.A.E., starting from a position with xyz aligned with XYZ, are, in sequence:

(1) A yaw rotation ψ about the z-axis,

(2) A pitch rotation θ about the y-axis,

(3) A roll rotation ϕ about the x-axis.

Note that the rotations are taken about the vehicle-fixed axes. (A nomenclature list for each chapter appears near the end of the book.)

The translational velocity of the vehicle is taken as the velocity of its centre of mass G, at the origin of xyz, measured in the system XYZ. For convenience the velocity is resolved into components along each of the xyz axes, Figure 1.3.2. There is:

(1) A longitudinal velocity u along the x-axis,

(2) A side velocity v along the y-axis,

(3) A normal velocity w along the z-axis.

Because the x and y axes are not generally exactly parallel to the ground plane, other terms are also defined as follows. Forward velocity is the velocity component in the ground plane perpendicular to y: essentially this is the longitudinal velocity resolved into the ground plane. Lateral velocity is the velocity component in the ground plane perpendicular to the x-axis: essentially this is the side velocity resolved into the ground plane. The total velocity in the horizontal plane is the velocity tangential to the path of the vehicle centre of mass. It is also convenient to define the heave velocity as the velocity component perpendicular to the ground plane, positive away from the ground ($-V_Z$). Special care is required with definition of terms if the ground plane is not horizontal.

Because the *xyz* axes are attached to the vehicle, the position of G is constant and it has zero velocity and acceleration in these axes. The vehicle's actual acceleration in the *XYZ* system is again resolved into components parallel to the *xyz* axes, there being a longitudinal acceleration $\dot{u}$ along x, a side acceleration $\dot{v}$ along y, and a normal acceleration $\dot{w}$ along z. There are also lateral, forward and heave linear accelerations. Figure 1.3.3, with exaggerated attitude angle, shows how the total horizontal acceleration may be resolved into forward and lateral components, or into tangential and centripetal components. The centripetal acceleration is the component parallel to the road plane, and perpendicular to the vehicle path, i.e. directed towards the path centre of curvature.

Figure 1.3.3. Acceleration components.

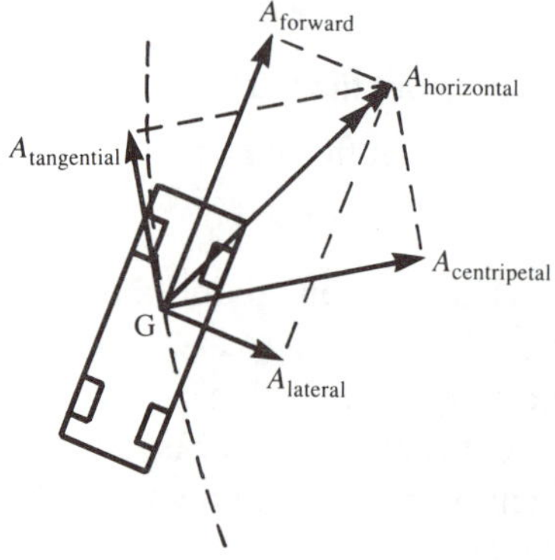

The vehicle's angular velocities are naturally measured relative to the inertial *XYZ* system, but are resolved about the *xyz* axes for convenience, to give the roll angular speed p, the pitch angular speed q, and the yaw angular speed r, expressed as rad/s or deg/s. In rotation there are roll ($\dot{p}$), pitch ($\dot{q}$) and yaw ($\dot{r}$) angular accelerations.

Referring again to Figure 1.3.1, looking down on the horizontal *XY* plane, the angular position ψ (psi) of the projected x-axis relative to the X-axis is called the heading angle, positive clockwise. The angular position β (beta) of the forward velocity, relative to the projected x-axis, is called the attitude angle (sometimes sideslip angle), again positive clockwise. The angular position of the velocity vector, which is tangential to the path and denoted by ν (nu), is called the course angle.

Consequently

$$v = \psi + \beta$$

The terminology introduced here only becomes familiar with regular use. Its complexity reflects the fact that the vehicle motion is complex. The complete system of notation is only called for in the general approach to vehicle dynamics. Happily, many problems can be treated without the full generality, and therefore in a more easily understood form; this latter approach is the one that will be adopted throughout most of this book.

1.4 Vehicle forces and notation

The motion of the vehicle is governed by the total resultant force acting on the vehicle and by the inertial properties of the vehicle. Expressed in its most general form we have two vector equations that can be written for the vehicle motion; force equals rate of change of linear momentum, and moment about the centre of mass equals rate of change of angular momentum:

$$\mathbf{F} = \dot{\mathbf{G}}$$

$$\mathbf{M} = \dot{\mathbf{H}}$$

These are the basic laws of motion as expressed by Euler. Newton's three laws of motion can all be deduced from the first of these. In practice in three-dimensional mechanics the first of Euler's two equations is much easier to apply than the second. Considering $\mathbf{F} = \dot{\mathbf{G}}$ first, by approximating the vehicle as having constant mass, this can be expressed as $\mathbf{F} = m\mathbf{A}$, or as applied in practice in a coordinate system, in three equations of the form $F_x = mA_x$ where F_x is the sum of all force components in the x-coordinate direction. Thus if all the forces on the vehicle are known with reasonable accuracy, the vehicle acceleration can be determined. In practice, in most cases the centre of mass can be taken as having a fixed location relative to the vehicle body. However, this is not exact, as discussed in Section 1.6 on body inertia. The total force acting is a combination of the forces exerted by the ground on the tyres, detailed in Chapter 2, of the forces exerted by the air on the body, Chapter 3, and of the weight force exerted by gravity.

The response of the vehicle to the total moment acting about the centre of mass is much more difficult to determine. This is because of several factors. By neglecting the fact that there are rotating wheels, engine and drive train, i.e. neglecting associated gyroscopic effects (which is not necessarily a good approximation), and by using the same approximations as for the translational analysis, the vehicle can be treated as a rigid body. However, the detailed equations of motion are still complex, not least because the principal axes of inertia of the vehicle do not in general coincide with the body-fixed axes. Fortunately, most practical handling problems are adequately analysed by dealing with simplified cases, for example by treating the vehicle as being in approximately plane motion parallel to the ground plane, so that the full generality of the equations is not required. For those readers interested in the general approach, the bibliography at the end of this chapter gives references.

Notation for forces on the vehicle follows a similar pattern to kinematic notation, including the use of the various subscripts for axis directions, and terms such as longitudinal force and sideforce. Notation for forces in the ground plane follows the acceleration notation of Figure 1.3.3. The centripetal force F_C gives the centripetal acceleration that causes path curvature. The tangential force F_T controls the acceleration along the path.

Figure 1.4.1. Plan-view free-body diagram: (a) in *XYZ*, (b) in *xyz*.

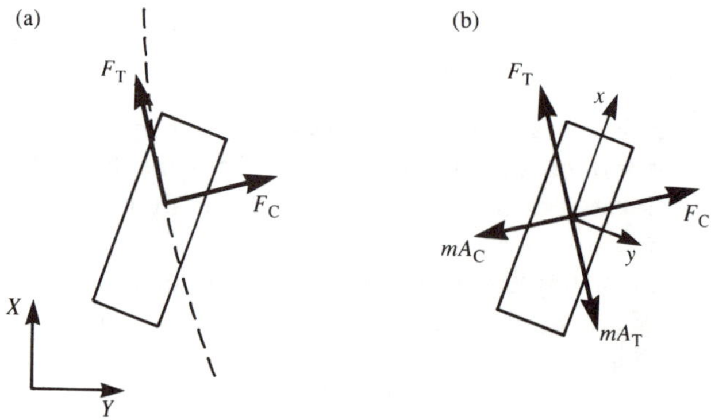

Figure 1.4.1(a) shows the free-body diagram of the vehicle in the ground plane, viewed in the *XYZ* inertial coordinate axes. The free-body diagram shows the chosen free body with the relevant forces that act upon it. As a result of the net forces in Figure 1.4.1(a) the vehicle experiences accelerations A_T and A_C according to $\mathbf{F} = m\mathbf{A}$ in the inertial *XYZ* system. For example, the equation of motion perpendicular to the path is

$$F_C = mA_C$$

If we now view the vehicle in a non-inertial coordinate system, having an acceleration relative to *XYZ*, the measured vehicle acceleration in this system will be different. Thus the acceleration calculated from $\mathbf{F} = m\mathbf{A}$ will be wrong – Newton's second law fails in an accelerating coordinate system. This difficulty may be overcome by 'adjusting' the free-body diagram, by adding compensation forces, or fictitious forces or d'Alembert forces as they are sometimes known, to bring the value of $\mathbf{F}$ into agreement with the measured $m\mathbf{A}$ in the accelerating coordinate system. The value of the compensation force needed equals the mass of the body times the acceleration of the coordinate system seen in non-accelerating coordinate axes. The compensation force must be added to the free-body diagram acting in the opposite direction to the true acceleration of the free body.

This method is shown applied in Figure 1.4.1(b), for the special case of the body-fixed axes. The true centripetal acceleration requires that we add the force mA_C opposing the true acceleration. Similarly mA_T is added. Because xyz is a body-fixed system, the vehicle has no acceleration in this system. From this free-body diagram the equation of motion perpendicular to the path is

$$F_C - mA_C = 0$$

because there is no acceleration. Comparing this with the equation from the XYZ system free body, we see it to be correct: the acceleration in inertial axes is $A_C = F_C/m$.

The force mA_C, balancing the centripetal acceleration, is called the centrifugal force because it acts outwards from the centre of curvature. This centrifugal force is a compensation force, and appears only in order to make Newton's second law valid in the accelerated reference frame. Unfortunately, this special nature of compensation forces is often overlooked, and they are believed to act even in inertial coordinates such as XYZ, which leads to great confusion.

Figure 1.4.2. Rear-view free-body diagram: (a) in XYZ, (b) in xyz.

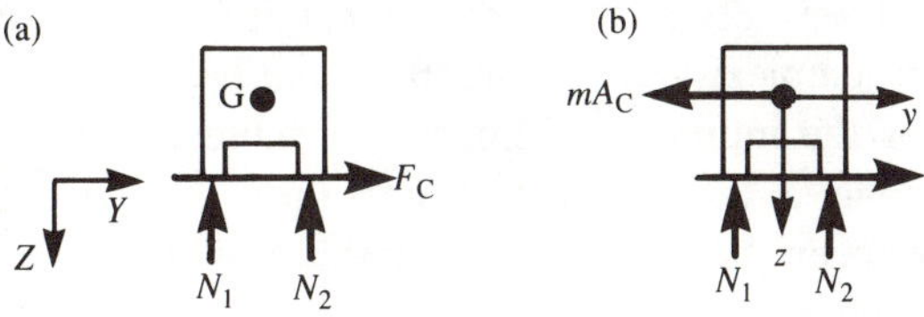

Figure 1.4.2(a) shows the rear elevation of the same vehicle in the inertial axes XYZ, simplified to one axle. Here F_C gives the real centripetal acceleration A_C. Also there is lateral load transfer: N_1 exceeds N_2 to balance the moment of F_C about G because there is no

angular acceleration in roll in the steady state. Figure 1.4.2(b) shows the corresponding free-body diagram viewed in the accelerating body-fixed axes, and therefore including the compensation centrifugal force mA_C. By definition, in this axis system the vehicle has no acceleration, i.e. it is in equilibrium.

Both inertial and body-fixed accelerating coordinate systems will be used here, according to which is most convenient at the time. In accordance with the usual conventions, the inertial coordinate system *XYZ* is in use unless otherwise stated.

1.5 Body stiffness

From a handling perspective, a high stiffness of the body or chassis is nearly always preferred, so that handling control may be realised by the design of the suspension and tyres. The body is loaded by the suspension at the transverse planes of the axles, so it is the effective stiffness between these planes that is significant. The principal loads are the vertical ones, applied primarily through the springs, dampers and suspension links, and essentially equal to the vertical tyre forces. If the sums of the vertical forces on the two vehicle diagonals are not equal then the body is in torsion. For cornering, the suspension is generally designed to create such an effect, for example by anti-roll bars, so that the tyre normal forces influence the lateral forces in a desirable way. Because of this, the body torsional stiffness can influence handling behaviour, in particular if the stiffness is inadequate, by preventing the suspension acting in the desired manner.

The torsional stiffness of the body or chassis is measured by applying appropriate forces and torques in the vertical transverse planes of the two suspensions. The most satisfactory way to do this is perhaps to support each axle wheel pair on a transverse beam that can be inclined laterally, so that the torsion-causing forces are fed into the body in a realistic way, and to measure the body distortion with dial gauges.

The torsional stiffness of the body or chassis is not easily calculated because of the complexity of the structure, and because of the difficulty of allowing for the influence of imperfect connection between panels, which may have a large effect. Since the 1950s most passenger vehicles have been based on an integral monocoque stressed-skin structure, as opposed to the previous method of having a chassis with a separate and relatively unstressed body, as is still used on trucks. Because of the greater enclosed cross-sectional area of a body, compared with that of a chassis, torsional stiffness is much greater, e.g. 10 kN m/deg, and modern car bodies can generally be treated as torsionally rigid for handling analysis. For example, a vehicle of 12 kN weight in the most extreme possible steady-state case would have vertical forces of 6 kN at each end of one diagonal, giving a torque on the body of about 4 kN m, but the angular deflection would be only about 0.4°. In normal use it is even less. Although the stiffness is measured between the suspension planes, the bodywork beyond the suspensions can also contribute significantly to the stiffness.

Where a separate chassis is used, as on most trucks, the torsional compliance is substantial, and certainly enough to influence handling. Many trucks and tractor units tend to instability because of the stiff rear springs that are required, and because there is insufficient torsional stiffness in many current designs to allow this to be compensated effectively by front anti-roll bars. The typical truck chassis construction uses a ladder-like assembly – a pair of side rails with connecting cross members. The rails are typically steel or aluminium channels with the flanges pointing inwards, typically 200 mm × 70 mm, 6 mm thick. Such open channels have much poorer torsional rigidity than closed tubes or boxes. Because they pass just over the axle, and hence between the wheels, they are usually rather close together, especially where double rear wheels are used, and this further limits the rigidity that is achieved.

Truck chassis side rails also bend as beams in side view; in steady state this has little effect on handling, although it can influence the steering mechanism. In plan view there may be bending or lozenging; this may influence the geometric relationship between the wheels and hence alter handling to some extent, although probably less than torsional effects.

1.6 Body inertia

The inertia properties of a vehicle are usually represented by:

(1) The mass, m.

(2) The position of the centre of mass, G.

(3) The second moment of mass about each axis.

(4) The cross products of inertia.

In practice, because of near symmetry, the cross products of inertia I_{xy} and I_{yz} are usually taken as zero; I_{xz} is also small but sometimes considered. This is equivalent to saying that the principal inertia axes are close to the xyz axes, although the longitudinal principal axis is often slightly inclined, usually down at the front.

The inertia properties depend on the loading condition, which can best be dealt with by establishing values for the unloaded vehicle, and then incorporating any particular loading condition as required. In general the vehicle is treated as a rigid body, but in some cases it is desirable to include load shifting effects, for example for large fluid loads.

The mass and centre of mass may be found theoretically if the masses and positions of the centres of mass of the component parts are known. In view of the large number of parts this may be inconvenient, even if computerised, but is the only way available at the design stage. In this method a coordinate system is defined, typically measuring x back from the front suspension plane, y laterally from the centre plane, and z vertically upwards from ground level. The total mass moment for each axis may then be calculated by summing the contributions, e.g. $\Sigma x_k m_k$ where m_k is the mass of the kth component. The centre-of-mass position is then

$$X_G = \frac{\Sigma x_k m_k}{m}$$

where m is the total mass. The second moment of mass can then be found for axes through the centre of mass. Considering now the standard body-fixed axes xyz, using the parallel axes theorem gives, for example,

$$I_{xx} = \Sigma\,(I_k + m_k x_k^2)$$

where m_k is the mass of each component and I_k is the second moment of mass of the component about an axis through its own centre of mass parallel to, in this case, the x-axis. The second moment of mass of a uniform rectangular solid of dimensions a by b by c about an axis through its centre of mass perpendicular to the side a by b is

$$I = \frac{m}{12}(a^2 + b^2)$$

The products of inertia can also be found, and then the principal axes and principal moments of inertia follow from the usual transformation equations (e.g. McLean & Nelson, 1962). This can all be rather laborious unless computerised. However, steady-state handling problems depend only on the total mass and the position of the centre of mass.

If the vehicle already exists, the mass and centre of mass are perhaps most easily found by weighing. Each wheel vertical reaction can be measured, for example the front-left reaction denoted F_{VfL}. The front axle reaction is then

$$N_f = F_{VfL} + F_{VfR}$$

Figure 1.6.1 shows the weight and axle reactions. For ground vehicles the centre of gravity (the point at which the weight force acts) and the centre of mass (the mean position of the mass elements) can be taken as coincident, although this is not always an adequate approximation, for space satellites for example. This effectively common point will be denoted G.

Figure 1.6.1. Side view static free-body diagram.

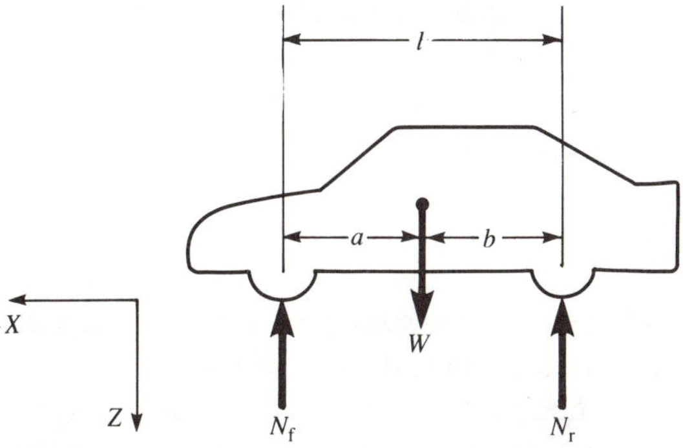

From the free body of Figure 1.6.1, which is in equilibrium, it follows that $W = N_f + N_r$. Of course, this is with zero vertical aerodynamic force. In the International System of Units (S.I.), the reaction forces should be expressed in newtons, giving a weight force in newtons. The mass is then $m = W/g$ where g, although varying slightly over the Earth at ground level, is often standardised as 9.81 m/s^2, and m is in kilograms. In the most commonly used version of the Imperial system, the forces are in pounds force (lbf, equals the weight of one pound mass in the standard gravitational field), the mass is in slugs (1 slug equals 32.2 pounds mass), and the standard gravitation is 32.2 ft/s^2. Commercial spring-balance type weighing devices may be calibrated in 'kg', intended to indicate mass directly, a liberty that is allowable because they are used in a gravitational field that is close to constant. Really they indicate forces in units of 'kilograms force' (kiloponds), the weight of one kilogram in the standard gravity, so in a scientific context the force value in newtons should be found by multiplying by 9.81 N/kg.

As a matter of interest, Figure 1.6.1 is, strictly, incomplete because it neglects the atmospheric buoyancy force, which acts vertically upwards at approximately the centre of volume. If we include this, taking it to act at approximately the centre of mass, then we see that the sum of the measured reaction forces is actually

$$N_f + N_r = W - B$$
$$W = N_f + N_r + B$$

Thus the weight is really slightly greater than $N_f + N_r$. The buoyancy force is $B = \rho Vg$ where ρ is the air density and V is the volume. Considering the enclosed air to be part of the vehicle, the total enclosed volume for a large car is about 8 m^3, giving $B \approx 100$ N. Thus the mass calculated in the ordinary way from the apparent weight $N_f + N_r$ is too small by about 10 kg. This is under 1%, although possibly more in the case of large enclosed-volume light vehicles. Normally this is neglected.

Returning to the simplified Figure 1.6.1, the longitudinal position of G follows by taking moments, for example about the front wheel contacts, neglecting tyre rolling resistance moment, the total moment being

$$\Sigma M_f = aW - lN_r = 0$$
$$a = \frac{N_r l}{W}$$

The lateral position of G follows similarly from the left and right totals of the vertical tyre forces on an end-view free body.

In the determination of the vertical forces it is not essential, although it is convenient, to have four load cells. One pair, adjustable for the track, is suitable. A single cell is possible, but less convenient. At the time of measurement, all wheels must be acting against a flat plane, i.e. the load cell must be recessed into the effective floor plane, or the suspension and tyre stiffnesses will combine with the wheel deflection to distort the results. The vehicle must be in its standard loading condition, for example dry or fuelled, with or without passengers, as appropriate.

Determination of the height of G is more difficult. One method is to hoist one end of the vehicle, for example as in Figure 1.6.2, with a force F at point P, to an angle θ. The vehicle may be rested on its rear wheels on columns mounted on ground-level load cells, as an alternative to a hoist with a load cell. Because the suspension forces change from their values in the horizontal position, the suspension should be locked in its normal standing position before hoisting, for example by pinned-through dampers or screwed rods replacing the normal dampers. Applying moment equilibrium to the free body of Figure 1.6.2 gives

$$Wc - F(c + d) = 0$$
$$Wd - N_f(c + d) = 0$$

Figure 1.6.2. Centre-of-mass height determination.

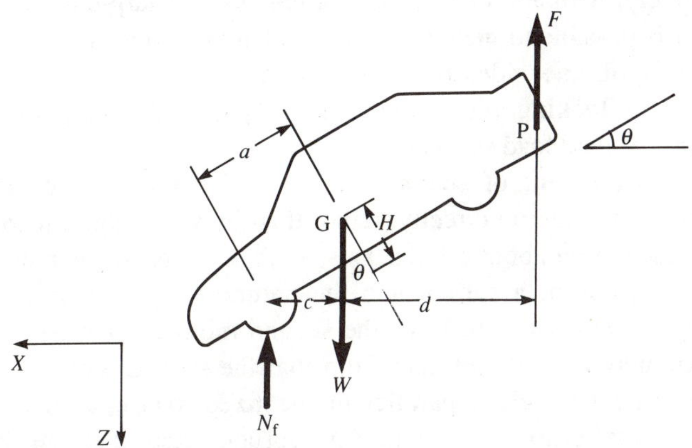

Thus dimension c, from the front wheel centre to the line of action of W, may be determined, depending on the actual parameters measured. In conjunction with the angle θ the line of W is determined, and in conjunction with the already known value of a the height of G may be found using

$$c + H\sin\theta = a\cos\theta + R_L\sin\theta$$

where R_L is the loaded wheel radius, and H is the normal height of the centre of mass above the ground plane. This gives

$$H = R_L + \frac{a\cos\theta - c}{\sin\theta}$$

so in this test the forward load transfer really indicates the height of G above the wheel radius. Hence, for some racing cars the load transfer in this test will be small, or even negative.

The accuracy of this method depends upon achieving an adequate value of θ, preferably about 30°. Great care is required to achieve reasonable accuracy in H. Because of the difficulty of doing this, particularly with large trucks, other methods are sometimes adopted. In some cases (e.g. trucks without their rear bodies) it is feasible to rest the vehicle on knife edges, at a point directly above G by an unknown amount, the dimension a already being known, at which point the vehicle hangs level. The addition of a known load at the front or rear will cause a measurable pitch position change, from which the vertical centre of mass position is easily found. Alternatively, if the knife edges are below the centre of mass, for example under the chassis rails, then in a pitched position there is a destabilising moment from W, which can be resisted by a load cell or balanced at the other end by a weight. Alternatively, with appropriate precautions, using sharp-edged solid disc wheels it is possible to incline the vehicle laterally until it is balanced on the wheels of one side. In all of these methods, caution must be exercised in locking the suspension, and in preventing or making allowance for fluid load shifting.

The measurement of second moment of mass must be done by investigating the inertia directly, rather than by weighing. The vehicle is put into oscillation about a fixed axis, with a restoring moment due to the weight force or a spring, and the natural frequency is measured. Knowing the restoring stiffness, the second moment of mass about the fixed axis may be deduced, and from that the second moment of mass about an axis through G parallel to the fixed axis can be calculated using the parallel axes theorem. Appropriate fixed axes are therefore chosen to investigate the inertia in yaw, pitch and roll oscillations.

Figure 1.6.3. Trifilar pendulum.

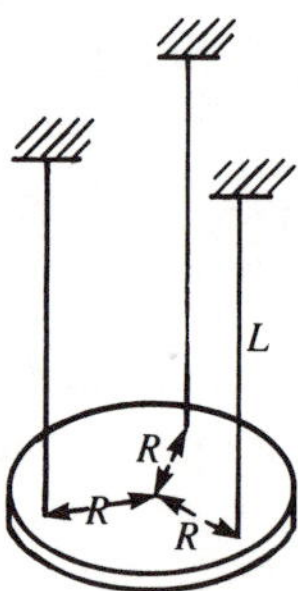

For yaw, using the standard trifilar pendulum, Figure 1.6.3, the vehicle is supported on a platform suspended by three equally-spaced wires of length L; the vehicle G must be directly over the platform G, which must be centrally positioned. Each wire has a tension $W/3$, where W is the total weight, so when the platform is rotated by θ there is a restoring moment, because of wire inclination, of

$$M = 3 \times \frac{W}{3} \times \frac{R\theta}{L} \times R$$

so the torsional stiffness is

$$\frac{\mathrm{d}M}{\mathrm{d}\theta} = \frac{WR^2}{L}$$

It is usually acceptable to neglect the wire torsional stiffness in this method, although a correction is easily added if desired. The natural frequency of torsional vibration is therefore

$$f = \frac{1}{2\pi}\sqrt{\frac{WR^2}{IL}}$$

By observing the natural frequency, first for the platform alone, then for the vehicle plus platform, I may be found for each case, and hence deduced for the vehicle.

The advantage of the trifilar pendulum over the quadrifilar pendulum is that the individual tensions are statically determinate. However it may be more convenient to use a quadrifilar pendulum. For this case, or for a trifilar pendulum with non-centralised load,

$$\frac{dM}{d\theta} = \sum \frac{T_k R_k}{L}$$

where T_k is the tension of the wire at radius R_k from the vertical line of the centre of mass. If the vehicle is centralised so that all the radii are the same, then the indeterminacy does not matter, i.e. it is not necessary to know each tension individually. One practical advantage of the quadrifilar pendulum is that each wire can be attached directly to each end of an axle; the effect of the centre of mass position on the various R values and tensions must then be allowed for. The quadrifilar pendulum's natural frequency may be found from the general formula

$$f = \frac{1}{2\pi}\sqrt{\frac{dM/d\theta}{I}}$$

Pendulum methods require a strong and rigid overhead structure, which may be problematic for very large or massive vehicles. An alternative method is to support the vehicle from below on a spherical air bearing, a hydrostatic oil bearing, or a small steel ball, of say 20 mm diameter. By placing this support a small distance d behind the vertical line of G, a small supporting reaction is required at the front wheels, met by a plate resting on balls or an air bearing. Thus the vehicle is constrained to rotate about a vertical axis just behind G. A horizontal spring with moment arm R and stiffness K provides a restoring moment at deflection θ of

$$M = KR^2\theta$$

The natural frequency is therefore

$$f = \frac{1}{2\pi}\sqrt{\frac{KR^2}{I_0}}$$

where I_0 is the second moment of mass about the fixed axis, which by the parallel axes theorem is

$$I_0 = I_Z + mx^2$$

where x is the distance of the axis from G, allowing I_Z to be deduced. A correction may be made for the inertia contribution of the plates under the front wheels.

Figure 1.6.4. Conventional pendulum.

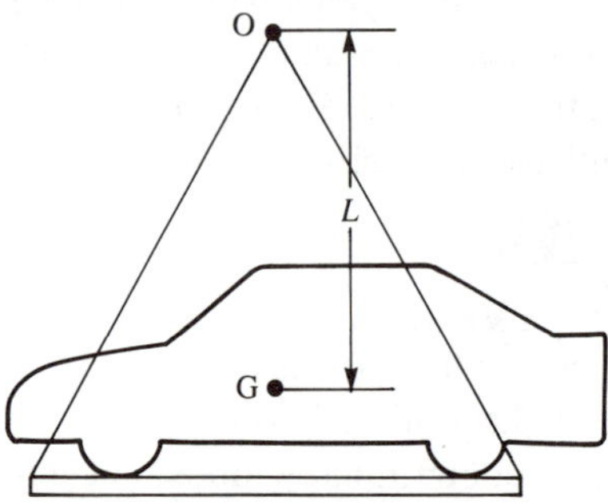

Experimental determination of the pitch or roll second moments of mass is usually performed by holding the vehicle in a cradle to form a pendulum, shown for pitch in Figure 1.6.4. The restoring moment at deflection θ is

$$M = WL\theta$$

where W is the total weight and L is the distance from the axis to the total centre of mass. By the parallel axes theorem the observed second moment of mass is

$$I_0 = I_P + m_V L_V^2 + I_C$$

where I_P is the desired vehicle pitch inertia, I_C is the cradle inertia about the pivot, and L_V is the pendulum length to the vehicle centre of mass. The natural frequency of pendulum motion is

$$f = \frac{1}{2\pi}\sqrt{\frac{WL}{I_0}}$$

The cradle inertia is found by testing the cradle alone; then $I_P + m_V L_V^2$, and hence I_P, may be deduced. It may be easier to pivot the cradle on knife edges lower down nearer to the centre of mass, in which case additional restoring stiffness may be provided by a suitable spring. This may also be more accurate, because the $m_V L_V^2$ term is much smaller.

As a matter of interest, by finding the pitch frequency for two pendulum lengths, both I_P and the height of G may be deduced. However this requires great precision, and it is necessary to allow for inertia of the vehicle's added and internal air masses, and for buoyancy forces.

Because of approximate lateral symmetry, the only product of inertia that is significant is I_{xz}. This may be measured in the following way. The vehicle is again tested for yaw inertia, but this time with the x-axis inclined upwards at the angle θ, say 30°, giving a result I_r. Then

$$I_{xz} = \frac{I_r - I_x \sin^2\theta - I_z \cos^2\theta}{\sin 2\theta}$$

The inclination of the principal axes from the xz axes is θ_p where

$$\theta_p = \tfrac{1}{2}\tan^{-1}\left(\frac{2I_{xz}}{I_x - I_z}\right)$$

Denoting the second moment of mass about a vertical axis through G, i.e. the yaw inertia, as I_Y, the radius of gyration for yaw is

$$k_Y = \sqrt{\frac{I_Y}{m}}$$

and has a typical value of about 1.4 m. Similar radii of gyration are defined for the other axes.

In analysis of transient dynamics, it is found that the significance of the yaw radius of gyration depends on how it compares with the distance of the axles from the centre of mass. Specifically, where a is the distance from the front axle to the vertical line of G, and b is the distance of the rear axle (Figure 1.6.1), the ratio

$$\frac{I_Y}{mab} = \frac{k_Y^2}{ab}$$

is called the yaw dynamic index, or just the dynamic index.

The corresponding pitch dynamic index is

$$\frac{I_P}{mab} = \frac{k_P^2}{ab}$$

where k_P is the pitch radius of gyration.

The roll dynamic index may be defined as

$$\frac{I_R}{mT^2} = \frac{k_R^2}{T^2}$$

where k_R is the roll radius of gyration and T is the mean track.

The influence of dynamic index will be discussed in Chapter 7 on unsteady-state handling, but briefly a high dynamic index, which tends to be associated with large cars because of their greater front and rear overhang, tends to lengthen the time constants of the response.

The part of the vehicle on the road side of the suspension is called the unsprung mass, discussed in more detail in Section 4.14. It is usually summarised as front and rear total unsprung masses designated m_{Uf} and m_{Ur}, basically found by weighing, with proportions of the partly moving elements, for example links. The unsprung masses may be a significant proportion of the total mass, especially for solid axles, so they are generally treated separately. The basic measured data are total mass at height H positioned a distance a behind the front axle and b in front of the rear axle on wheelbase $l = a + b$, and also the unsprung masses m_{Uf} and m_{Ur} at heights H_{Uf} and H_{Ur} usually taken as equal to the wheel radius. From these may be deduced the front and rear end-masses:

$$m_f = \frac{m\,b}{l}$$

$$m_r = \frac{m\,a}{l}$$

The total, front and rear sprung end-masses are

$$m_S = m_{Sf} + m_{Sr} = m - m_{Uf} - m_{Ur}$$

$$m_{Sf} = m_f - m_{Uf}$$

$$m_{Sr} = m_r - m_{Ur}$$

The sprung centre of mass distance behind the front axle is a_S, giving

$$a_S = \frac{m_{Sr}}{m_S}\,l$$

$$b_S = \frac{m_{Sf}}{m_S}\,l$$

$$l = a + b = a_S + b_S$$

The sprung centre of mass height follows by mass moments from ground level, giving

$$mH = m_S H_S + m_{Uf} H_{Uf} + m_{Ur} H_{Ur}$$

In rolling and pitching it is often the angular inertia of the sprung mass alone that is of interest, which may be deduced by standard techniques found in mechanics textbooks, mainly involving transformation to new axes and addition or deletion of components.

1.7 Loads

The loading of even a normal passenger car may have a substantial effect on the mass, the centre of mass, and the second moments of mass.

The process of finding the loaded properties is the same for all axes. The mass is simply the sum of the masses, of course. The new centre of mass is the point about which the first moment of mass of the vehicle plus load is zero. The second moment of mass about axes through the new centre of mass is the sum of contributions from the vehicle and the load, each found by the parallel axes theorem,

$$I_0 = I + mx^2$$

The second moment of mass of a uniform rectangular load, a by b, about its own centre of mass, is $m(a^2 + b^2)/12$.

For a typical saloon (sedan) the driver and front passenger are near to the longitudinal centre of mass and so have relatively little effect on the longitudinal properties. Two rear seat passengers have a total first moment of mass about the centre of mass of about 140 kg m, moving the centre of mass back about 100 mm. A full fuel tank may move G by 70 mm. A loaded boot (trunk) may move G by 100 mm rearwards. The total movement of 250–300 mm is about 10% of the wheelbase, and of considerable significance in influencing behaviour.

Loading of a small or large commercial vehicle may have an even greater effect on the inertia properties. Because of the large effect, each case must be considered individually. The usual effect of adding load is to move G backwards, and slightly upwards.

The repositioning of even supposedly minor components may be significant. Moving a battery from the engine bay to the boot may move the centre of mass by 30 mm, changing the front–rear load distribution by 1%. On front-drive vehicles, which are very light at the rear, this may be a useful improvement. Alternatively, for rear-engine vehicles, which are often light at the front, a useful improvement may be achieved by moving components forward, or even by adding dead load as trim as far forward as possible, for example lead in the bumpers in one well known case.

The influence of fuel load may be a considerable problem on certain specialised vehicles. For example, the standard layout of racing cars before 1959 was a front engine, followed by the driver, with the fuel tank placed over and behind the rear axle. Because of the small total mass and the large fuel-tank mass, there was a marked change of centre of mass position during a race. One Indianapolis racer of the early 1950s carried 380 kg (115 gallons) of fuel, moving the centre of mass by almost 20% of the wheelbase. In an attempt to overcome this problem, one European manufacturer put the fuel in outboard tanks filling the longitudinal space between the wheels, also reducing wheel drag. This placed the fuel load close to the longitudinal position of G, but the method proved inferior to the standard layout of the period, apparently because it greatly increased the roll inertia, which reduced the roll natural frequency and spoiled the transient dynamics.

With the change to rear engines in racing in 1959, the fuel was placed in various tanks, including at the sides, but this time close in to the flanks. With the ground-effect aerodynamics revolution of 1978, it became essential to minimise the width of the true body, in order to maximise the width of the ground effect tunnels on each side. For this reason the fuel was placed in a single central tank between the driver and the rear engine. This was also ideal for minimising the shift of centre of mass, especially with the large fuel loads, up to 200 kg of fuel with a dry vehicle mass of 600 kg or less.

The lateral position of the centre of mass is usually taken in the longitudinal plane of symmetry, because usually the vehicle itself has only minor asymmetries when unloaded and dry. However loading may offset G significantly. Often the fuel tank is on one side, and this may offset the centre of mass by 30 mm, 2.5% of the track. Each driver or passenger may cause an offset of 20 kg m, giving 20 mm offset (1.5% of the track). On some vehicles the tank and driver are on the same side, so an offset of 50 mm may commonly arise. A worst case would be with one passenger on the same side in the rear, giving 70 mm offset, 5% of the track.

The centre of mass is considered to be a fixed point in the sprung body in most cases, for a given loading condition, and usually on the centre plane, and the total mass is treated as constant. The constant total mass approximation is a good one because the mass rate of change is small, e.g. fuel consumption. The constant position approximation is less accurate because of the possibility of shifting of the load or passengers, i.e. there are inaccuracies in the rigid-body model, although again these are usually small. For example, the leaning of four passengers in severe cornering moves their centre of mass typically 15 mm, giving a total centre of mass movement of about 3 mm. Shifting of fuel in the tank of a typical saloon car may also give a centre of mass movement of about 3 mm, so when combined with the passengers the total movement is about 6 mm or 0.5% of the track, and normally negligible.

For some specialised vehicles the shiftable load may be a much larger proportion of the total mass, tankers for liquids providing an obvious extreme example, and in such cases special precautions are required, for example baffling of the tank. Load shifting is at its worst for a range of part loads around 50%, depending on the tank shape and the lateral acceleration. In all cases of fluid load shifting, in steady-state cornering the fluid surface is at an angle to the horizontal of $\theta = \tan^{-1}(A/g)$, where A is the lateral acceleration, which in the vehicle-fixed axes we may explain as the consequence of centrifugal force on the liquid. Such vehicles are also characterised by a high ratio of centre-of-mass height to track, and when tyre and suspension compliance are included, the steady-state roll-over lateral acceleration is as low as 4 m/s^2 (0.4g). When dynamic response of the liquid, and the behaviour of trailer units, is included, then the limit may be only 2 m/s^2 (0.2g). Hence it is understandable that roll-overs are a significant contributor to accidents for such vehicles.

1.8 Engine and brakes

The engine and brakes are the means by which tyre longitudinal forces are controlled, the longitudinal force being the force in the direction in which the tyre points. Because of the steering on front-drive vehicles, and because of the vehicle attitude on rear-drive vehicles, in general the longitudinal tyre forces have components contributing to, or detracting from, the vehicle centripetal acceleration, but primarily the engine and brakes control the acceleration tangential to the path.

In steady-state cornering on a level surface, the engine must provide the net power to overcome the aerodynamic drag and tyre drag forces. Also the longitudinal forces on the driven tyres influence the lateral forces at a given slip angle, because of the tyre characteristics, and hence have an effect on the handling.

Figure 1.8.1. Plan view of unpowered cornering tyre.

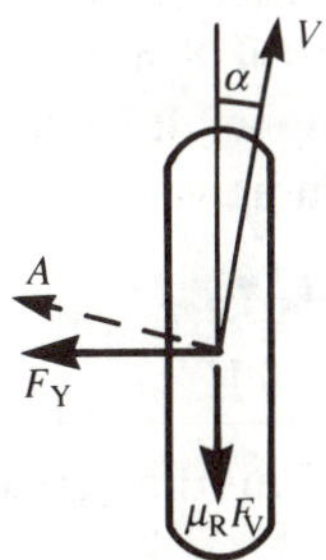

Approximate values for the cornering thrust and power requirements can be found by considering the vehicle to be concentrated at a single tyre, with a mean slip angle. Figure 1.8.1 shows an unpowered tyre with slip angle α, and with the forces exerted by the road on the tyre. These are the lateral force F_Y, and the rolling resistance force $\mu_R F_V$, where F_V is the tyre vertical force. To provide the vehicle centripetal acceleration, and neglecting aerodynamic forces, at path radius R

$$F_Y \cos\alpha \; - \; \mu_R F_V \sin\alpha \; = \; mA \; = \; mV^2/R$$

Because μ_R is small ($\mu_R < 0.02$ typically)

$$F_Y \approx \frac{mV^2}{R \cos \alpha}$$

The tyre drag is

$$D_T = F_Y \sin \alpha + \mu_R F_V \cos \alpha$$

Because α is reasonably small

$$D_T \approx F_Y \sin \alpha + \mu_R F_V$$

Substituting for F_Y,

$$D_T = (mV^2/R) \tan \alpha + \mu_R mg$$

The aerodynamic drag (Chapter 3) is given by

$$D_A = \tfrac{1}{2} \rho V^2 A_D$$

where A_D is the drag area, typically 0.7 m^2 for a European saloon, and actually influenced somewhat by the attitude angle.

The thrust power required to overcome these resistances is

$$P = (D_T + D_A) V$$

On level ground, the engine must provide this, plus losses in the transmission.

Steady-state testing is normally performed at a sufficiently small radius for a moderate power requirement, for example 32.9 m (108 ft). For a centripetal acceleration of 8 m/s^2 the speed is 16.22 m/s (36.5 mile/hr). With a mean tyre slip angle of 8°, a vehicle mass of 1200 kg, and $\mu_R = 0.017$, the total tyre drag is

$$D_T = (mV^2/R) \tan \alpha + \mu_R mg$$

$$= 1349 + 200 = 1549 \text{ N}$$

The aerodynamic drag, for 0.7 m^2 drag area, is

$$D_A = \tfrac{1}{2} \rho V^2 A_D = 113 \text{ N}$$

This is much less than the tyre drag. The thrust power required at the wheels is

$$P = (D_T + D_A) V = 27 \text{ kW} \quad (36 \text{ b.h.p.})$$

This is well within the capabilities of the average vehicle. On the other hand, if the test radius is increased to 100 m, for the same lateral acceleration the speed is 28.3 m/s (63.3 mile/hr). With the same slip angle, which will be approximately true, the tyre drag is still 1549 N, but the aerodynamic drag is up to 343 N, and, mostly because of the increased speed, the total thrust power is now 53.5 kW (71 b.h.p.). Not all vehicles could achieve this. This also implies a power dissipation of 11 kW at each tyre, with corresponding temperature increases.

From the above example we may conclude that in typical steady-state testing the power requirement depends primarily on the tyres rather than on the aerodynamics, and that the power output is usually adequate for testing at a moderate radius, but that at larger radii the engine power may be a limiting factor. In other words, the engine power defines a limit to the steady-state cornering performance envelope. The maximum steady-state centripetal acceleration at any given speed is shown in Figure 1.8.2. Section AB of the curve is limited by steering lock, section BC by tyre friction limits, generally declining slightly with speed, and section CD by engine power. At D all the engine power is required to sustain forward speed in a straight line, and steady-state centripetal acceleration is not possible.

Figure 1.8.2. Cornering performance envelope.

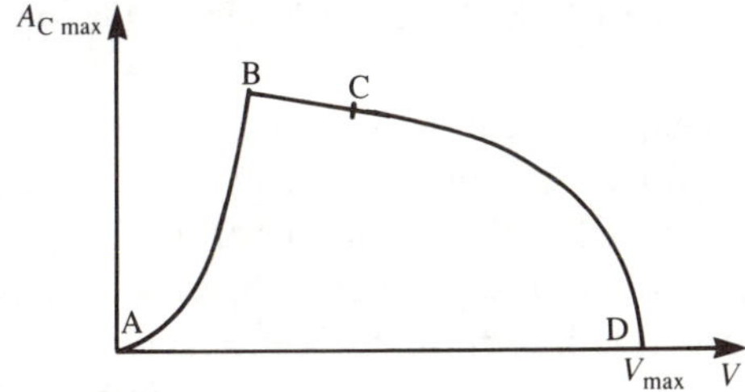

It is sometimes stated that the total tyre drag on a vehicle is greater on a small test radius, because of the increased steer angle. This is incorrect. The vehicle is travelling in a slightly different direction at the two axles, and part of the steer angle compensates for this (Section 5.15 discusses turning geometry). The drag and power dissipation depend on the tyre slip angles, not on the steer angle.

For an uphill corner the steady-state power requirement is increased because of the rearward component of the weight force, and for a downhill corner the power is reduced. For example, on the 32.9 m radius condition discussed above, a downhill spiral slope of about 1 in 7 would eliminate the need for engine power, although there would then also be other effects such as forward load transfer.

Clearly then, for transient manoeuvres in acceleration the engine power limit may be significant. Most modern cars have brakes capable of locking the wheels if required, so braking capability is not generally a limitation, although the front-to-rear balance is important.

An interesting common case is that of engine over-run without braking, which applies a torque to the 'driven' wheels. This torque depends upon the gear and the engine rotational speed. The actual engine resistance is zero at a vehicle speed giving tickover r.p.m., and increases in a way that is sensitive to the tickover setting, i.e. a small increase in tickover speed setting may considerably reduce the engine over-run braking effect at higher vehicle speeds.

1.9 Differentials

The basic types of differential are:

(1) Free
(2) Solid
(3) Locking
(4) Limited-slip

Most road vehicles are fitted with a free differential. Some utility vehicles, for example tractors, have a free differential that can be manually locked when required, as distinct from the automatic locking type above.

The free, or open, differential is a mechanical force balancer that applies equal torques to the two driveshafts whilst allowing them to rotate at different speeds. This is the usual equipment of most road vehicles. The speed difference is required because in a corner the outer wheel is on a larger radius from the centre of curvature, and must travel further, and it therefore wishes to rotate faster. The ratio of the outer speed to inner speed, with zero longitudinal slip, is

$$\frac{\Omega_o}{\Omega_i} = \frac{R + T/2}{R - T/2} \approx 1 + \frac{T}{R}$$

where R is the mean path radius and T is the track. Hence the proportional speed difference is most marked for small-radius corners, and may be as great as 30%. The disadvantage of a free differential is that if one wheel can withstand only a small torque, then the other wheel is also restricted to this torque. This can badly affect the traction in patchy or low-friction conditions. It also occurs in cornering, where the load transfer reduces the vertical force on the inner wheel, and hence reduces the limiting friction force. Although the driveshaft torques are equal, the tyre longitudinal forces may not be exactly so; this can occur if one wheel is spinning and accelerating, when the acceleration torque $I\alpha$ on that wheel will be matched by a thrust at the other tyre. For example, with $I = 0.4$ kg m^2 and $\alpha = 100$ rad/s^2, at $r = 0.32$ m the opposite longitudinal force from this effect will be 125 N, which is small.

Because the longitudinal forces produced by a free differential are normally equal, these forces do not have a net moment in plan view about the centre of mass, and in this sense the free differential is neutral in its influence on handling.

The opposite of a free differential is a solid differential. This is interesting as a limiting case of differential types, and although not found on road vehicles it is not unusual in racing, where it may be preferred over a limited-slip differential because of strength, reliability, availability or cost, and over a free differential because of performance. In the locked differential the driveshafts are connected rigidly so that they are constrained to have the same angular speed. Evidently in a corner there must be some elastic slip or actual sliding of the tyres on the ground. The total discrepancy in the advance of the two tyres, over a corner arc of θ, is

$$R_{\mathrm{o}}\theta - R_{\mathrm{i}}\theta = T\theta$$

If this is distributed evenly on each tyre, then each tyre has a longitudinal slip of approximately

$$\frac{T\theta/2}{R\theta} = \frac{T}{2R}$$

This slip gives rise to longitudinal tyre forces, and evidently is of increasing severity as the corner radius becomes smaller. Because the inner wheel wants to go faster, and the outer slower, the longitudinal forces on the tyres are forward on the inner and backward on the outer wheel, hence trying to turn the vehicle out of the corner, a so-called understeering effect. Such an effect will certainly be significant at a longitudinal slip of 1% (see Chapter 2), which for a typical vehicle track occurs at a radius of 60 m. The steer effect and the high rate of tyre wear caused by the slip mean that the solid differential is unsuitable for normal road use.

The solid differential does sometimes find application in racing, particularly and understandably on those circuits with only large radius corners or superelevated banking where the disadvantages are minimal: this means that it appears on the large oval circuits of America rather than the intricate road circuits of Europe. Because the ovals have turns of only one direction the solid differential is used in conjunction with a slightly larger-diameter outer tyre ('tyre stagger') alleviating the slip discrepancy on the corners at the cost of introducing some acceptable steering effect on the straight.

The locking type of differential has an over-run device on each driveshaft, ensuring that each is free on the over-run, but when power is applied the differential becomes solid. In corners, because the outer wheel has a greater angular speed, under power all the tractive force occurs at the inner wheel, steering the vehicle out of the corner, an understeer effect. Also, if one wheel slips because of a low-friction area of road, all the torque is applied to the other wheel.

Limited-slip differentials, also known as torque-bias and torque-proportioning differentials, act as free differentials on the over-run, or when moderate driving torques are applied, but partially lock up when one wheel tries to spin, so that the torque is directed to the other wheel. In this way the total longitudinal force at the driven wheels is limited by the total friction force at the tyres, rather than by that at the worse side only. In a corner, the outer wheel tries to run faster, which the limited-slip differential interprets as wheel spin and therefore partially locks, therefore giving more tractive force on the inside wheel; this is a similar effect to the locking differential under these circumstances, but less severe.

There are various designs of limited-slip differential, including the cam and the pawl, the friction clutch and viscous coupling types. In the cam and pawl type, power is applied to the pawls which wedge between ramps on the 'cams' that drive the drive-shafts. The mechanical friction of the wedging action, produced by the input torque, means that even when one drive shaft runs free, some torque can be applied to the other one. Without input torque, the differential is virtually free. The degree of locking that is achieved under power, usually about 75%, depends on the design angles and dimensions. This type is subject to wear, particularly when heavily loaded, and the locking action reduces as the critical dimensions are changed by the wear pattern.

In the clutch type, the two driveshafts are connected together through clutch plates, the clutches generally having some spring pre-load, and are further loaded by wedging action of the input torque. Thus if one driveshaft slips, the clutch pack returns the torque to the other driveshaft. The clutch loading can also be done hydraulically, so that the limited slip action can be controlled electronically, for example with wheel speed sensors, giving any desired characteristic.

The viscous coupling type transfers torque from the faster to the slower driveshaft simply by the viscous effect with the speed difference, and is not controlled by the input torque.

One disadvantage of limited-slip differentials is that during locking operation there is power dissipation, which may lead to heating problems in continuous operation. Actually, a limited-slip effect may be achieved by any system that introduces inefficiency if slipping occurs.

Thus there are other types based on various arrangements of gears, including worm-gear systems.

One difficulty that may be experienced with locking and limited-slip differentials is that although the design may, under given conditions, have an equilibrium state with the desirable transmission of torque, the equilibrium may be dynamically unstable, so that the differential locks and unlocks in rapid sequence, leading to an oscillating torque application to the wheels, which may lead to an erratic response from the vehicle.

Figure 1.9.1 illustrates differential characteristics; the tractive force $T_1 + T_2$ is shown as a function of the tractive force limit at the low-friction side, T_1. For a free differential $T_2 = T_1$, up to the point where both are limited by the input torque. For a solid or locked differential, T_2 is independent of T_1. A limited-slip differential has some kind of intermediate characteristic, being virtually free at small T_1 but becoming virtually solid at higher values of T_1.

Figure 1.9.1. Differential traction characteristics.

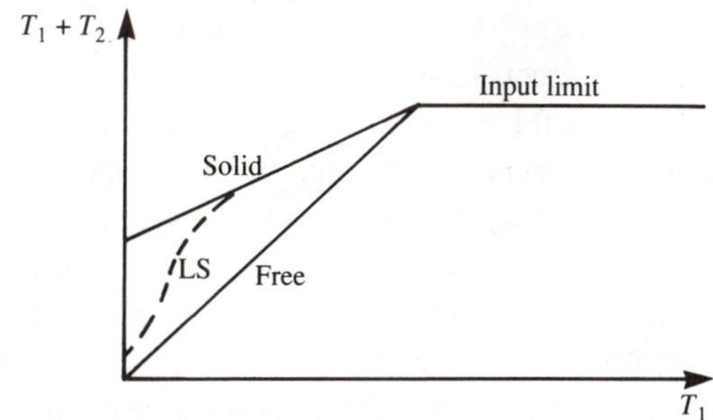

Limited-slip differentials are quite commonly fitted to more expensive or sporting vehicles, although they are a significant extra cost. From the point of view of conventional motoring, their value is in discouraging wheelspin in adverse conditions in plain longitudinal motion. In cornering, they may unfortunately cause less progressive breakaway. On front-drive vehicles the difference between the torque in the driveshafts means that there can be considerable steering wheel reaction ('wheelfight') when the locking comes into play. In good road conditions and normal driving, the free differential seems to be quite adequate. In racing, some form of locking or limited-slip differential is essential for realising the maximum performance.

Four-wheel-drive systems are common on off-road, mixed-use and military vehicles, and are recently becoming more common on passenger cars. This requires some sort of central differential to distribute torque between front and rear axles. Typical mixed-use vehicles have an open centre differential that can be manually locked, sometimes combined with a locking or limited-slip differential for the rear axle. An open differential is generally used for the steered front axle, except when the most difficult conditions will be encountered, because of adverse effects on the steering feel, although the viscous coupling type is attractive here. Some military vehicles have four-wheel-drive with a solid centre differential. Although this is obviously useful for obtaining maximum traction in poor conditions, it may have serious adverse effects on the behaviour on surfaced roads, especially if the tyre diameters are slightly different front to rear, because of wear.

1.10 Wheels

The term 'wheel' may be used in a wide sense to include the whole of the rotating element including the tyre, or in a narrow sense to mean the part that connects the tyre to the hub. The wheel probably originated at least 5000 years ago. The first wheels were presumably developed from the older idea of using a tree trunk as a roller, and possibly made by cutting discs from a trunk; such wheels are weak because of the grain direction. Fabricated wooden wheels are probably almost as old, and would have been very satisfactory on agricultural carts. With higher performance demands from military chariots, it may well have been the military engineers who devised the spoked wheel, which was certainly in common use at least two thousand years ago. The wooden-spoked wheel with wooden felloes and steel rim became very highly developed by the end of the 19th century. However, the arrival of the bicycle, the motor vehicle and the aeroplane all created a demand for improved wheels.

In fact, today there is a wide variety of wheel types for cars and commercial vehicles, the main types being:

(1) Steel disc
(2) Spoked
(3) Light alloy
(4) Flat-base and wide-base
(5) Divided
(6) Modular

The steel disc wheel is the standard fitting on most cars today. The wheel size is usually specified by the rim diameter and rim width, and a letter defining the rim section shape, for example 14×5J. The wheel rim dimensions must of course match those of the tyre. The outset is the distance from the mounting surface, on the central disc, out to the rim/tyre centre-line. Inset is the negative of outset, outset being more common. The tyre bead seats against the flanges, which are curved outwards to provide a smooth edge and a progressive contact surface for the tyre as it distorts because of bumps or cornering. The inside of the wheel must be clear of the brakes and suspension components. To improve brake cooling, holes are usually provided in the central disc, and so this standard wheel is often known as a ventilated disc wheel.

The steel disc wheel is constructed from two components – the rim and the disc or dish. The rim is formed from steel strip by cutting, rolling and butt-welding. The disc is pressed to a dished shape, typically from an octagonal blank, to give the inset, and is contoured to improve stiffness compared with a flat disc, and then usually spot-welded to the rim. This is a very economical method of wheel manufacture.

Although the spoked or wire wheel of early sports cars offered at that time a lower unsprung mass, nowadays it is selected for appearance only. The rim and a small hub are connected by about seventy spokes. These are arranged at various angles to enable the driving and cornering forces to be transmitted. The setting up of such a wheel with an even distribution of spoke tension and with the rim accurately located requires a systematic procedure and can be time-consuming. Unfortunately the wire wheel requires a good deal of attention to maintain its appearance and adjustment, and so is now virtually obsolete.

The light-alloy wheel has, for quite a long period, enjoyed success as an aircraft wheel, and also for car rallying and racing. More recently it has enjoyed a considerable vogue on passenger cars, essentially for appearance, and is becoming a common option rather than just the province of after-market suppliers. Either aluminium or magnesium alloys are used, cast or forged. The best strength is achieved by forging, although this is much more expensive. Magnesium alloys, of density about 1800 kg/m^3, are potentially lighter than aluminium alloys, of density about 2800 kg/m^3, and although requiring larger sections are usually preferred where performance is paramount. Most road vehicle light-alloy wheels are made of aluminium, and are hardly any lighter than a good steel wheel. In all cases precise alloy selection is critical for strength, fatigue resistance and corrosion resistance.

Flat-base and wide-base wheels are variations on the steel wheel, for use on commercial vehicles. For cars, the tyre is fairly flexible, and the rim well is adequate to allow assembly. The tyres of trucks and commercial vehicles are much more rigid, so in some cases it is necessary to completely remove one flange; the removable flange has a joint so that it can be sprung into position after the tyre is slid onto the rim. The well is then no longer required. There are several methods of locating the removable flange. On some light trucks, with tyres of intermediate stiffness, a rim is used which combines a removable flange with a shallow well.

In the divided wheel, the flanges are integral with the rim, but the rim is split circumferentially at its centre, where it has additional turned-in flanges to allow the whole wheel to be bolted together. This method is favoured for military vehicles using run-flat tyres. Although normally run with air pressure, these have very stiff side-walls which are trapped between the flanges and a spacer between the beads, so that in the event of tyre damage the side-walls act essentially as cantilevers, supporting the remains of the tread, if any, and in any case allowing some mobility until the tyre can be replaced.

In the modular wheel, which has become popular in racing, there is a cast or forged light alloy centre or spider, which is bolted in between two spun or pressed aluminium rim halves. This has various advantages: it is light, it has good crash energy absorption, various total rim widths and offsets can be made up, minor rim damage is often repairable unlike a completely cast wheel, and even if not, then only a part of the wheel may need replacing. This type has not so far found significant application on road vehicles.

In addition there are innumerable special wheel designs for specialised vehicles, for example agricultural equipment such as tractors, and construction equipment; solid tyred wheels are used for fork-lift trucks; and so on. One interesting development is on land speed record vehicles, where the very high speeds, about 300 m/s, mean that it has become very difficult to produce tyres that will withstand the centrifugal forces arising from wheel rotation, even when these are minimised with large-radius wheels. For example, at 300 m/s, a wheel of radius 1 m has a centripetal acceleration at the edge of 90 km/s^2, or about 9000g. Consequently the current jet reaction vehicles use solid metal wheels. These vehicles run on salt flats or mud playas, rather than hard metalled surfaces, so the wheel is harder than the road.

One method of mounting road vehicle wheels to the hub is by several studs, typically four on cars. To improve vibration resistance it is desirable to have some compliance of the wheel at the place where the stud nuts bear against it. Therefore on a steel wheel these parts of the wheel are stood forward from the plane of the mounting face. The nuts are tapered on the bearing face, so that the wheel locates on the nuts rather than the studs. For normal use it is more accurate to locate the wheel on its centre bore, and to use the nuts for retention only.

The alternative method of wheel retention is the centre-lock. The wheel is trapped and centralised between two tapers, one on the hub and one on a single large nut that screws onto the protruding hub centre. In order to provide self-tightening, it is usual to have left-hand threads on the left of the car. This method is more expensive than the multi-stud method, although aesthetically more appealing. It was often associated with spoked wheels, and similarly is becoming obsolete. One advantage is that rapid wheel changes are possible, and for this reason it still finds regular application in racing.

The manufacturing process naturally leaves the wheel with dimensional imperfections. The important feature is how the wheel locates the tyre relative to the hub, so dimensional tolerances are specified for radial and axial run-out of the rim at the tyre bead locating point. Cast and machined alloy wheels tend to be more accurate than fabricated steel wheels. Additional inaccuracy can arise because of inconsistent location of the wheel on the hub.

The wheel is generally slightly out of balance both statically and dynamically, i.e. the centre of mass may be off-axis and the principal axes may be misaligned, and this is usually dealt with by balancing the combined wheel and tyre by adding appropriate counter-masses.

The wheel is subject to quite a complex load distribution in transmitting the tyre forces to the hub. Although these can be analysed, wheels are largely designed by experience. Mounting the tyre on the rim puts the rim transverse section into bending. Inflation results in a tension of the rim, and also some bending stresses. The standing vehicle load is transmitted to the rim by redistribution of the compressive force of the bead on the rim. When the vehicle runs, this gives a fatiguing stress. Braking and driving apply an axial moment to the wheel, and tend to make the tyre slip around the wheel. Hence substantial normal forces are required at the bead to provide adequate frictional forces. In cornering, the centre dish or spider is put into substantial bending, the whole wheel being effectively subject to a bending moment about a diametral axis, of value approximately equal to the product of lateral force and tyre loaded radius, $F_Y R_l$, with a typical value up to 1000 N m. This gives substantial fatiguing stresses on the dish as the wheel rotates. The wheel must have sufficient fatigue strength to withstand these, and so, because the fatigue strength of light alloy is lower than that of steel, larger sections must be used.

The applied forces also cause wheel deflections. Although there does not seem to have been any directly observed measurement of the influence of wheel flexibility on handling, there is some anecdotal evidence that it can be significant in some circumstances, especially on the race track where racing tyres produce large lateral forces, wheels are designed for lightness, wide tyres are camber-sensitive, and vehicles are finely balanced.

The maximum moment about an axis on a vertical diameter is the maximum self-aligning torque, about 50 N m, much less than the moment of 1000 N m found about a horizontal axis. Hence, although the tyre lateral force is much more sensitive to slip angle than camber angle, the wheel deflection in camber is significant. Considering a wheel diametral torsional stiffness of 500 N m/deg, 1000 N m will give a camber deflection of 2°, which is not negligible. Because it is necessary to provide adequate stiffness, there is little incentive to use high-strength steel in wheels, because the moduli of elasticity of various steels are virtually the same, and so a section reduction allowable because of high-strength material would result in excessive flexibility unless an improved shape could be found. Plastic wheels, with their low modulus, are also subject to this problem, even if they are strong enough. Fibre-reinforced plastic wheels would be expensive for production road vehicles, although their properties might be acceptable; however, they have not found application so far. Plastics are generally prone to fatigue. The mass of a typical car wheel is approximately 10 kg.

Alternative wheel concepts have been proposed at various times, for example the now defunct Dunlop Denovo 'run flat' system. Currently in limited use are the Michelin/Dunlop TD system and the Pirelli/Goodyear asymmetric-hump tyre-retention rim. Various other new systems are under development.

1.11 Roads

The ground surface is naturally a very important element in the environment of a vehicle. In some cases vehicles are required to be able to traverse a very wide range of terrain, for example military vehicles, and to a lesser extent agricultural, construction and mining vehicles. However, a great majority of vehicles are designed to operate almost exclusively on specially prepared roads. As a minimum, the surface shape is controlled, basically by smoothing, and usually a gravel substructure (for strength) and a hard-wearing surface are provided, for example tarmacadam or concrete. The provision of roads is more or less as old as civilisation itself.

Road density correlates well with industrial development and population density. The world average is about 0.14 km/km^2 of land area. Assuming an average road width of 10 m suggests that 0.14% of the world's land surface is road. The figures are highest in Western Europe, with an average of about 1.2 km/km^2, 1.2% of the land surface. A road density of 2 km/km^2 would be equalled by a square mesh of roads at 1 km intervals. Such high road densities imply a large number of road junctions, so vehicles must often negotiate the sharp corners that are characteristic of junctions.

Car ownership in Europe is approximately 0.2 cars per person, against a world average of about 0.05. This means that in Europe there are about 50 vehicles per kilometre of road, or the amount of road is 20 m per car. Fortunately, at any one time most cars are off the road, but nevertheless this low figure does mean that the driving environment is often characterised by the presence of other vehicles, which may greatly constrain a driver's manoeuvring options. In a direct sense, the presence of the other vehicles only impinges on a vehicle when there is a collision. Happily this is relatively rare, European insurance statistics indicating that the average car and driver has an accident claim at intervals of about eight years. Adapting a concept from statistical thermodynamics, we could say that the average car has a mean free path of order 10^8 m between collisions.

The kinematic design of a road depends on the service that it is required to provide. The number of lanes depends on the traffic density; one lane can handle an uninterrupted flow of about 1200 vehicles per hour, although this can be influenced by many factors including slope, curvature, sight distances, width, frequency of interruptions such as junctions or bus stops, lighting, speed limits, weather, etc.

The mechanical design of a road depends on the size and frequency of the imposed traffic loads. A road may be likened to a loaded metal component, having a fatigue life. For loads below some threshold, the life is virtually infinite, but larger loads rapidly cause damage and consume the fatigue life. The important load is not the total vehicle weight but the individual tyre loads. In practice, in road design lateral load transfer is neglected, and the axle load is considered as the vital parameter. Tyre lateral forces certainly increase the surface wear rate, but the road structure is designed around the distribution of vertical forces to be accommodated with an acceptable life. Any undulations of the road considerably increase the peak dynamic vertical forces above the mean value, and this must be allowed for.

From the civil engineering point of view, the basic structure of a road is considered to be either flexible or rigid. In either case, there is a sublayer of typically 250 mm of gravel placed over the compacted subsoil. For a flexible pavement the top layer is about 70 mm of tarmacadam or similar tar-coated gravel. A rigid pavement has a top layer of about 250 mm of concrete, often reinforced; this is in sections with expansion joints. The rigid type has substantial bending strength, and distributes the load more widely, although both types may accommodate heavy loads with appropriate thickness of material. Even the flexible type has only a small deflection under the actual loads, and for vehicle dynamic analysis on road surfaces it is usual to treat the road as completely rigid; this is a very helpful simplification, which would be quite inappropriate for many off-road analyses. There is of course a wide variety of other surfaces, rarely laid new today, such as cobbles and pavé.

For handling analysis, the surface character of the road must be represented in some way, i.e. we must consider deviation of the road from the simplest possible case of perfectly smooth, horizontal and straight. Much of the character of the road can be represented by considering a spectral analysis of the road surface deviations. Table 1.11.1 shows one possible way of partitioning the wavelengths, and how they bear upon handling analysis.

Table 1.11.1. *Road spectral analysis*

Characteristic		Scale	Influence on:
Slopes		$50\text{ m} < \lambda$	Pseudo-static
Undulations		$1\text{ m} < \lambda < 50\text{ m}$	Dynamic
Roughness		$10\text{ mm} < \lambda < 1\text{ m}$	Dynamic
Texture	Macro	$1\text{ mm} < \lambda < 10\text{ mm}$	Friction
	Micro	$10\ \mu\text{m} < \lambda < 1\text{ mm}$	Friction
Material		Molecular	Friction

For wavelengths exceeding 50 m, i.e. hills and slopes, at an example speed of 20 m/s the forcing period exceeds 2.5 s, which is slow enough for the vehicle to be in effective equilibrium on its supension. Thus in this range the effect may be dealt with by a static analysis; in other words, to the vehicle a hill is essentially a sequence of steady-state conditions, each of constant slope.

Wavelengths from 1 m to 50 m, i.e. periods of 50 ms to 2.5 s, called 'undulations' here, are dealt with primarily by the suspension, and stimulate a dynamic response. This includes the basic sprung mass resonance at about 1 Hz, which must be controlled by the damping. Wavelengths of 10 mm to 1 m, with periods 0.5 ms to 50 ms, called 'roughness' here, are dealt with largely by tyre deflection, and again require a dynamic analysis of the vehicle response. Some roads, for example cobbles and pavé, exhibit a large spectral peak at a wavelength corresponding to the dimension of the constructing elements, around 100 mm.

For wavelengths under 10 mm there is little dynamic response from the vehicle, the forcing period being under 0.5 ms. The main effect is on the frictional behaviour of the tyre. The division into macrotexture and microtexture is a result of analysis of actual roads and their influence on friction, where it has been found that the road can be well represented by giving it roughness values on the macrotexture and microtexture scales, or simply defining it as rough or smooth on each scale, Figure 1.11.1. Macrotexture is described as rough or smooth, whereas microtexture is described as harsh or polished. Macrotexture corresponds essentially to the size of gravel used in the tarmacadam, and in a sense measures the extent to which the gravel protrudes or has been rolled-in flat with the tar, whilst the microtexture corresponds to the surface finish of the gravel elements. Their influence on wet roads is somewhat similar to the influence of tyre tread design, in that a rough macrotexture assists water clearance in the tyre footprint like the tread grooves, whilst a rough microtexture, like sipes, gives high local pressures to clear the final film. In wet conditions the porosity of the upper layer is important in providing additional drainage paths below the tyre, whereas on a smooth impervious surface such as concrete, tyre footprint clearance must be achieved exclusively by the tyre tread design. This topic is explored more extensively in Section 2.19. Attempts have been made to improve concrete road drainage below the tyre by introducing lateral grooves in the road surface; this is effective but causes substantial tyre noise.

At a molecular scale, the actual road material has a considerable

Figure 1.11.1. Macrotexture and microtexture of road surface.

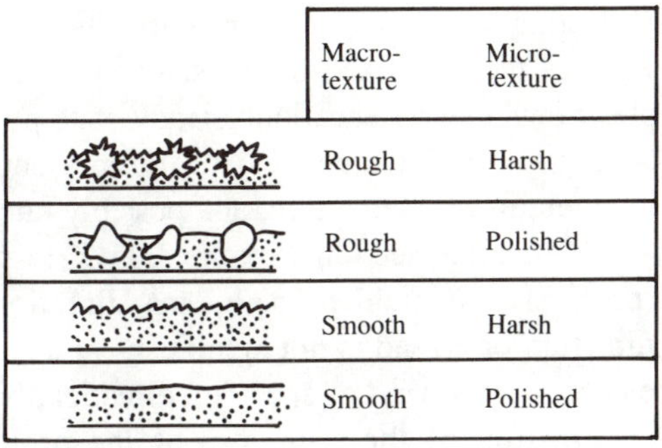

	Macro-texture	Micro-texture
	Rough	Harsh
	Rough	Polished
	Smooth	Harsh
	Smooth	Polished

influence on the frictional characteristics. Also, the variation of frictional coefficient with temperature is quite different for tarmacadam and concrete (Section 2.16).

The above spectral analysis of road deviations can be performed for three directions, i.e. we can analyse the road shape in plan view, in lateral section and in longitudinal section. The comments made above are mainly concerned with the last of these. In plan view the analysis would only be significant for wavelengths of undulation size or larger; i.e. in this case the class 'slopes' corresponds to the direction of the road, and the class 'undulations' corresponds to curves or corners. Smaller deviations are not meaningful in this case. In lateral section the terminology used for longitudinal sections is generally appropriate. However lateral slope and undulations are normally given the special names of 'road camber' and 'road camber curvature'. Lateral section roughness influences the vehicle in a similar way to longitudinal roughness, but through tyre camber effects. It is possible in principle for lateral section texture to be different from longitudinal texture, for example because of fine lateral grooves in concrete, or because of the longitudinal rolling process in road manufacture, but in practice such differences are usually neglected.

As far as handling analysis is concerned, road factors of type texture and material are dealt with as part of the tyre properties for a particular surface; frictional properties are always a property of the combination of surfaces, so the expression tyre properties should really be considered as an abbreviation for tyre–road properties.

Road factors of type roughness and undulations stimulate a dynamic vehicle response, and are generally regarded as constituting part of the field of unsteady-state response. The mean response of a cornering vehicle to a systematic form of roughness might, however, in some cases be considered to be an effectively steady-state response. Lateral undulations, i.e. successive corners, will generally constitute a continuous unsteady state. However the special case of a constant-radius corner, neglecting entry and exit transients, obviously constitutes a steady state, as does the true fixed-radius standard test. Road slopes, or direction in plan view, clearly constitute steady-state problems.

For handling analysis it is often convenient to represent the road at a given point by its direction and curvature, and possibly rate of change of curvature, in each of its three sections, Figure 1.11.2. In plan view the road, or more precisely the vehicle path, has direction and corner curvature. The direction of a road is not significant in a handling sense, other than in the context of a wind of specified direction, but as stated in Section 1.3, the direction of the path, i.e. of the projected vehicle velocity, is measured from the X-axis of an arbitrarily selected inertial XYZ system, clockwise positive in plan view, and represented by ν (nu). A positive radius of corner curvature will be considered to have a centre in positive y of the vehicle-fixed axes in normal running, i.e. Figure 1.11.2(a) shows positive curvature. Positive curvature, and a positive radius, correspond to a right-hand turn.

Figure 1.11.2. Road slope and curvature.

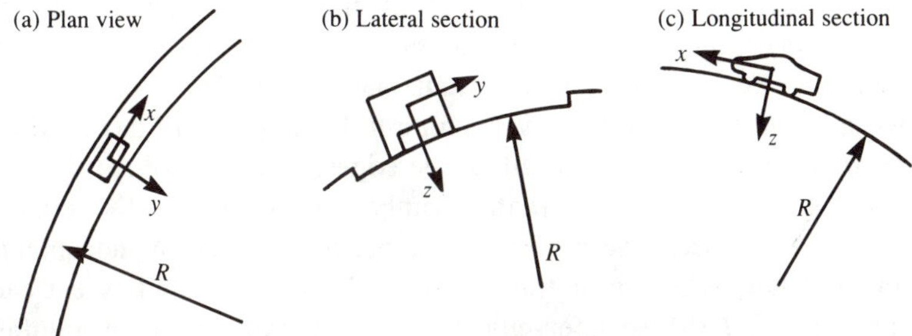

Positive camber will be increasing height towards positive y; camber is expressed in radians or degrees, or as a gradient, typically 2° or a gradient of 0.035 or 3.5%. For large gradients, the distinction between the use of $\sin\theta$ and $\tan\theta$ becomes significant; $\sin\theta$ will be used here. Positive camber radius will be with the centre of curvature in positive z. Figure 1.11.2(b) shows positive camber slope and positive camber curvature.

Positive longitudinal slope will be increasing height towards positive x, i.e. uphill, expressed as radians or degrees or as a gradient. Positive longitudinal radius will be with the centre of curvature in positive z. Thus a hump-back bridge has positive curvature and radius at the centre, and negative curvature and radius at its entry and exit. Figure 1.11.2(c) shows positive slope and positive curvature.

The influence of these road shapes on handling will be discussed in the chapters on handling. Briefly, however, corner curvature evidently demands lateral forces to provide the centripetal acceleration. Camber slope demands a lateral force on a straight road, to balance the side component of the weight force, and also causes lateral load transfer. Camber curvature introduces camber angles to the tyres. Longitudinal slope influences the thrust requirement in steady state, and causes longitudinal load transfer. Longitudinal curvature influences the tyre normal forces, which in the vehicle-fixed axes can be explained as the consequence of a vertical centrifugal (compensation) force. The influence of slopes is particularly significant in low-friction conditions.

For example, with a camber gradient of 5% and an ice-covered road with a tyre friction coefficient of 0.1, half of the available friction force is required simply to drive straight ahead.

In practice these road deviations are often combined, for example camber in a corner, or a corner on a crest, posing interesting handling problems.

The surface of a road is often at variance with its basic design condition, through damage, e.g. pot holes, or contamination, e.g. rain, ice or fallen leaves. Rain is dealt with by having a camber, the slope of which depends on the water thickness that is tolerable for a given rainfall rate. For example, at a rainfall rate of 20 mm/h, which is 5.5 g/m^2s, a 10 m wide carriageway draining to both sides will have a flow of 28 g/s for each metre of length at the edges, and proportionately less towards the centre. Thus greater camber is needed at the edges; typically 2% is used. The necessary camber depends on the acceptable depth of water, which is a function of vehicle speed and tyre tread design (Section 2.19), so safe vehicle speed is directly related to rainfall rate.

Ice is a serious hazard because of its low friction properties. It is white when broken, or frozen unevenly, for example frost, because of refraction and scattering of light by the facets of the ice fragments. So-called black ice is water that has frozen *in situ*, and is therefore transparent, appearing to be the colour of the underlying road. This is particularly dangerous because it is difficult to see.

1.12 Drivers

In any particular situation, the motion of the car depends on the performance of the complete car–driver system. Figure 1.2.1 showed the principal inter-relationships. The driver accepts feedback information, mainly from the vehicle motion, including position on the road, and steering feel, and hence this is a closed-loop system. A large part of handling theory is concerned with expressing the vehicle transfer function in terms of its detailed design, for example tyre and suspension parameters. This is conventional engineering, and considerable progress has been made in this area (Chapters 6 and 7), although the process is not yet complete.

When we consider the closed-loop system, which includes the driver, many new questions arise. What is the optimum vehicle transfer function to enable the driver–vehicle system to achieve optimum performance? Indeed, does such an optimum exist, or does it vary considerably between various drivers and roads? If there is no optimum, can limits be set on acceptable vehicle open-loop characteristics? In the interests of safety, should these be the basis of legal requirements? These have proved to be very awkward questions – the identification and objective specification of good handling qualities are extremely difficult. In practice a vehicle must be suitable for a range of activities and conditions. Another reason for difficulty lies with the driver, because the driver transfer function is highly complex, variable between drivers, and with time for a given driver; the driver is a highly adaptable control system and to a considerable degree the driver adjusts according to the vehicle characteristics to give a satisfactory overall system characteristic. Also, it is difficult to keep constant the experimental conditions such as mental and physical state and skill.

The driver uses most of his senses in driving; in some respects the human is far superior to the machine in processing data at present, for example in sight and visual field processing, i.e. pattern recognition, but inferior in other respects, for example the inner ear for inertial navigation. Sight is used primarily to examine the road shape ahead including obstacles such as other vehicles, and to assess some of the vehicle motion variables, particularly vehicle position on the road, path curvature, speed, yaw speed and roll angle. Tactility, i.e. external feeling over the body, is used to detect linear accelerations, as perhaps is feel at the hands indicating steering torque, and at the feet for pedal forces. Kinaesthetics, i.e. the internal feeling of muscular effort, is used to assess steering torque and pedal loads, and also lateral acceleration in those cases where the seat does not adequately locate the body. The inner ear also contributes to acceleration assessment. Hearing gives occasional information, including speed cues such as wind and engine noise and tyre squeal, but is perhaps most important as a warning channel, for example other vehicles' horns. In certain military and competition activities, hearing is important as the communication channel with a navigator. With the advent of the talking microprocessor, this channel can be used to advise the driver of road hazards or of inadequate distances from other vehicles, detected by suitable sensors.

The data are transmitted to the brain and processed, and then suitable muscles are activated, thus introducing a time delay. In a simple tracking problem this is typically 0.15 s, and associated largely with the transmission speed in the motor neurones from brain to hand. This compares with a total time of about 1.0 s for a realistic braking or swerving situation, requiring about 0.5 s for recognition of the need for action and about 0.5 s to activate the muscles. Thus human response time constants are typically 0.15 to 1.0 s, provided that deliberation is not required. This compares with typically 0.2 s for the time constant of a car in yaw.

The driver judges much of the vehicle's behaviour through the seat. Experiments have been performed with the seat pivoted about a vertical axis so that any lateral acceleration of the vehicle causes a slight rotation of the seat. This strongly influences the driver's perception of the vehicle's behaviour, with a forward pivot making the vehicle seem unstable and giving the driver a low confidence, and a rearward pivot giving a good stability feeling and possibly overconfidence. A forward pivot results in the driver's body having an increased attitude angle, so this is consistent with the general result that it is subjectively good for attitude angles to be small. For those commercial vehicles with isolated cabs there may be significant motion of the cab relative to the chassis, so if the driver's perception is being considered then it is necessary to instrument the cab in addition to the chassis.

On the whole it seems that the driver is remarkably sensitive to minor cues regarding the vehicle's behaviour. In open-loop handling tests the path of the vehicle can be quite well predicted, even from steady-state data, but the reaction of a driver to a vehicle, i.e. favourable or unfavourable, is much more difficult to predict. This is important; minor changes that would normally be regarded as negligible from the engineering perspective and which hardly affect the vehicle response at all may have a substantial effect on the driver's perception of the vehicle, and may make the difference between a vehicle that is subjectively good or bad. Thus in answering the question of what design features lead to good handling, it is essential to distinguish clearly between the problem of achieving a good vehicle response, i.e. a fast stable response to control inputs, and the problem of achieving a good 'driver feel'.

Attempts to find specific vehicle parameters that give favourable subjective assessment have met with only limited success, with different and sometimes opposite results from various studies, and with a wide variety of driver preferences and ability to discriminate between different vehicles. Typically, drivers can discriminate approximately 10% changes in steering gear ratio G, in the understeer gradient k, and in the yaw response time τ, when driving vehicles at intervals of a few minutes. Chapters 6 and 7 give information on the best values for design variables.

Actual driver behaviour on the road, in terms of frequency of lateral acceleration demand, varies considerably between roads and drivers. Figure 1.12.1 shows percentage distance travelled against lateral acceleration for an average driver. For example, from this figure we can read that on rural roads an average driver exceeds $0.25g$ lateral acceleration for 1% of distance travelled, and that in all cases about 90% of distance travelled is covered at less than $0.07g$. Note that speed alone is not a good indicator; motorways, with their absence of small-radius bends, have high speeds but a small lateral acceleration demand. Rural roads have the highest lateral acceleration demands. Trunk and urban roads are similar, although mean speed on the former was twice that on the latter. As an example of the differences between drivers, on rural roads relatively slow, medium and fast drivers produced results with a similar pattern to Figure 1.12.1, but the demands for $0.3g$ were zero, 0.2% and 1.0% respectively.

Figure 1.12.1. Average driver behaviour.

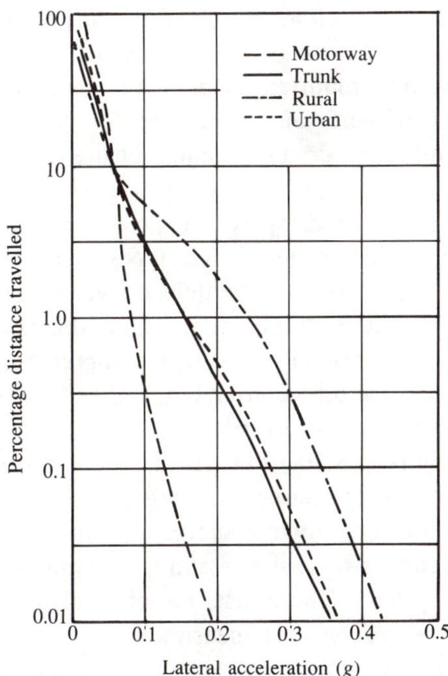

1.13 Testing

Vehicle testing is performed for various purposes, for example to confirm that a proposed vehicle design is acceptable, or to test proposed changes, or to test theory in the hope of gaining improved understanding, which should in turn lead to better vehicle behaviour. Referring back to Figure 1.2.1, which defines the open-loop system as the vehicle response only, and the closed-loop system as the vehicle plus driver, the tests can be subdivided into:

(1) Open loop
- (a) Position control
- (b) Force control
- (c) Disturbance response

(2) Closed loop
- (a) Task performance
- (b) Subjective evaluation
- (c) Disturbance response.

Position control means that the control positions are specified, for example a ramp-step or sinusoidal steering input. Force control tests are unusual for road vehicles, although common for aircraft in the special case of zero force – known as the 'stick-free test'. Disturbance response includes tests such as side-wind gusts. For vehicle plus driver closed-loop tests, specific tasks may be set, for example to perform a manoeuvre at the maximum speed or in the shortest time, giving an objective measure of performance in speed or time. Alternatively, in subjective assessment the driver's opinion of the vehicle's behaviour is solicited.

The ultimate test of handling is to expose a vehicle to experienced test drivers over a wide range of road types and weather conditions. The drivers can then score the vehicle on various qualities, and in particular can point out any problem areas, i.e. this is a subjective test. The advantage of this method is that it is comprehensive and realistic. However, it may also be time-consuming, and because it is subjective it is necessary to have several test drivers.

Although it may be possible to correlate a test driver's ratings with objective performance measures and with the vehicle's detail design, this is not easy and not always successful. In order to understand handling and to relate it to design features it is normal to use various standard open-loop tests. These tests are of a relatively simple nature so that the vehicle response is measurable objectively, and can be directly and causally related to design. With the present state of knowledge, there is a substantial gap between the idealised tests and the broader road behaviour of the vehicle, i.e. it is not yet possible to define a series of standard tests that adequately encompass road behaviour, although there has been some progress in this area.

The basic steady-state test is to operate the vehicle in steady cornering, i.e. constant speed and constant radius. Other steady-state test conditions include operation on a cambered, i.e. laterally sloping, road, and operation in a side wind. The most fundamental unsteady-state test is perhaps one of constant speed with varying path curvature, which for a given curvature can be compared with the corresponding steady state; this is a test at a point condition. In practice, manufacturers and independent testers use various more specific tests, for example capability of a lane change, or maximum speed through a chicane, i.e. a double lane change, specific corner entry and exit tests, and speed through a slalom; these are task performance tests, and are closed-loop tests over a given path, as opposed to tests at a given point condition. Open-loop tests measure response to a given steer input, for example step, ramp, sinusoidal or random input. In addition there are unsteady wind response tests, for example exposure to a gust or to a step change of side wind. Amongst given-path unsteady-state tests there is little standardisation of the details, e.g. of chicane dimensions, although there are I.S.O. drafts for a number of open-loop tests.

The steady-state test has been subject to some standardisation, for example I.S.O. 4138. The purpose is to explore the relationship between each of steering wheel angle, steering wheel torque, sideslip angle and roll angle with the lateral acceleration; the primary result is that for the steering wheel angle. The test is performed on a fairly smooth area of a hard high-friction surface, for example asphalt or concrete, with a radial slope of up to 2% for drainage. The radius should be at least 30 m; for example the Motor Industry Research Association (M.I.R.A.) pad course is 32.9 m (108 ft). A large radius means a higher speed for a given lateral acceleration, so for a larger radius aerodynamic forces may become significant. At 30 m radius, 8 m/s^2 lateral acceleration is 15.5 m/s (34.7 mile/hr); typically $C_DA < 1$ m^2 and $C_LA \approx 1$ m^2 giving drag and lift forces of about 150 N, about 1% of the weight, and just about negligible. At 300 m radius the required speed is 49 m/s, giving drag and lift of about 1.5 kN, about 10% of the weight. The side forces will generally be of the same order. Thus the test radius can be chosen to include or exclude aerodynamic effects, within limitations of engine power and track facilities. Lift and pitch are basically proportional to speed squared for constant coefficients, and hence to lateral acceleration at a given radius. On the other hand, because yaw angle grows with lateral acceleration, sideforce and yaw moment are basically proportional to speed to the fourth power, and to lateral acceleration squared. Engine power demands were discussed in Section 1.8. This can be a particular limitation for commercial vehicles. For large radii, only partial circuits are used, but sufficient to allow steady state to be established; this should be for at least 3 s, with a path deviation not exceeding 0.3 m.

Wind speed should not exceed 7 m/s for 30 m radius, and less for larger radii where the vehicle speed will be higher and wind sensitivity greater. Tyres are warmed up for 500 m at 3 m/s^2. Data points are taken at lateral acceleration intervals of no more than 0.5 m/s^2, and less if the results are sensitive to lateral acceleration, which may be so for small values, e.g. under 1 m/s^2, and is always so for large values, e.g. over 6 m/s^2. Increasing values of lateral acceleration are used until steady state can no longer be maintained. Data are taken and presented for both directions, because in some cases this makes a considerable difference.

The actual speed, yaw rate and centripetal acceleration can be measured and calculated from each other in various ways, since $V = rR$ and $A = r^2R$. Data may be taken from:

(1) A y-aligned accelerometer,

(2) A yaw-rate gyro measuring r,

(3) An inertial velocity transducer,

(4) Elapsed time over a given path, for example one lap.

Table 1.13.1 shows suggested ranges and maximum errors for the complete measurement system employed, and gives some idea of the accuracy that can realistically be achieved. The frequency bandwidth should be at least 3 Hz for steady-state measurements, to indicate fluctuations and to allow automatic sensing of when an acceptable duration of effective steady state has been achieved. For vehicles with significant chassis torsional compliance, for example commercial vehicles, it is necessary to measure the roll angle at at least two points, for example at the axles. The use of vehicle transducers means that the data are measured in the vehicle-fixed axes, so appropriate transformations must be applied to obtain the desired plotting values, for example acceleration from a transducer aligned with the y-axis will need consideration of both roll and sideslip angles.

Instead of using transducers in the vehicle, some parameters can be measured by using a small trailer, e.g. the M.I.R.A. roll/slip trolley. This is a small trolley of mass 18 kg, with two small cycle wheels, usually attached to the rear of the vehicle. Being light and balanced, it corners with small roll and sideslip, thereby providing a reference from which the vehicle roll and sideslip can be measured, at the connection to the vehicle. A geometrical allowance must be made for the trolley position relative to the vehicle centre of mass. This method has proved to be more suitable for steady-state response investigations. Laser distance meters can be used for ride height and roll angle measurement, and laser-doppler devices for accurate speed measurement.

Table 1.13.1. *Transducer and recorder specifications*

Variable	Symbol	Min. range	Max. error
Steering wheel angle	δ_{sw}	±360°	±2° (< 180°) ±4° (>180°)
Lateral acceleration	A	±15 m/s^2	±0.15 m/s^2
Yaw angular speed	r	±50°/s	±0.5°/s
Speed	V	> 50 m/s	±0.5 m/s
Attitude angle	β	±15°	±0.5°
Steering wheel torque	T_{sw}	±30 N m	±0.3 N m
Roll angle	ϕ	±15°	±0.15°

An unusual alternative form of steady-state testing developed at M.I.R.A., now no longer used, was tethered testing, where the vehicle was firmly displacement-located, at a point close to its centre of mass, from a rigid arm cantilevered sideways from a large truck. The test vehicle could rotate freely, and hence adopt steady-state yaw, pitch and roll angles according to the steer and throttle positions. Extensive data acquisition and processing equipment could be carried in the truck, avoiding the problem of loading the vehicle.

In steady-state testing a human driver is normally used, attempting to maintain constant radius, constant steering and constant accelerator position. Near to the maximum lateral acceleration it proves increasingly difficult to maintain steady conditions. Thus an open-loop test of the vehicle only becomes to some extent a closed-loop test.

When studying the unsteady response to a specified steering input, it is very difficult for a driver to control the wheel position accurately, for example over a ramp change at a specified rate to a particular value. In some cases, therefore, steering wheel control has been automated, driven by a stepper motor. This gives much better repeatability.

The random steer input test is typically performed over a straight of at least 1 km at several speeds, e.g. 50, 80 and 120 km/hr. After analysis, the basic result is a transfer function for the vehicle.

The results of testing can be expressed in many different ways; these are discussed in Chapters 6 and 7. Typically, the steady cornering result is a plot of the various parameters against lateral acceleration. Although the centripetal acceleration should really be used, it has become common practice to use the lateral acceleration. The difference is usually small because it is a cosine difference in β, e.g. 1% at 8°. Often the steering wheel angle is divided by the effective steering gear ratio between the hand wheel and the road wheels to give the reference steer angle δ_{ref}, and then this and the sideslip and roll angles are all plotted on one figure, as for example in Figure 1.13.1. The transient response is often summarised by plotting the steering response gain and the phase against frequency.

Figure 1.13.1. Example steering pad result.

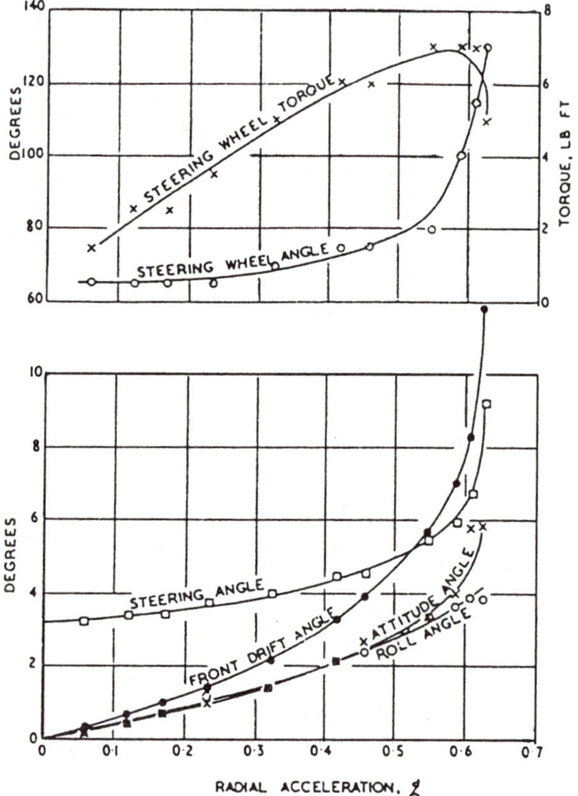

Figure 1.13.2 shows the Dunlop–M.I.R.A. handling and stability circuit, designed to provide preliminary assessment of handling qualities of vehicles or tyres before more extensive road testing. A variety of radii are provided, each long enough for the steady state to be established, so that all phases of cornering can be studied. There is also an area for chicanes and small-radius tests. Part of the circuit passes in front of the M.I.R.A. wind generator, for wind response tests.

M.I.R.A. also has a ride and handling circuit for suspension evaluation, including pavé, corrugated track, long-wave pitching track, single bump, 1.5 inch dip, manholes, broken edges, corrugated bend, level crossing, hills and dips, adverse camber, road intersection, spoon drain and Australian Creek, and other 'standard' challenges to the suspension, some rarely used nowadays.

Figure 1.13.2. Dunlop–M.I.R.A. handling and stability circuit.

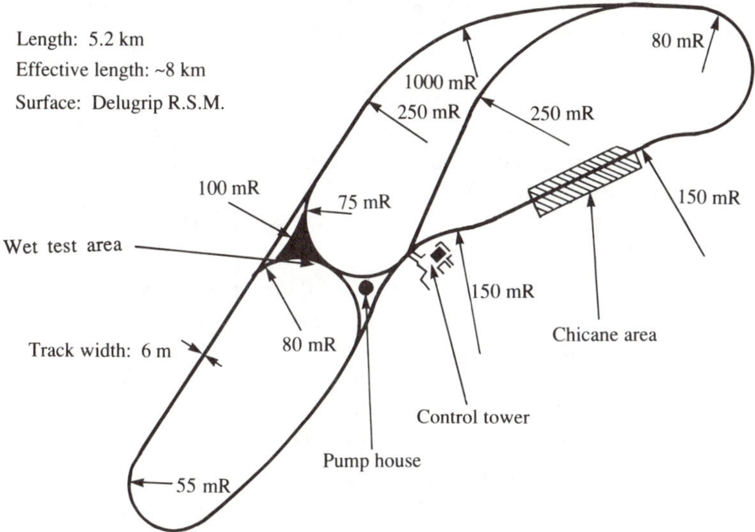

1.14 Problems

Q 1.1.1 Discuss the meaning of the terms cornering, road-holding and handling.

Q 1.2.1 Review briefly the history of the theory of handling.

Q 1.3.1 Explain the differences between longitudinal, side, normal, forward, lateral, heave, tangential and centripetal accelerations.

Q 1.3.2 Explain the relationship between the heading angle, sideslip angle, attitude angle and course angle.

Q 1.4.1 Discuss the validity of treating the vehicle as a rigid body, with regard to wheel rotation.

Q 1.4.2 Explain what a compensation force is.

Q 1.5.1 Describe in some detail how you would measure the torsional rigidity of a chassis.

Q 1.6.1 Four wheel reactions are measured as F_{VfL} = 4.1 kN, F_{VfR} = 4.9 kN, F_{VrL} = 4.1 kN and F_{VrR} = 4.3 kN on a wheelbase of 3.20 m and track of 1.40 m. Where is the centre of mass?

Q 1.6.2 On the Moon (g = 1.60 m/s^2) a spring balance intended for use on the Earth is used, and indicates 4.300 kg. What are the weight and mass of the object?

Q 1.6.3 A 3.50 m wheelbase vehicle has a wheel radius of 0.30 m, and has $N_f = 9.20$ kN and $N_r = 7.80$ kN when level. When inclined at 30°, still resting on its wheels, $N_r = 6.3$ kN. Find the longitudinal position and height of G.

Q 1.6.4 In a trifilar pendulum test, with wires of length 3 m at radius 2 m, the platform alone has a mass of 220 kg and a frequency of 0.270 Hz, and the platform plus vehicle 0.256 Hz. The vehicle mass is 1640 kg. What is the vehicle yaw inertia?

Q 1.6.5 A vehicle of wheelbase 2.500 m with mass 1540 kg and a =1.100 m is suspended on a quadrifilar pendulum, supported directly at each axle at a width of 2.100 m, on wires of length 2.500 m. The natural yaw frequency is 0.290 Hz. Calculate the yaw inertia.

Q 1.6.6 A 1760 kg vehicle of wheelbase 3.00 m (a = 1.50 m) is mounted on a hydrostatic pivot bearing 0.52 m behind G, and put into fixed-axis oscillation about a vertical axis. The restoring moment is from a 20.0 kN/m spring acting with a moment arm of 1.620 m. The natural frequency is 0.411 Hz. What is the vehicle yaw inertia, and corresponding radius of gyration and dynamic index?

Q 1.6.7 A vehicle of mass 1330 kg and centre of mass height 0.580 m rests on a pendulum plate with upper surface 3.00 m below the pivot. The plate centre of mass is 3.05 m below the pivot, and the plate mass is 210 kg. The plate alone has a pendulum frequency of 0.239 Hz, the plate plus vehicle 0.270 Hz. What is the vehicle pitch inertia and corresponding radius of gyration?

Q 1.6.8 Obtain an expression for the lateral distance of G from the centre-line in terms of the static wheel reactions.

Q 1.6.9 For Figure 1.6.2, by drawing a suitable diagram, show that

$$c + H \sin \theta = a \cos \theta + r \sin \theta$$

Q 1.6.10 The centre-of-mass height and lateral position are to be investigated by fitting a vehicle with thin disc wheels and inclining it laterally until balance is achieved. The angles from the horizontal are found to be θ_l and θ_r for left and right wheels on the ground respectively. Obtain equations for the desired values.

Q 1.6.11 A complete vehicle has mass 1750 kg with G at height 720 mm, 1.420 m back on a 2.920 m wheelbase, and has front and rear unsprung masses of 130 kg and 270 kg both at height 340 mm. Find the front and rear end-masses, and sprung G position behind front axle and height.

Q 1.7.1 An unloaded vehicle has mass 1800 kg, with G at a point 1.422 m behind the front axle and 0.630 m above ground level. I_P = 3400 kg m^2. A uniform load of mass 340 kg is added, with its centre of mass 2.413 m behind the front axle, and at a height of 1.104 m. The load dimensions in side view are length 2.142 m and height 1.010 m. Find the combined centre-of-mass position and the pitch inertia.

Q 1.8.1 Describe how the engine power requirement depends on test radius for a given lateral acceleration, and on lateral acceleration at a given radius.

Q 1.9.1 Explain the functional difference between open, solid, locking and limited-slip differentials.

Q 1.10.1 Describe briefly the different types of wheels.

Q 1.10.2 What influence does the wheel have on handling?

Q 1.11.1 Describe how the longitudinal fluctuation of road shape may be considered in terms of a spectral analysis, and how the wavelengths may be grouped.

Q 1.11.2 Describe how road texture influences tyre characteristics in wet road conditions.

Q 1.11.3 Sketch the three cross-sections of a road which has positive slope and curvature in each case. What part does each play in handling?

Q 1.12.1 Discuss the reasons for the difficulty in determining vehicle design features that optimise closed-loop system performance, and that give good subjective assessment of vehicle dynamics.

Q 1.12.2 Describe how a driver perceives vehicle motion.

Q 1.12.3 With what percentage probability, based on distance, does an average driver call for a lateral acceleration exceeding 2 m/s^2 on typical rural roads and trunk roads in Europe?

Q 1.13.1 Discuss the relative merits of subjective, task performance and vehicle response tests.

Q 1.13.2 Describe the most common forms of unsteady-state tests.

Q 1.13.3 Sketch a typical graphical result of a steady-state circular test, and briefly describe its main features.

Q 1.13.4 Describe the various ways in which speed, yaw velocity and lateral acceleration can be measured and calculated from each other for a circular pad test.

1.15 Bibliography

There are only a few books on vehicle dynamics, mentioned as appropriate below and in other Chapter bibliographies. There is, however, extensive research literature, of which it is possible to mention only a limited number of the more salient and useful papers.

Most of the useful papers appear in the following publications: *Automobiltechnische Zeitschrift* (in German), *Bulletin of the Japanese Society of Mechanical Engineers* (in English), *Proceedings of the*

Institution of Mechanical Engineers, Society of Automotive Engineers technical paper series, *Transactions of the S.A.E.*, *Transactions of the American Society of Mechanical Engineers*, and the journals *Vehicle System Dynamics* and the *International Journal of Vehicle Design.* There have also been many relevant conference proceedings, mostly sponsored by the above bodies, including, for example, the long-running I.A.V.S.D. symposium series 'Dynamics of Vehicles on Roads and Tracks'.

To put road vehicles into context, the broader background of ground vehicles in general may be found in Bekker (1956, 1960, 1969) and Wong (1978). (References can be found in the references section near the end of the book.) A good general introduction to the mechanical design of vehicles is Korff (1980). A useful general introduction to the analysis of vehicles is that by Artamonov *et al.* (1976).

For a useful general perspective on road-vehicle handling and suspension design, see Bastow (1980 or 1987) and Giles (1968). Another general work on suspension and handling is that by Steeds (1960). A largely practically orientated overview of the subject is the useful popular work by Puhn (1976, 1981). For a theoretical approach in a mathematical vein, see Ellis (1969), now out of print, or more recently Ellis (1989). The latter has some overlap with this book (perhaps 30%), but is generally different in approach and coverage, and is recommended as a 'second opinion' on those areas that are in common. Extra material there includes more extensive linear analysis, ride, steering oscillations, and articulated vehicles. Smith (1978) gives an interesting discussion of optimising racing car handling. Truck handling is dealt with extensively in Segel, Ervin & Fancher (1981). This also discusses load shifting.

The history of early handling, with extensive references, is given in Milliken & Whitcomb (1956). Details of the historical papers can be traced there. Figure 1.1.1 comes from Olley (1934). Figure 1.1.2 comes from Evans (1935). For more details on measurement of body inertia, see Winkler (1973) and Goran & Hurlong (1973). For details of the macrotexture and microtexture of roads see Sabey (1969). Driver behaviour on the road is examined in Smith & Smith (1967). For a description of limited-slip differentials, see Lewis & O'Brien (1959), Hall (1986) and Garrett (1987). The standard for wheels is BS AU 50: Part 2 (1979).

2

The tyre

2.1 Introduction

This chapter presents information on the handling properties of tyres. To put this in context, some general background information is also given. The essential function of a tyre is to interact with the road in order to produce the forces necessary for support and movement of the vehicle body. Forces must be created to cause forward acceleration, braking, and cornering. The tyre is also required to cushion the vehicle against road irregularities and to operate for many miles with great reliability. We shall be concerned here only with the vehicle control forces produced on the tyre by the road; this is not to suggest that other factors, such as vibration isolation, are unimportant in the wider view.

In the production of the forces required to give the vehicle its desired kinematic behaviour, it is the contact patch, the 'footprint', that is the focus of attention. To quote the first paper on tyre cornering properties, by Evans in 1935, "The areas of contact between tyres and road ... are the very front line trenches in the furious battle between space and time." The rubber-carcase pneumatic tyre is uniquely well adapted to this evolutionary niche. During the serious shortages of rubber during the Second World War, various alternative ideas were reappraised – some even tried on the road – but none had any real success. The post-war years have not seen significant commercial success for any alternative. This dominance by the rubber tyre is the result of a remarkable combination of properties that enables the tyre to provide support and control with good durability in difficult conditions, whilst being highly adaptable to specific applications. Nevertheless, there is a continuing effort to improve its properties.

For a given tyre and road surface, the lateral force produced by road contact depends upon many factors, especially the angular position of the tyre relative to its direction of travel (the slip angle, sometimes called drift angle), its angle of lean to the vertical (the camber angle), the vertical force, the inflation pressure, and the rolling angular speed (associated with any tractive or braking forces also demanded). Many other factors have a secondary influence, including for example the speed of travel. When various tyre designs and various road surfaces are considered, then a whole host of further factors are introduced. It is impossible to predict with high accuracy the tyre force that will result in any real conditions. This is so for several reasons, including tyre production variations, tyre wear, road surface variation, and so on. Even in the laboratory, testing causes such rapid wear that a tyre is likely to change its characteristics very quickly, and a tyre will certainly wear out before a truly comprehensive set of results can be obtained. Nevertheless, characteristics of engineering accuracy can be achieved. These are reasonably accurate in relation to natural road variations. This chapter attempts to provide a qualitative, and to some extent a quantitative, understanding of the behaviour of the tyre and factors that influence it, and to provide some insight into the modelling of tyre behaviour for vehicle dynamic simulations.

2.2 Construction

The feature peculiar to a road vehicle, in contrast to a fluid-borne vehicle, such as a ship or an aircraft, is the wheel. Since there is no fluid to provide the compliance necessary for both comfort and control, then compliance must be engineered into the vehicle. In practice this is provided by the tyre, the suspension and the seat.

The essential constructional feature of the modern pneumatic tyre is the carcase – a moulding of rubber reinforced by several layers of cords or fabric, each layer called a ply. The carcase makes contact with the wheel at the bead. Multiple beads are sometimes used if there are more than eight plies, for example on trucks or aircraft. The bead is an interference fit on the rim which has a taper of about 5°. Sometimes using a thin inner tube for sealing, the carcase is inflated with air, thus tensioning the carcase. Pressures are typically 120 to 200 kPa for cars, 300 to 600 kPa for trucks. The reinforcing cords, typically nylon, rayon or terylene, have a higher modulus of elasticity than the rubber, and their creep is less, so they carry the tension whilst the rubber acts essentially as a gas sealant. The alignment of the cords is an important feature affecting the vehicle handling behaviour. In the radial-ply tyre, the cords arc radial in side view, running directly across the tread area (the crown) at 90° to the centre-line. This is described as a crown angle of 90°. On the bias-ply (also called diagonal-ply or cross-ply) tyre, the cord crown angle is typically 40°, alternate plies being angled left and right from the centre-line, and they are no longer simply radial in side view.

The rubber of the carcase is fairly soft to give it good fatigue resistance. Various materials have been used for the cords. Dunlop's original tyre used flax. Cotton was dominant up to 1945, then rayon up to 1960, and then nylon. Since about 1975 polyester has been most common. Steel, glass fibres and aramid fibres have also gained ground since 1970. These changes are directly related to the mechanical properties of the candidate materials. Brass-plated steel wire is favoured for commercial vehicles and off-road applications. Laying up the cords is an expensive manual operation, but cordless tyres have not yet come to commercial fruition. The density of the reinforcement varies considerably, e.g. nylon 1100 kg/m^3, glass fibre 2500 kg/m^3, steel 7800 kg/m^3.

On radial-ply tyres, the crown has a circumferential reinforcing belt of nylon or steel cords. Some bias-ply (diagonal-ply) tyres have been manufactured with belts (the belted-bias type) but these are no longer produced. All radial-ply tyres are belted, so the simple term radial always implies this. The cord direction in a belt is relatively close to circumferential, i.e. the belt itself has a small crown angle of typically 20°. Sometimes a breaker is used on a bias-ply tyre – this is like a belt but with a crown angle close to that of the main carcase cords. A belt is sometimes known as a rigid breaker. The carcase construction so far described plays a dominant role in determining the tyre characteristics and vehicle dynamics as far as normal driving is concerned, i.e. for lateral and longitudinal accelerations up to about 3 m/s^2 (0.3g).

On the outside of the crown, the perimeter of the tyre, is the tread – the wearing course of rubber that actually contacts the road. This is patterned with grooves, slots and sipes (cuts), typically to a depth of about 8 mm when new (12 to 14 mm on truck tyres), to encourage drainage in wet conditions. The tread pattern also assists cooling in dry conditions. When large accelerations in any direction are required of the car, then the frictional properties of the tread become most important. The tread rubber must be quite hard to give good wear characteristics, and therefore in choosing the rubber blend there is a conflict between wear and grip. Sometimes a softer cushion layer is used beneath the tread.

Tyre size is normally specified by two dimensions: the wheel rim diameter, and the section width, Figure 2.2.1. The section width is specified in millimetres for radial-ply tyres. The safe load-carrying

Figure 2.2.1. Tyre cross-section.

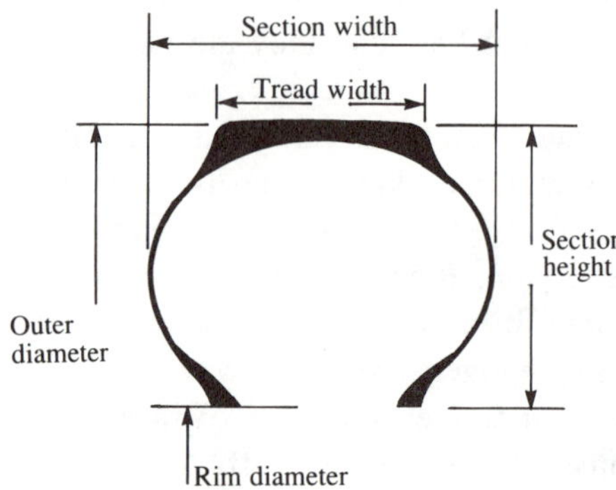

capability of a tyre is primarily dependent on its size. The outer diameter is independently variable from the rim diameter, giving rise to various cross-section proportions normally expressed by the aspect ratio or profile, which is the section height divided by the section width. The tendency is towards greater width and less height, giving a lower aspect ratio, led by the extremely low profiles in racing. Modern passenger car tyres have aspect ratio values of 50, 60, 70, 78 and 82%; in racing it may be as low as 30%. Low aspect ratios (65%) are now also used on trucks. For a bias-ply tyre the natural aspect ratio depends on the cord crown angle, whereas for the radial it depends on the restricting effect of the belt. The structural design and materials selection for a tyre is a delicate balance, and although it is known how to improve each separate property of a tyre, the problem is to achieve this without an even greater sacrifice in some other property.

Although a design load may be quoted for a tyre, its load-bearing capabilities depend in practice upon service conditions. The combination of load and inflation pressure results in a deflection which causes fatigue stresses whilst running. Hence the optimum pressure is load-dependent. This also implies that in severe cornering the interests of the dominant outside tyres are better served by a higher pressure than is ideal for straight running. The acceptable deflection, and hence load, depends upon the load-time duration, so that a load acceptable briefly in extreme cornering is not acceptable on a continuous basis. The sidewalls are normally expected to outlive the tread, although at very low wear rates embrittlement and sidewall cracking may occur first. The effect of operation at excessively high material temperatures (unacceptable combination of load, speed and environment) typically manifests itself as blistering (delamination), tread chunking (loss of small pieces of tread), tread stripping (loss of large pieces of tread), and roughness caused by loss of homogeneity.

The choice of a tubed or tubeless tyre depends to some extent on the service conditions, although for road use it is largely a matter of local custom, with habits varying greatly between different countries. The tubeless tyre effects some initial economy, and is more puncture-resistant; both types have a typical road distance between carcase penetrations of 30000 km, but the tubeless type requires a roadside wheel change in only about one penetration in five. In rough conditions such as rallying or other off-road activities, the tubeless type may lift from the rim under severe impact, with immediate deflation. For a normal road puncture, however, it deflates more slowly than a tubed type. It is also a little lighter.

Charged with producing a new tyre for a new vehicle, the tyre designer is usually faced with basic dimensional restrictions – sufficient wheel diameter to accommodate suitable brakes, plus maximum outer diameter and section width for reasons of body clearance. The load-carrying ability is governed by the dimensions and the inflation pressure. Once the diameters and section width are determined, the cross-sectional shape must be decided. For an unbelted tyre this is a function of the angle of the carcase cords. There being little bending stiffness, the cord tension must provide equilibrium against the inflation pressure and only one crown angle is compatible with a given set of values of rim radius, rim width, tyre radius and section width. The cord crown angle of an unbelted tyre governs many of the basic performance properties. For a larger crown angle (i.e. nearer to radial) the section width is lower, tread radius lower, rolling resistance lower, shear stresses lower, bead stresses higher, cord tension lower, loaded deflection greater, contact length greater, lateral stiffness less, tread wear worse, sidewall cracking worse, tread cracking less, cornering stiffness less, ride softer, and bursting strength higher.

The carcase constructional details are much more complex and subtle than may appear from this terse description, and the properties of the tyre can be markedly changed by quite small design alterations, especially of cord angles.

2.3 Rubber

The density of commercial rubber is 1100 to 1200 kg/m^3. Tyre rubber includes other constituents such as carbon black and oil, and has an average density of about 1200 kg/m^3. Rubber, whether natural or synthetic, is a visco-elastic material. Essentially this means that when a piece of rubber is distorted, it will resist with a force, but the rubber relaxes because of the viscous effects, and the force diminishes. Figure 2.3.1 shows a very simple model that exhibits this kind of behaviour. If a sinusoidally varying displacement is applied to the top point A, then the consequent force amplitude at A will depend on the frequency. At low frequency the damper will exert negligible effect, and the apparent stiffness will be that of the two springs in series. At high frequency the damper will have a large resistance and hardly move at all, so the stiffness will be greater, being that of the upper spring only. At both these frequency extremes, the damper will dissipate little energy. However there is an intermediate frequency at which the energy dissipation is at a maximum.

Figure 2.3.1. Mechanical model of rubber.

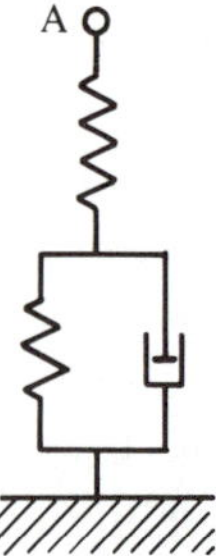

If a pure rubber is given a laboratory friction test against glass in clean conditions, then the frictional coefficient is found to depend on the sliding speed and the temperature, Figure 2.3.2. The peak friction coefficient observed may be very high. The curves for various temperatures can be reduced to a single one, the master curve, Figure 2.3.3, by the Williams–Landel–Ferry (WLF) transformation, which is based on a visco-elastic model of rubber. This very strong correlation between the equations of visco-elastic behaviour and of rubber friction suggests that at least the phenomena have a common origin, and possibly that visco-elasticity is the cause of rubber friction.

Figure 2.3.2. Rubber friction against sliding speed.

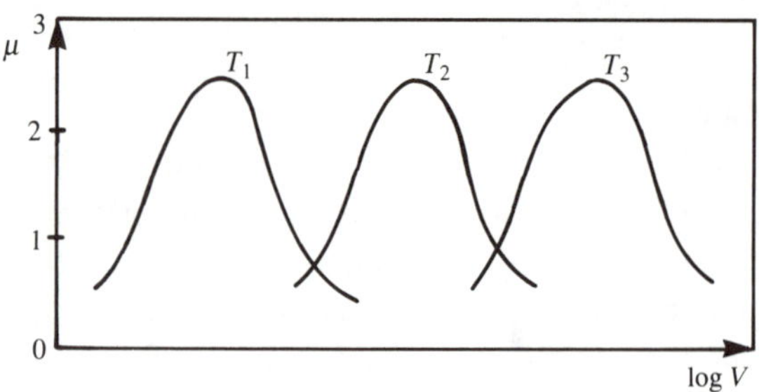

Figure 2.3.3. Friction of rubber on glass and silicon carbide.

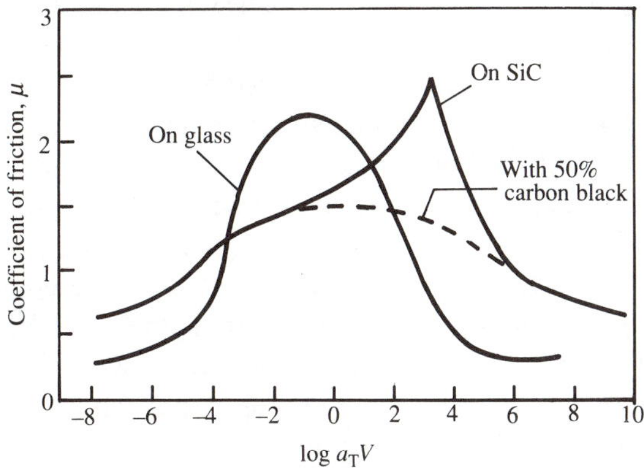

If the rubber is tested instead against a rough surface such as silicon carbide, then the WLF transformation is still successful, but the master curve is now shaped differently, Figure 2.3.3. The new and rather sharp peak is due to rubber distortion around the rough asperities, whilst the previous much smaller scale molecular effects are of diminished importance.

The addition of carbon black, to improve wear resistance, smooths out the curve, and lowers the average value, giving the more practical representative friction values of normal tyres. This is part of the reason for the temperature sensitivity of racing tyres, where the maximum frictional coefficient is sought. There are some thirty basic types of carbon black with various properties, the particle size and the chemical interaction with the rubber being important – it is not just a filler. Most widely used is high-abrasion furnace black (HAF black). In addition to wear resistance, benefits include improved tensile strength and tear strength. The addition of carbon black was first proposed by Mote in 1904. It was first used by Pirelli in 1907, when it was found to improve tread life by a factor of two or three. The hardening of rubber by vulcanisation was invented by Goodyear, and first performed in 1839.

Another constituent of tread rubber is oil; the result of including oil is to increase friction, shifting the master curve upwards, and towards lower sliding speeds. There is, however, a region of reversal, so that under particular conditions of temperature or skidding speed an oil-extended rubber can be inferior.

When tested on ice, rubber friction is found to undergo rather sudden changes. Very cold ice is just a smooth surface, with the rubber behaving normally. High friction can be achieved by using a rubber that is at the peak of its master curve at, say, –10°C. In laboratory conditions, ice can give higher friction than glass. Just below the melting point of ice, however, there is a depression in the curve of friction coefficient versus temperature, presumably due to some melting of the ice caused by frictional heating and pressure.

Figure 2.3.4 represents a rubber slider on an idealised rough road. If the surface is lubricated and offers no shear force, there can still be a friction force because of different pressures normal to the inclined surfaces arising from rubber hysteresis, i.e. the visco-elastic effect. Increasing the vertical load will increase the area of contact between rubber and road, although in a non-linear manner. Hence the friction force would increase, but the friction coefficient would decrease, as is found in practice, typically in proportion to the mean contact pressure to the power –0.15. Detailed analysis of the model of Figure 2.3.4 shows that the friction coefficient does not depend upon the scale of the serrations, but does depend on their angle. From Figure 2.3.5, this is because the element contributes a friction force $P_h A \sin\theta$, where P_h is the hysteresis pressure difference. Evidently the angle is important, and a surface with a microscale roughness of larger average angle can produce better friction. This outline explanation gives a possible illustration of why the coefficient of friction of rubber is load-dependent, and hence why a greater rubber area for a given load can result in a higher limiting friction force.

Figure 2.3.4. Idealised tyre-to-road contact.

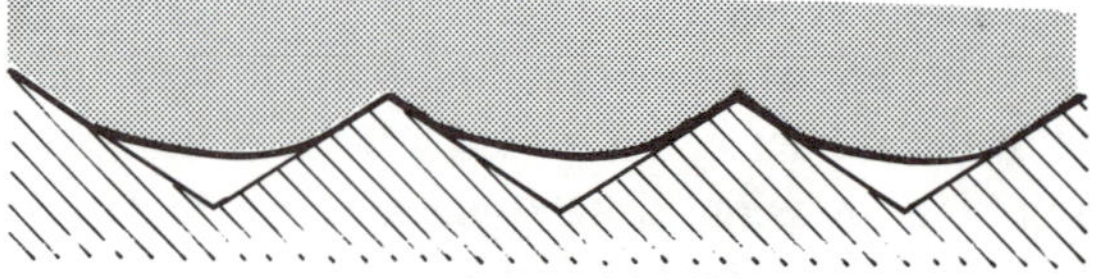

Figure 2.3.5. Idealised tyre-to-road contact (detail).

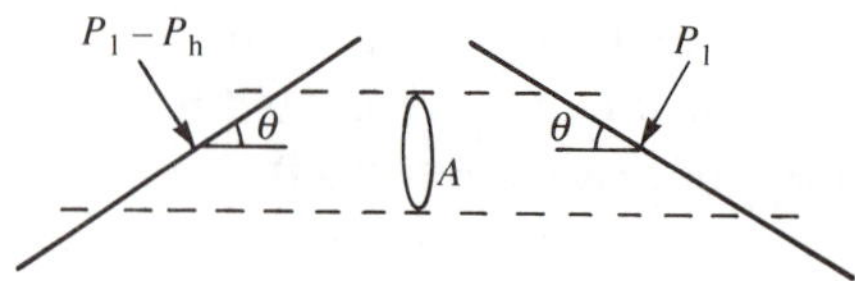

Equally important is the adhesive component of the total friction. This is due to molecular bonding between tyre and road. The formation of the bond does not provide useful energy, but the breaking of the bond requires an input of energy, hence causing energy dissipation in sliding or rolling. This adhesion component is most important in dry conditions, whereas the hysteresis component is more important in wet conditions.

A complete tyre contains several different blends of rubber, each optimised for properties and cost according to its specific application, e.g. tread, carcase wall, bead filler, inner liner. One important parameter is hardness, governed largely by the quantity of carbon black and the degree of vulcanisation. This is usually measured by a dial gauge pressing a conical tip a fixed distance into the rubber (Shore A Durometer) giving a hardness reading from 0 to 100. A typical tread hardness value is about 60.

A modern passenger car tyre has a mass of about 12 kg, comprising about 4 kg of rubber, 2 kg of carbon black, 2 kg of oil extenders, 3 kg of steel and 1 kg of rayon.

2.4 Axes and notation

Figure 2.4.1 shows the S.A.E. axis system. It is expected that the I.S.O. will soon publish a standard, also using a right-hand set of axes, and with X forward, but with Y to the left and Z upwards.

Figure 2.4.1. S.A.E. tyre axes and terminology.

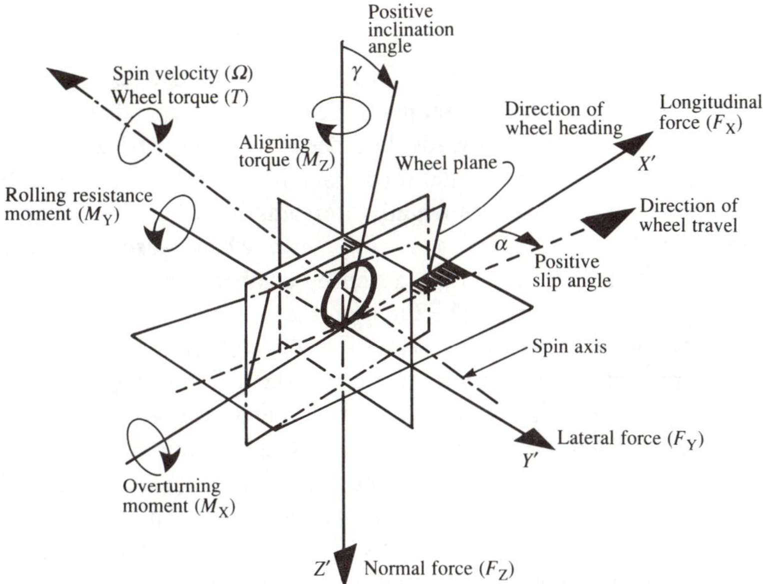

In its simplest state, the wheel stands vertically and rolls in its plane of symmetry. If the direction of travel is other than in the plane of symmetry then there is a non-zero slip angle α, with an associated cornering force, Figure 2.4.2(a). Regardless of slip angle, the wheel may also be inclined, Figure 2.4.2(b). Inclination is measured positive for right-hand rotation about the X'-axis. In keeping with common use, S.A.E. camber angle equals inclination in magnitude, but takes a positive value when the top of the wheel leans outwards from the car centre-line. This results in a camber force, directed towards the low axis side. It is sometimes known as the camber thrust.

Figure 2.4.2. Tyre angles: (a) slip, (b) camber.

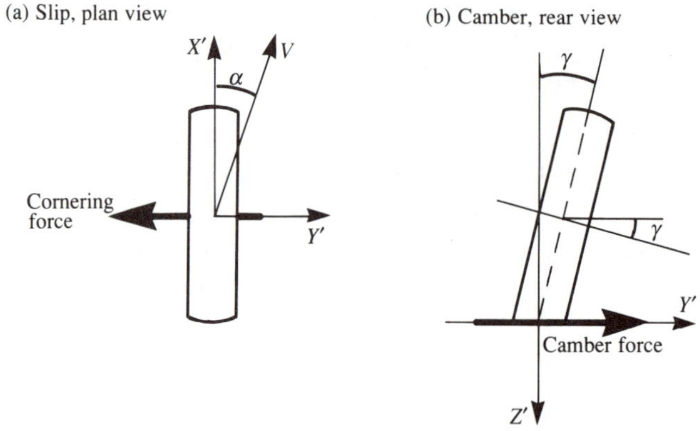

Figure 2.4.3 shows combined slip and camber, plus axes and forces. The $X'Y'$ axes are in the ground plane, parallel and perpendicular to the wheel. The Z'-axis is downwards, in order to form a right-hand set. The force exerted on the tyre by the road along X', denoted by F_X, is called the longitudinal force, and is negative for braking. The force along Y', denoted by F_Y, is called the cornering force when caused by the slip angle only, camber force when caused by camber angle only, or more generally the lateral force. The force on the tyre along Z', F_Z, is called the normal force; this is negative, so for convenience the upward force on the tyre is called the vertical force, F_V. The aligning moment M_Z and the overturning moment M_X act in the right-hand positive sense relative to their associated axes. It is best to treat the rolling resistance as a force acting at spin axis height, opposing motion (Section 2.7). All forces and moments mentioned so far are exerted by the road. In addition there are forces exerted by the suspension components, the driveshaft and the brakes, which will be considered later.

The term 'tyre force' generally means the total force exerted by the ground on the tyre. Where the context makes it clear, it is also used to mean the total force parallel to the ground plane, Figure 2.4.4(a). Here the tyre force is resolved into lateral and longitudinal components. Note that the $X'Y'$ axes rotate when the wheel is steered, so the force along

Figure 2.4.3. Tyre forces.

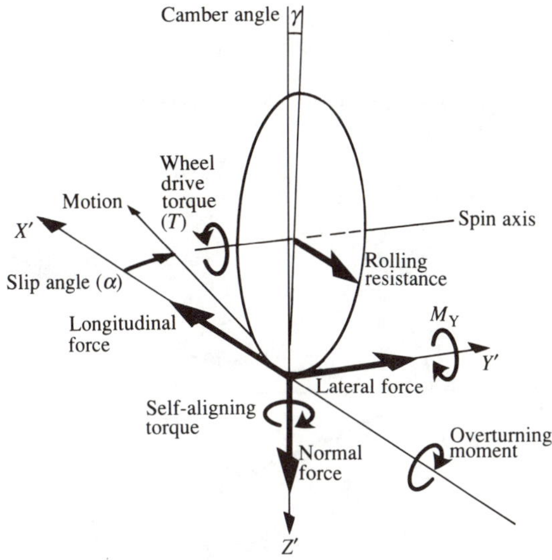

X' is the result of engine or brake action, together with the rolling resistance if significant. The force along Y' is the lateral force, which is the cornering force because of slip angle plus the camber force because of camber angle.

The term 'lateral force' was used differently in early S.A.E. papers, to mean the force perpendicular to motion, which is now known as the central force. For the analysis of the vehicle, rather than the tyre, it is in fact more useful to know the forces along and perpendicular to the vehicle motion. This is not quite the same as the tyre motion because of other effects such as vehicle rotation. The XYZ (unprimed) axes are used for the vehicle. The $X''Y''Z''$ axes are the axes aligned with the motion of the individual wheel, Figure 2.4.4(b). They are usually fairly close in direction to the vehicle XY. If the tyre forces are resolved into X'' and Y'' components, then these are called the tractive force and the central force. In the absence of engine or brake action, neglecting rolling resistance the longitudinal force is zero and the tyre force is perpendicular to the wheel, so the tractive force is then negative, and called the tyre drag force.

It is essentially the central force that provides the cornering centripetal acceleration of the vehicle. As Figure 2.4.4(b) illustrates, it is the tractive force that determines whether the vehicle rounds a curve at constant speed or not, i.e. the tractive force governs the tangential acceleration. However, for a freely-rolling tyre the longitudinal force is small, so tyre cornering force data are most simply and conveniently expressed in terms of the lateral force.

Figure 2.4.4. Tyre force components: (a) in $X'Y'$, (b) in $X''Y''$.

(a) Tyre force is lateral force plus longitudinal force

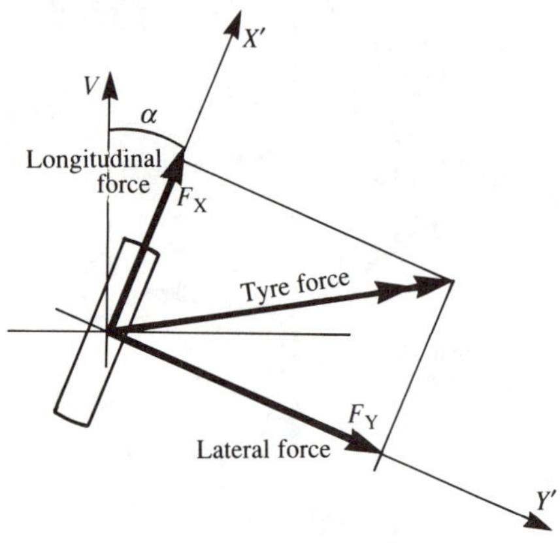

(b) Tyre force is central force plus tractive force

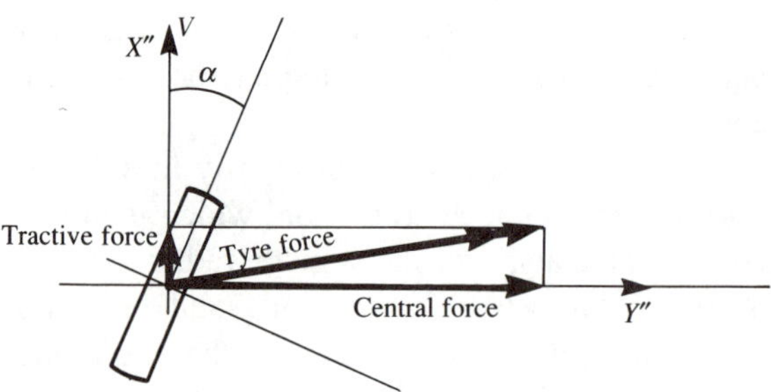

2.5 Tyre radius

Three types of tyre radius are generally recognised: the unloaded radius R_u, the loaded radius R_l, which is the wheel centre height under normal conditions, and the effective rolling radius R_e. The last of these is defined as the translation velocity of the wheel axis divided by the wheel angular speed:

$$R_e = V/\Omega$$

Note that for a locked sliding wheel R_e is infinite, whilst for a spinning non-translating wheel R_e is zero. In normal running R_e adopts a value rather closer to R_u than R_l. Also it depends a little upon speed, although less so for the radial tyre because of the restricting belt.

The loaded radius R_l depends on the vertical deflection under vertical load, which is typically 18 to 24% of the section height. The footprint length is typically about $\frac{1}{2}R$ (150 mm) for an average car tyre, subtending about 30° at the wheel centre. Therefore the crown undergoes a fairly sharp bend of 15° at each end of the footprint. Vertical deflection is about $0.035R$ (10 mm) and is quite closely proportional to vertical force, so it is meaningful to refer to the vertical stiffness. This depends upon size, construction and inflation pressure. A typical value is 250 N/mm. However, because of hysteresis, when the speed is low there is an extra deflection of about 3 mm for decreasing load. Over 80% of the total stiffness is due to the inflation pressure (over 90% for aircraft tyres) so the pressure has a large effect on stiffness, the unpressurised carcase having relatively little stiffness. Vertical stiffness generally increases with load capacity, rim width, and decreasing cord crown angle. There is a variation of typically ±10%, for a given size and pressure, from various manufacturers. Increasing speed increases the stiffness at about 0.4% per m/s. When large cornering forces are induced, the vertical stiffness may reduce by about 20%.

Damping in a tyre is non-linear, being a function of amplitude, but it is small and for many purposes it may be neglected. It is generally insensitive to inflation pressure because it arises from carcase hysteresis and ground friction. One published result for a typical car tyre gave a damping coefficient of 200 N s/m at 6 mm amplitude, varying as amplitude to the power −0.64.

2.6 Speed limitations

Aircraft speed is often characterised by the Mach number, relating the vehicle speed to the speed of sound waves in the air, arising from the elasticity of that medium. The road vehicle is also supported by a flexible medium, namely the tyre, which is therefore capable of supporting wave phenomena. Such were in fact first studied by Gardner & Worswick (1951), beginning at a speed of around 40 m/s. Instead of leaving the ground and smoothly taking up its normal circular form, the crown goes through a wave behind the footprint. At constant speed the wave is static, i.e. it does not pass around the wheel. The wave amplitude may be 10 mm or more, and the wavelength is typically 100 mm; the waves may extend around one quarter or more of the tyre circumference with gradually reducing amplitude.

A successful theory of such waves in tyre treads was provided by Turner in 1954, who showed that the wave speed was

$$C_W = \sqrt{\frac{t_C}{\rho_C}}$$

where t_C is the crown circumferential tension per unit width, itself depending on inflation pressure and speed, and ρ_C is the crown density per unit area. Once the wheel peripheral speed reaches this value then the tyre, before arriving at the ground, can no longer receive any advance warning of the impending impact by elastic transmission. Also, the tread leaving the ground with a radial speed component carries a great deal of energy which is dissipated by the wave action behind the contact patch. The predicted wavelength is theoretically dependent on speed:

$$\lambda = \text{constant} \times \sqrt{\left(\frac{V}{C_W}\right)^2 - 1}$$

which agrees with the experimental data. The analogy of this wave with the Mach number for fluids seems apt, giving the Turner number for tyres:

$$N_T = \frac{V}{C_W} = \frac{\Omega R_u}{\sqrt{(t_C / \rho_C)}}$$

where V is the peripheral speed of the tread (ΩR_u); this should be correct for both slipping (spinning or locked) or non-slipping wheels.

The presence of such waves dissipates energy, increases the rolling resistance, and raises the temperature of the carcase, soon causing damage. Because of carcase flexing and tread scrubbing, even at lower speeds there may be a substantial temperature rise in the tyre, so for sustained operation this sets acceptable limits to speed and load combinations. A thick tread may cause heat dissipation problems, although the tread pattern itself helps cooling. Also, at high speed there are significant centrifugal stresses on the outer tread elements. Therefore, for high-speed tyres the crown density per unit area must be small. For land speed record vehicles the total crown thickness has been as little as 0.4 mm.

2.7 Rolling resistance

Considering a notional isolated wheel rolling at constant speed down a ramp in a vacuum, forces and moments on the wheel must be in equilibrium. This determines the nature and value of the rolling resistance force and moment. There must be zero moment about the wheel axis (centre of mass) to maintain zero angular acceleration, so the total rolling resistance must act through that axis. At the footprint it must comprise a drag force plus a moment, or a drag plus a forward shift of the normal force, which actually arises from changes in the footprint pressure distribution. Figure 2.7.1 shows some possible representations.

Figure 2.7.1. Alternative representations of rolling resistance with vertical force.

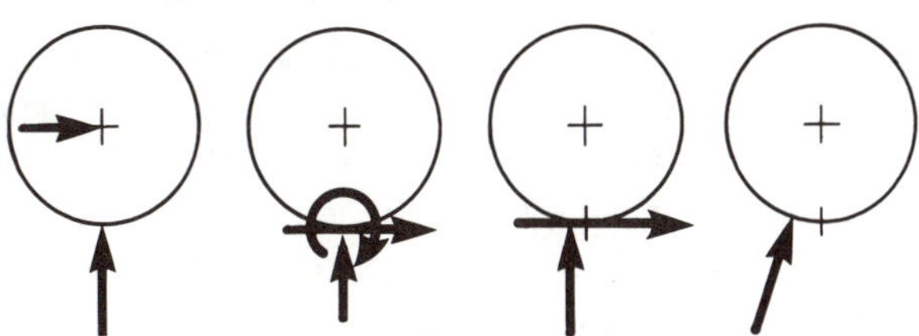

Fortunately the rolling resistance is a relatively small force and not of decisive importance in most handling problems; it will be treated as acting at the axis if it is included at all. Its main effect in the context of handling is on steering feel. Practical measurements of rolling resistance show it to be reasonably constant with speed, possibly increasing slightly, up to a Turner number of about 0.8, beyond which it increases rapidly because of the wave energy loss. These losses effectively limit the permissible speed of tyres because of the danger of overheating and tread separation. Tyres are speed rated; for example SR not to exceed 180 km/h, TR 190 km/h, HR 210 km/h. For high-speed running, the Turner number can be minimised by increasing the tread tension which results from an increase in inflation pressure. In steady state the cavity air temperature and pressure themselves depend upon speed. There is a rise of typically 35°C and 40 kPa in steady state at 30 m/s compared with cold static conditions. Provided the Turner number is less than about 0.8, speed does not normally play a primary role in tyre handling effects, although there are some secondary effects through visco-elasticity, changing of the tension in the carcase, and changes of the friction of the tread.

For a static or freely rolling tyre with a perfectly compliant tread and carcase, the footprint contact pressure would equal the inflation pressure (plus a negligible tread weight contribution). This first estimate is a better approximation for the radial-ply than for the bias-ply tyre. For a radial tyre, the normal force distribution along the footprint is correspondingly roughly constant. For the bias-ply tyre it is trapezoidal, some one-quarter of the footprint length being required for the increase, and one-quarter for the decrease. The normal force is slightly greater over the front half, associated with the rolling resistance. In tyre modelling, rectangular, parabolic and elliptical distributions have been used as analytical models. The longitudinal shear force distribution, associated with tread shear strains, has small fore–aft asymmetries that are manifestations of the rolling resistance. The lateral distribution is non-uniform, especially for the bias-ply type, i.e. separate thin strips parallel to the centre-line can have longitudinal force distributions differing markedly from each other. Even for a freely rolling tyre, the average longitudinal shear stress at the road surface is typically 30 kPa, and peak values often exceed 100 kPa. These are significant in the context of a mean contact pressure of about 200 kPa. The normal pressure distribution is, in fact, similarly complex, peaking at typically 800 kPa (five times the inflation pressure) for bias ply, and 300 kPa for radial ply. Lateral shear stresses are likely to exceed 400 kPa locally. These figures, generally more extreme for bias-ply than radial, suggest that some local sliding will occur even for a freely rolling tyre, and are the principal reason for the radial-ply type exhibiting less wear and lower rolling resistance.

The rolling resistance coefficient μ_R is defined by the equation

$$F_R = \mu_R F_V$$

where F_R is the rolling resistance force and F_V is the vertical force. The total rolling resistance coefficient ranges from 0.01 to 0.025 for cars. For trucks the range is somewhat lower at 0.005 to 0.012, because of higher inflation pressures and harder tread compounds. The total rolling resistance of a good tyre is apportioned typically as 90% material hysteresis, 8% surface friction, and 2% air friction, at moderate speed.

2.8 Tyre models

In vehicle handling analysis, the tyres are generally represented by empirical models; these are non-phenomenological, that is, the tyre behaviour is not derived from the material properties and structure of the tyre. The material of a tyre is a multi-layered, non-uniform, anisotropic, cord–rubber composite, so to understand tyre behaviour there is a pressing need for simplification. There are three principal models used to understand tyre forces and deflections in cornering and to give some insight into footprint behaviour: the elastic foundation, the string, and the beam.

In the elastic foundation model, each small element of the contact patch surface is considered to act independently; if forced by the ground it can be displaced from its null position relative to the foundation and resists with a given stiffness. Figure 2.8.1(a) represents a plan view of a tyre during cornering, showing the lateral deflection of the tyre centre-line in the footprint. Figure 2.8.1(b) is a representation of this, with the various elements each constrained by a foundation stiffness spring, attempting to restore the element to its central position. This is the simplest model. In the string model, lateral displacement of each element is also resisted by tension between the elements, because of changes in the displacement slope. In the beam model, each element exerts bending moments on its neighbours. The foundation stiffness model allows a discontinuous distribution of displacement and slope of the centre-line. The string model allows discontinuous changes of slope, but not of deflection. The beam model does not allow discontinuities of either. The string and beam models are sometimes combined. None of these models reflects directly, in a physical sense, the true complexity of a real tyre, and the various stiffness values are selected empirically to obtain realistic results. For radial-ply or belted bias-ply tyres, the beam model is generally found to be superior. The implication is of course that the bending stiffness is largely associated with the belt.

Figure 2.8.1. (a) Tyre displacement, (b) foundation stiffness model.

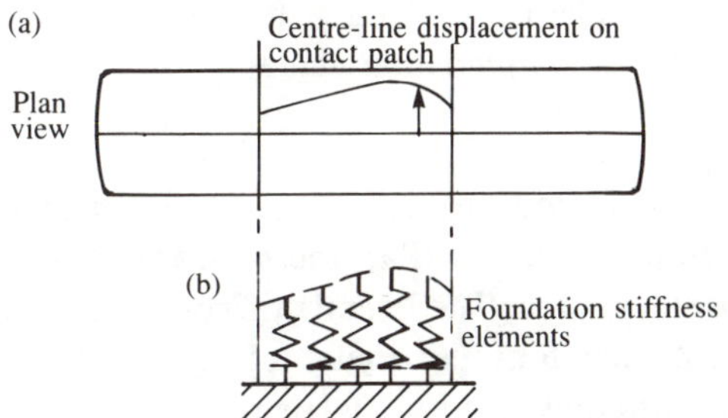

There have been various attempts to use finite-element techniques to model tyres. These do go some way towards predicting tyre characteristics, but they are generally unsuitable for vehicle handling simulations because of their complexity, i.e. because of the large number of parameters or degrees of freedom, which means that the program is too slow, except perhaps for the very fastest computers.

Even the simplest model, the elastic foundation alone, produces many of the interesting characteristics of a real tyre, and this model will be used here to ‘predict’ various properties of tyres and to illustrate frictional behaviour in the footprint. For simplicity, a displacement graph can be constructed to show only the tyre centre-line, not the whole tread width; Figure 2.8.2 illustrates this, with a view down through a ‘transparent’ idealised tyre with foundation stiffness only, and its corresponding centre-line displacement graph. Because of the foundation stiffness, this distortion of the tyre requires a force to be exerted on the tyre by the road; this is the lateral force to be determined.

Figure 2.8.2. Model tyre and displacement.

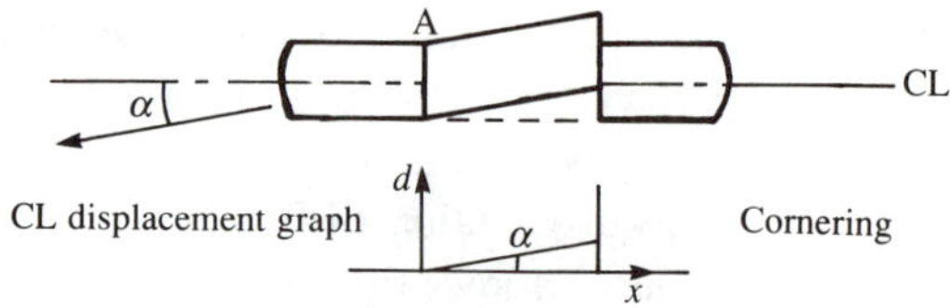

Figure 2.8.3. Model tyre centre-line displacement.

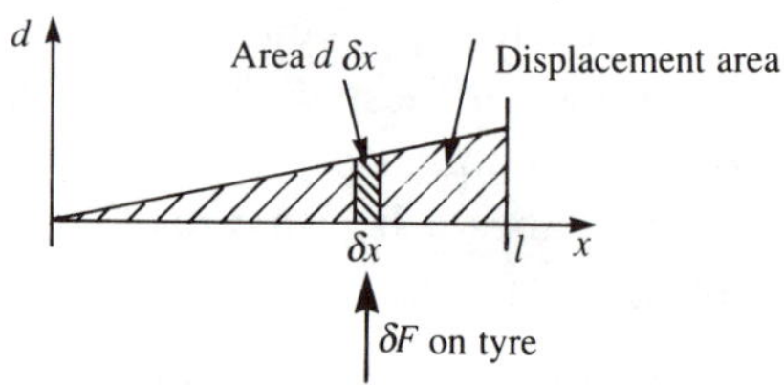

In Figure 2.8.3, the element of length δx and centre-line displacement d, according to the elastic foundation model, gives a force on the tyre of

$$\delta F = cd\,\delta x$$

where c is the foundation stiffness. Hence the effective foundation stiffness c has units of pascals (Pa), the total force being the stiffness times the displacement area. The lateral stiffness of a particular element of length δx is $c\,\delta x$ newtons per metre of lateral displacement. The lateral force concentration, in newtons/metre, is

$$\frac{\mathrm{d}F}{\mathrm{d}x} = cd$$

For any given distribution of centre-line displacement, the total force magnitude is

$$F = \int_0^l cd\,\mathrm{d}x$$

and the moment about the contact patch centre-point is

$$M = \int_0^l cd\,(x - \tfrac{1}{2}l)\,\mathrm{d}x$$

For realistic results, the foundation stiffness c must be given a suitable empirical value, not simply that measured from the actual overall lateral stiffness of the tyre tread relative to the wheel.

The purpose of this exercise is simply to give some physical insight into tyre characteristics, rather than to produce accurate real characteristics. However, with suitable choice of the effective foundation stiffness c and the friction coefficient, quite realistic performance curves are produced. For accurate numerical handling predictions it is necessary to measure the actual tyre characteristics.

2.9 Slip angle and cornering force

An example variation of cornering force with slip angle was shown in Figure 1.1.2. It is characterised by an initial approximately linear region with force proportional to angle, and a final friction-limited value. This section will show how such a characteristic results from the simple elastic foundation model.

Figure 2.9.1. Model tyre centre-line displacement: (a) small slip angle, (b) larger slip angle.

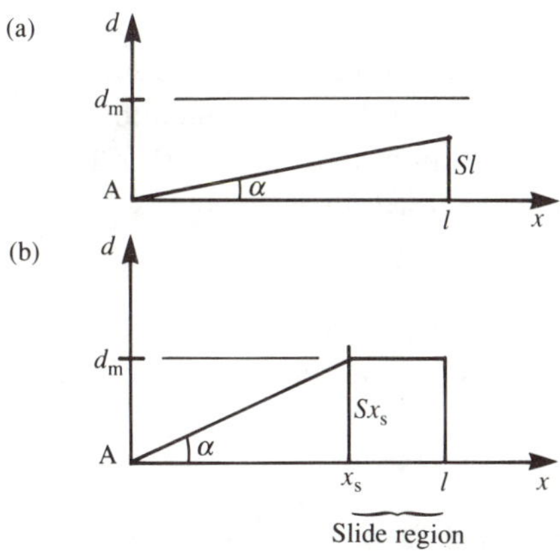

For a small slip angle, the centre-line deviation is that of Figure 2.9.1(a). The tyre contacts the ground at A. The tread is supported by the ground, which exerts the vertical force F_V. The normal force concentration, considered to be distributed uniformly, is F_V/l (newtons/metre). To avoid sliding, the highest lateral force concentration allowable is $\mu F_V/l$. The previous section showed that the lateral force concentration is cd where c is the effective foundation stiffness. Hence, for no sliding,

$$cd \leq \frac{\mu F_V}{l}$$

at all points along the footprint, so the maximum possible displacement for no sliding is therefore

$$d_M = \frac{\mu F_V}{cl}$$

Denoting the lateral slip as S,

$$S = \tan \alpha$$

For small slip angles, with α in radians,

$$S \approx \alpha$$

For small α and S, the largest d is at $x = l$, Figure 2.9.1(a), where

$$d = l \tan \alpha = Sl$$

and the condition for no sliding is that d must not exceed d_M, giving

$$S \le d_M/l$$

Substituting $d_M = \mu F_V/cl$ gives

$$S \le \frac{\mu F_V}{cl^2} = S_1$$

Realistic figures are: contact patch length 180 mm, normal force 5 kN, friction coefficient 1.0, and effective foundation stiffness 3 MPa. This gives a slide-free maximum slip $S_1 = 0.051$ ($\alpha_1 = 2.95°$). The corresponding maximum tread displacement is

$$d_M = \mu F_V/cl = 9.3 \text{ mm.}$$

In this small slip angle regime, the triangular displacement graph as in Figure 2.9.1(a) is valid. The total force magnitude is

$$F_Y = \int_0^l cd \, \mathrm{d}x$$

This is just the stiffness times the area of the displacement diagram,

$$F_Y = \tfrac{1}{2}cl^2S$$

so the force is proportional to the slip. For small α, with α in radians

$$F_Y \approx \tfrac{1}{2}cl^2\alpha$$

The cornering stiffness is

$$C_\alpha = \frac{\mathrm{d}F_Y}{\mathrm{d}\alpha} = \tfrac{1}{2}cl^2 \quad \text{(per radian)}$$

The maximum non-slide force is

$$\tfrac{1}{2}cd_M l = \tfrac{1}{2}c\, l\,(\mu F_V/cl) = \tfrac{1}{2}\mu F_V$$

Thus the model predicts a lateral force proportional to slip angle, for a force up to half of the maximum total friction limit force μF_V. For the example values, the cornering stiffness is 49 kN/rad (848 N/deg) and the maximum non-slide force is 1500 N.

The force acts through the triangle centroid, a distance $2l/3$ from the first contact, and therefore $l/6$ behind the mid-point. Hence the self-aligning torque is

$$M_Z = \frac{F_Y l}{6} = \frac{cl^3 S}{12}$$

which is proportional to the slip. The pneumatic trail (the moment arm) is

$$t = \frac{M_Z}{F_Y} = \frac{l}{6}$$

which is constant. In practice rather larger values are usually found at small slip angles, t/l being typically 0.2 to 0.35.

It is important to appreciate that the relationship between lateral force and slip, i.e. the slope of the F against S (or α) curve, depends not on the friction coefficient but on the foundation stiffness, i.e. the tyre compliance which is due to the carcase and tread compliance. The available friction does however define the limit of the non-slide linear operating regime.

If the slip angle is increased further, then there will no longer be sufficient friction to keep the rear of the contact patch fixed on the road. Considering a simple model in which the friction coefficient is independent of sliding speed, then at larger slip angles the rear of the contact patch will reach the friction limit displacement, Figure 2.9.1(b), and remain there until lifted from the road at $x = l$. The position at which sliding begins is denoted by x_s.

$$S = \tan\alpha = d_M / x_s$$

$$d_M = \frac{\mu F_V}{cl}$$

$$x_s = \frac{d_M}{S} = \frac{\mu F_V}{clS}$$

The lateral force is the stiffness times the displacement area, integrated from the displacement graph:

$$\begin{aligned} F_Y &= \int cd\,\mathrm{d}x \\ &= \tfrac{1}{2} c d_M x_s + c d_M (l - x_s) \\ &= c d_M (l - \tfrac{1}{2} x_s) \end{aligned}$$

Substituting $x_s = \mu F_V/clS$ and $d_M = \mu F_V/cl$ gives

$$F_Y = \mu F_V - \frac{\mu^2 F_V^2}{2cl^2 S}$$

Hence, as the lateral slip S continues to increase beyond initiation of sliding at the rear of the footprint, the lateral force approaches the limit μF_V asymptotically. Since $S = \tan \alpha$, the ultimate limit force is actually reached at a slip angle of 90° when the tyre is sliding fully sideways. Figure 2.9.2(a) shows the graph of lateral force against slip angle for the example values. Real tyres do indeed have a characteristic of this nature, as may be seen by comparison with Figure 1.1.2.

The self-aligning torque after sliding begins is:

$$M_Z = \int_0^l cd(x - \tfrac{1}{2}l)\,\mathrm{d}x$$

$$= \frac{cld_M x_s}{4} - \frac{cd_M x_s^2}{6}$$

Substituting $x_s = d_M/S$ gives

$$M_Z = \frac{cld_M^2}{4S} - \frac{cd_M^3}{6S^2}$$

Thus for a slip angle greater than the non-slide limit, the self-aligning torque eventually reduces. Substituting for d_M gives

$$M_Z = \frac{\mu^2 F_V^2}{4clS} - \frac{\mu^3 F_V^3}{6c^2 l^3 S^2}$$

The pneumatic trail is:

$$t = \frac{M_Z}{F_Y}$$

$$= \frac{3\mu F_V c l^2 S - 2\mu^2 F_V^2}{12c^2 l^3 S^2 - 6\mu F_V clS}$$

Physically, from Figure 2.9.1(b), we can see that increasing the slip angle reduces the length x_s, and therefore the pneumatic trail reduces once sliding begins at the rear, and goes to zero as the slip angle approaches 90°. Hence the self-aligning torque reduces even though the force increases. Figure 2.9.2(b) shows the self-aligning torque variation for this model. The reduction of aligning moment and hence of steering-wheel torque as the lateral force approaches its limit is a valuable form of feedback to the driver. It is especially important in giving warning of a poor friction surface.

The general form of the curves of both lateral force and self-aligning torque derived from this simple model are in accordance with experimental data, although in the case of experimental self-aligning moment negative values are often observed at high slip. This arises because of a decline of friction with sliding speed, because of sideways shifting of the rolling resistance caused by carcase distortion, and because of the fore–aft asymmetry of the vertical force distribution which allows greater friction in the front half.

Figure 2.9.2. (a) Tyre cornering force against slip angle, (b) Tyre self-aligning torque against slip angle.

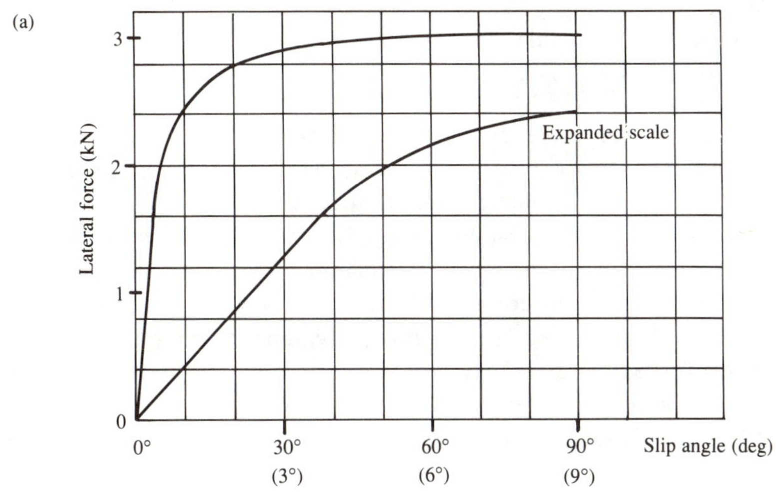

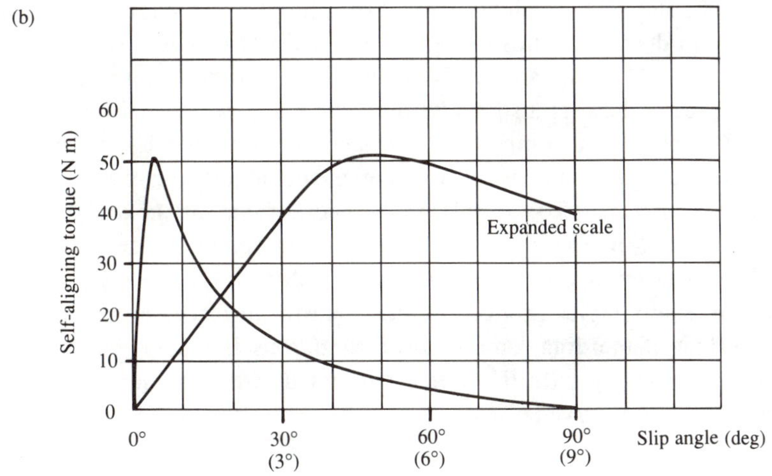

Since $\alpha = \arctan S$, the graphs of force and moment are not strictly linear even at first, although this error is negligible because of the small angles obtaining in this region. The initial slope of the force curve is denoted the cornering stiffness and given the symbol C_α. Synonyms for cornering stiffness are cornering rate and cornering power, but these terms are obsolete, and will not be used here. A typical value of cornering stiffness for a car tyre is 50 kN/rad (870 N/deg). To improve comparability between tyres of different load ratings it is often convenient to divide the cornering stiffness by the normal force F_V to give the cornering (stiffness) coefficient C_S (subscript S for slip angle), which has a typical value of 0.16/deg or 10/rad for radial-ply, and some 20% less for bias-ply.

For small α (non-slide regime),

$$F_Y = F_\alpha = C_\alpha \alpha = C_S F_V \alpha$$

The meaning of these equations is clear enough for experimental conditions, but in application to vehicle handling the vertical force is itself a variable, and some caution must be exercised.

It is of course essential to distinguish carefully between cornering force, cornering-force coefficient, cornering stiffness, cornering-stiffness coefficient, maximum cornering force, and maximum cornering-force coefficient, Figure 2.9.3. (In addition, if camber is not zero, there are camber force, camber-force coefficient, camber stiffness, camber-stiffness coefficient, maximum lateral force and maximum lateral force coefficient, Section 2.12.) In every case the word 'coefficient' is added where the variable has been normalised against F_V.

The force that provides the vehicle's centripetal acceleration is not actually the tyre cornering or lateral force. It is the tyre force perpendicular to the vehicle path; depending on the angle of the tyre relative to the vehicle this is approximately the tyre central force, Figure 2.9.4. Note that, for an undriven wheel, in addition to the desirable central force there is also a tyre drag force F_D, which will reduce the vehicle's speed. If compensating forces are provided by the engine, then the drag force may be overcome and a positive tractive force created, Figure 2.4.4(b). Resolving the lateral force into central and drag forces results in Figure 2.9.5. Note that the central force has a well defined maximum value, at a slip angle of 20° for this model, and that excessive slip angles produce inferior centripetal forces on the vehicle. The effect of the drag force in these conditions is readily observed by cornering sharply with the clutch depressed, when a marked loss of speed will be observed.

Figure 2.9.3. (a) F_Y against slip angle, (b) F_Y/F_V against slip angle.

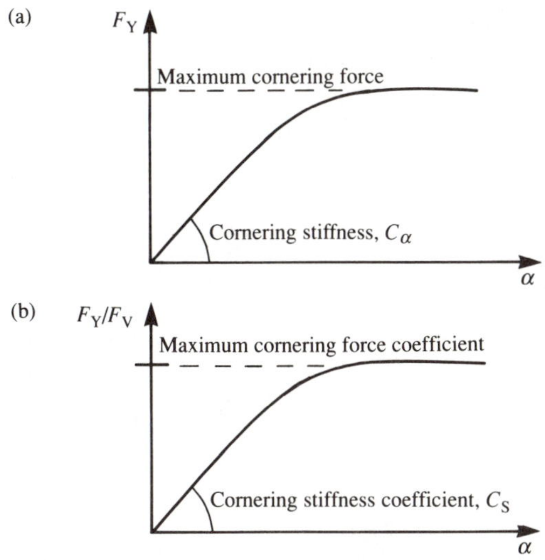

Figure 2.9.4. Force components for undriven tyre.

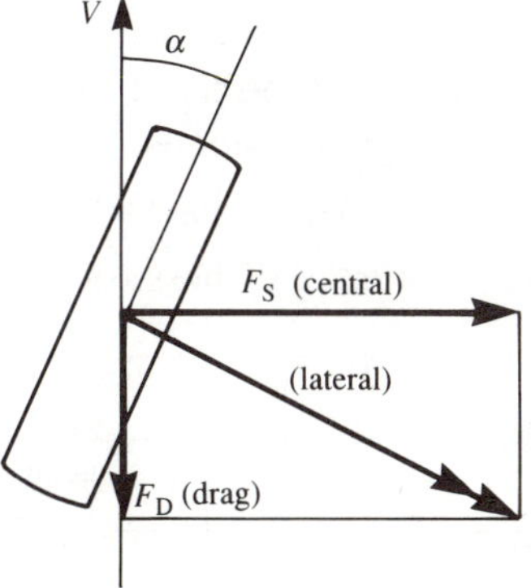

The central/drag force ratio is analogous to the lift/drag ratio for a wing. Neglecting rolling resistance this is simply cot α, which reduces continuously with slip angle. Higher cornering stiffness therefore has merit, especially for competition vehicles, because a given central force will be achieved at smaller slip angle, and therefore with a smaller tyre drag force. Including rolling resistance gives the central/drag force ratio, within the range of linear cornering force, as

$$\frac{F_S}{F_D} = \frac{C_\alpha \alpha \cos\alpha}{\mu_R F_V + C_\alpha \alpha \sin\alpha}$$

Figure 2.9.5. Tyre central and drag force against slip angle.

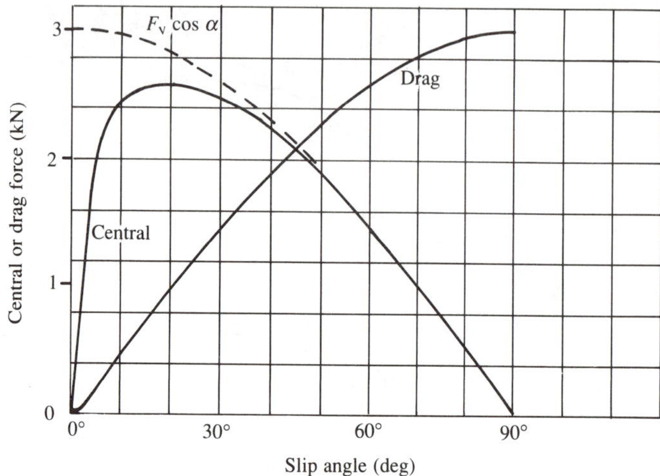

For realistic values this typically peaks at a value between 10 and 12, at 2° to 3° slip angle.

Figure 2.9.6 shows lateral force plotted against self-aligning torque for the simple foundation model; this is known as the Gough plot, and is of particular interest with regard to feel at the steering wheel. For real tyres, because the self-aligning moment goes negative, the curve actually goes off to the left beyond a slip angle of 20°.

Figure 2.9.6. Gough plot for simple model tyre.

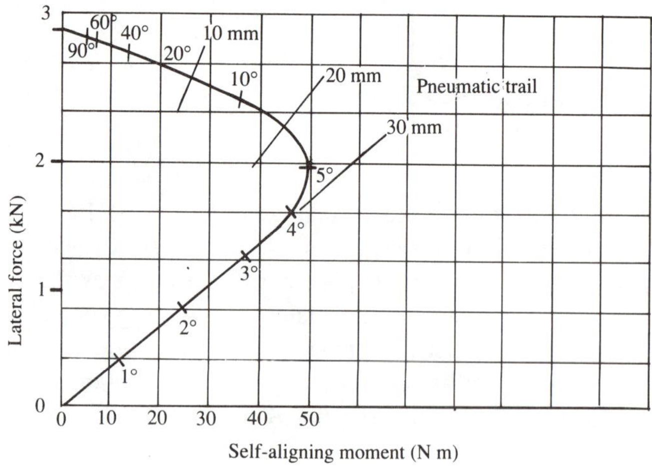

Cornering force causes tyre distortions that give an overturning moment. This can be conceived as a lateral shift y of the vertical force position, where

$$y = \frac{M_X}{F_V}$$

This has significant effects on lateral load transfer.

2.10 Non-dimensionalisation

It is often helpful to non-dimensionalise the tyre characteristics. The cornering force is easily non-dimensionalised by dividing by F_{Ymax}, in which case the force coefficient is

$$C_Y = \frac{F_Y}{F_{Ymax}} = \frac{F_Y}{\mu_C F_V} \qquad \text{(abnormal definition)}$$

When the lateral force has been non-dimensionalised like this, all graphs have a maximum value of 1.0. Alternatively, and more usually, the forces are non-dimensionalised by dividing by F_V. The cornering force coefficient is then

$$C_Y = \frac{F_Y}{F_V}$$

The maximum value is then the maximum cornering force coefficient μ_C.

The slip can be non-dimensionalised by defining a characteristic slip S^* (and characteristic slip angle α^*), and then plotting forces against S/S^* or α/α^*. The characteristic slip angle is the angle at which the tangent from the origin meets the lateral force value $F_{Y\,max}$. Thus

$$S^* = \frac{F_{Y\,max}}{C_\alpha} = \frac{\mu_C}{C_S}$$

which for this model is

$$S^* = \frac{2\mu F_V}{cl^2}$$

It has a typical value of 6° or 0.1 rad. Thus the slip coefficient (non-dimensionalised slip) is

$$s = \frac{S}{S^*}$$

In terms of angles,

$$\alpha^* = \arctan S^*$$

$$s = \frac{\tan \alpha}{\tan \alpha^*}$$

The non-dimensional slip angle is

$$\theta = \arctan s$$

$$\approx \frac{\alpha}{\alpha^*}$$

Now, for the simple tyre model, for no sliding ($S < S_1$)

$$s < \tfrac{1}{2}$$

The sideforce is

$$F_Y = \tfrac{1}{2}cl^2S$$

The sideforce coefficient is

$$C_Y = \frac{F_Y}{F_V} = \left(\frac{cl^2}{2\mu F_V}\right)\mu S$$

$$= \frac{\mu S}{S^*}$$

$$= \mu s$$

which introduces a rewarding simplicity.

Defining a self-aligning moment coefficient C_{MZ} as (not non-dimensional, but the usual definition),

$$C_{MZ} = \frac{M_Z}{F_V}$$

$$= \frac{cl^3S}{12F_V}$$

$$= \left(\frac{cl^2}{2\mu F_V}\right)\frac{\mu lS}{6}$$

$$= \frac{\mu lS}{6S^*}$$

$$= \frac{\mu ls}{6}$$

The non-dimensional pneumatic trail is t/l, which for this model is simply

$$\frac{t}{l} = \frac{1}{6}$$

Beyond the point of sliding

$$C_Y = \mu\left(1 - \frac{S^*}{4S}\right)$$

$$= \mu\left(1 - \frac{1}{4s}\right)$$

$$C_{MZ} = \mu l\left(\frac{S^*}{8S} - \frac{S^{*2}}{24S^2}\right)$$

$$= \mu l\left(\frac{1}{8s} - \frac{1}{24s^2}\right)$$

$$\frac{t}{l} = \frac{3s-1}{24s^2-6s}$$

Evidently, by using non-dimensionalisation, the equations become more manageable. More importantly, all the model tyres now have the same characteristic shape, with an initial gradient of 1 and maximum value of μ_C. Consequently, any particular example can be fully represented by only two parameters – μ_C and S^* – which define the actual scales of the two axes.

The process of non-dimensionalisation can usefully be applied to real tyre experimental data. Usually the cornering force is simply non-dimensionalised against vertical force, so peak values vary somewhat, equalling the maximum cornering force coefficient μ_C. There remain some other differences. The graph curvature in the transition region (slip around 6°) is the main one, radial-ply generally having a sharper knee to the curve than bias-ply, giving less progressive final handling. Also, there may be a decline of cornering force beyond a peak, especially at high speed.

2.11 Improved friction model

Tyre experiments show that friction coefficient depends on speed, but also depends on skid duration preceding measurement at a given speed. As described in Section 2.2, the coefficient of friction of any particular rubber on a given surface depends upon the sliding speed and temperature. Once the rear of the footprint is sliding, the rubber temperature will rise in a way that is rather hard to quantify. However the practical consequence is that there is a reduction of frictional coefficient because the temperature exceeds the optimum. As shown in Figure 2.3.3, the friction value actually rises and then falls with sliding speed. For realistic tyres and sliding speeds, friction is past the peak and therefore reduces with speed.

There are four commonly used analytic friction–speed models:

(1) Constant μ

(2) Different static and dynamic values μ_S and μ_D

(3) $\mu_D = \mu_S(1 - KV)$

(4) $\mu_D = \mu_S\, e^{-V/V_1}$

In model (2) $\mu_D/\mu_S \approx 0.8$. In model (3) $K \approx 0.012$ s/m. In model (4) $V_1 \approx$ 60 m/s. The low-speed friction coefficient μ_S can be as high as 1.2 for standard car tyres, but is much less for trucks, Figure 2.11.1, which shows results for locked-wheel braking tests. The lower value for trucks is due to harder tread compounds for low wear, and to high contact pressures. The high values for racing tyres are achieved by using special compounds and low contact pressures. The values are only roughly indicative, and may vary substantially according to the particular tyre, vertical force, etc.

Figure 2.11.1. Friction against speed for locked-wheel braking.

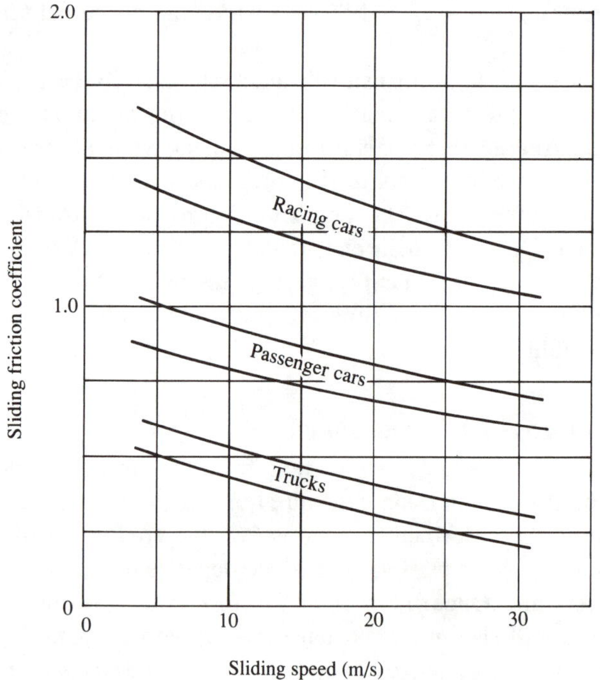

Although the more complex models are needed for accurate analysis, model (2) is sufficient to understand the effect of speed dependence of friction by defining one static coefficient of friction applicable to and limiting the non-sliding part of the footprint, plus a lower dynamic coefficient of friction applicable to the sliding part of the footprint.

The consequences of this on cornering force can be seen fairly easily. At high slip angles the sliding part of the footprint, at the rear, will have a reduced friction coefficient, so the cornering force will peak at a moderate value of slip angle and then decline. However, the decline of maximum cornering force with speed will be less dramatic than that of locked wheel braking, because the actual relative sliding speed is only $V \sin \alpha$, and the wheel continues to rotate so the local temperature rise is less. The reduced force behind the footprint centre contributes to the negative self-aligning moments that are observed in practice.

2.12 Camber angle and camber force

Camber is the inclination of the tyre in front or rear view, Figure 2.4.2(b). The S.A.E. defines positive rotation about the X-axis, as shown, as the inclination angle and defines camber as the modulus of the inclination, with a positive value if the top of the wheel is outward from the vehicle centre-line. Single-track two-wheeled vehicles operate at extreme camber angles in cornering; four-wheeled vehicles usually have angles less than 10°. Figure 2.12.1 represents a view down through a transparent cambered wheel showing the tread area distortion caused by the camber. The consequence is a lateral force. There is also a camber aligning moment about the Z-axis (associated with a camber trail), but this is small and usually neglected.

Figure 2.12.1. Tyre camber distortion.

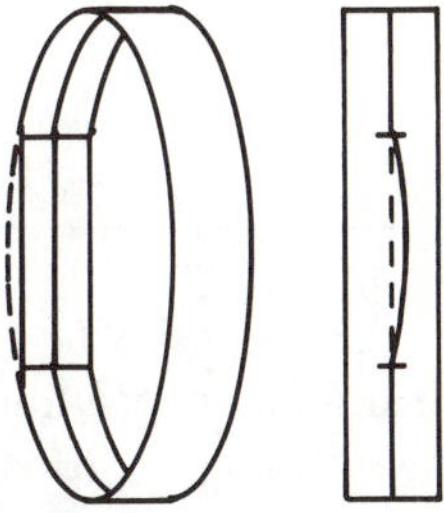

The force is called the camber force, sometimes the camber thrust (obsolescent); since camber angles are fairly small, friction effects are generally secondary, and the camber force is a function of the stiffness properties of the tyre, of the camber angle and of the vertical force. It acts towards the low axis side, and is typically found to be proportional to the camber angle; hence the camber stiffness C_γ is defined as the rate of change of camber force with camber angle. It has a typical value for a modern radial car tyre of 35 N/deg (2 kN/rad). The camber force is approximately proportional to the vertical force, so dividing by the vertical force gives the approximately constant camber stiffness coefficient C_C. (The analogous cornering stiffness coefficient was denoted C_S.) Hence, for zero slip angle

$$F_Y = F_\gamma = C_\gamma \gamma = C_C F_V \gamma$$

The value of C_C is typically 1.0 for a bias-ply tyre, and about 0.4 for a radial-ply tyre. Hence in the case of a bias-ply tyre, the normal force plus camber force acts approximately up the tyre centre-line, which is desirable for a two-wheeled vehicle. Radial-ply tyres for motor-cycles are given a large section radius to move F_V laterally to compensate for the low C_C.

The camber angles found on four-wheeled vehicles are typically 0 to 1° in the static position, increasing to a limit of ±10°, usually less, under extreme cornering. The forces caused by camber angle are therefore generally less than those caused by slip angle. However they can have a significant effect on handling behaviour because of camber changes during suspension movement, which may be different front and rear; for example many vehicles have independent suspension at the front with an axle at the rear.

Early tyre force experiments were interpreted as implying that camber angle affected the slip-angle cornering stiffness. However it is now generally accepted that for car analysis, for small angles, the two should be treated as independent, i.e.

$$\begin{aligned} F_Y &= F_\alpha + F_\gamma \\ &= C_\alpha \alpha + C_\gamma \gamma \\ &= C_S F_V \alpha + C_C F_V \gamma \\ &= F_V (C_S \alpha + C_C \gamma) \end{aligned}$$

The camber coefficient diminishes for large slip angles. This is necessarily the case because at high slip angles, near to peak F_Y, camber has only a relatively small effect on F_Y. Some experiments indicate that C_γ reduces significantly with speed beyond about 30 m/s. The effect of camber on force generation at high slip angles is discussed under tyre experimental data.

Lateral force is cornering force (from α) plus camber force (from γ). This is clear for small angles, but the distinction ceases to be clear in the non-linear regime, where the simple summation is no longer valid. At large camber angles, as achieved on motorcycles, the cornering stiffness is reduced significantly. The reduction in C_α is generally in the range 0.3 to 0.7% per degree of camber angle, and is virtually linear up to very large camber (60°).

The path curvature $\rho = 1/R$ also has an effect on tyre lateral force. The contribution to F_Y from path curvature alone, in the linear region, is

$$F_Y = F_\rho = C_\rho \rho$$

where C_ρ is the path curvature stiffness. Now a tyre of loaded radius R_l with camber γ rolling freely moves approximately as a cone following a path radius R, where

$$\gamma \approx \sin\gamma = \frac{R_l}{R} = \rho R_l$$

Hence

$$F_Y = C_\gamma \gamma + C_\rho \rho = 0$$

$$C_\rho = -\frac{\gamma}{\rho} C_\gamma$$

$$= -R_l C_\gamma$$

A representative value of C_ρ is 600 N m. The effect of path curvature on tyre force is usually neglected in ordinary handling problems, although it can be important in steering vibration problems.

2.13 Experimental measurements

The complexity of tyre construction, and the notorious variability of friction, means that it is not at present possible to predict tyre performance very successfully by direct theoretical means. However, the simple models already examined show that there are two principal parameters of tyre behaviour: the cornering stiffness coefficient and the maximum cornering-force coefficient. The former governs behaviour in normal driving, and depends primarily on the stiffness properties of the tyre. The latter governs the behaviour during extreme driving, and depends primarily upon the frictional properties of the tread on the relevant road surface. Even if conditions are restricted to dry hard road surfaces, the friction forces are rather variable. There are relatively few published measurements of maximum forces, although fairly extensive investigations have been made of the cornering stiffness and camber stiffness coefficients.

There are various possible ways to measure tyre characteristics. The most convenient is to operate the tyre on a drum, typically of 4 m diameter. Such results are good for comparative purposes, but must be treated with some caution because the drum curvature influences the results, especially the self-aligning torque. The inaccuracies are generally proportionately worse at small slip angles. Adjustments of the order of 10% can be made to correct the cornering force data to flat conditions although these are not really reliable; the cornering stiffness coefficient needs to be increased according to the tyre and drum diameters:

$$C_\alpha = C_{\alpha d}(1 + 0.4D_t/D_d)$$

Measurements have also been made inside drums and on discs. The best laboratory arrangement is the continuous belt, supported from below by an air bearing. A remaining limitation of this method is the problem of simulating the road surface. This can be overcome by running the tyre along a suitably prepared floor, but then only rather low speeds are practical. Each method therefore has its advantages and disadvantages.

An attractive alternative to laboratory measurement is to make mobile tests on the road – typically done by cantilevering the tyre out from the rear of a lorry. Whilst offering realistic results, this method obviously has problems of environmental and surface control, and repeatability.

The results of a tyre test depend upon the usage history of the tyre, especially carcase conditioning and tread wear. Changes can be especially rapid on a new tyre during its first few runs, seemingly from two effects. Firstly, flexing of the carcase results in some stiffness changes. Secondly, a new tread surface has a mould sheen, which quickly wears off. This is a surface unrepresentative of the rubber below, and it is also likely to be contaminated with mould lubricant. These effects are quite different from those of gross tread wear. Specific tests have been performed to investigate this running-in effect on the tyre. Generally, the effects on peak sideforce, which tends to increase, are

(1) A sharp and often erratic change in the first run due to wearing of mould sheen.

(2) An upward exponential decay, with a distance constant of typically 1 km eventually resulting in a total coefficient increase of about 0.08.

(3) A steady coefficient change of about 0.015/km caused by uniform wear of the tread during initial operation at high slip angle.

On the other hand, cornering stiffness coefficient is often reduced by conditioning. The effects are:

(1) A recoverable effect because of rise of carcase temperature (–0.0007/°C).

(2) Typically about 10% reduction because of visco-elastic effects in the carcase, of which 7% is permanent, and 3% recovered when resting with a time constant of about 1000 s.

The tread is likely to be worn away completely in 10 to 20 km of hard testing, reflecting the severe operating conditions. Even after initial transient effects, the steady variation can give a total change of 20% in cornering stiffness during the life of the tyre – sufficient to have a major effect on vehicle handling.

Even when using flat-bed belt testers with agreed conditioning procedures, there remain significant differences between cornering stiffness results from different machines, of the order of 5% or more.

Carcase flexing and initial tread wear often result in a marked increase of aligning torque stiffness, of typically 30%, with a modest increase during the remainder of the life. On the road, the full carcase conditioning period corresponds to typically 1000 km.

Conditioning effects are generally inconsistent, varying between tyre types. Some investigations have found little or no statistical significance for the results in relation to the random variation between tyre samples.

2.14 Stiffness measurements

Even if a new tyre is tested against a smooth (but frictional) surface, because of minor variations in structure such as cord angles and rubber thicknesses around the perimeter, there are small variations in lateral forces as the tyre rotates. These variations have an amplitude of the order of $0.01F_V$. Also, the average forces and moments are not zero for zero angles. In fact some tyres are produced with deliberate such asymmetry, of carcase or tread, or both. The force and moment values at zero slip and camber angles are called the residual cornering force and residual aligning moment.

The average lateral force at zero slip angle results from two factors; conicity and ply steer. These are more significant for radial-ply than for bias-ply tyres. Ply-steer forces result from the angle of the belt plies; for an equal number of plies of alternate angles each side of the centre-line, the outer ply exerts the dominant effect. The direction of the ply-steer lateral force depends upon the direction of rolling. This is also true of slip angle, so ply-steer is also known as pseudo-slip. A typical ply-steer force value is equivalent to 0.3°of slip angle, or about 250 N. For a given design of tyre, the ply-steer force is a fairly constant value, with small variations between samples.

Conicity force is so-called by analogy with a rolling cone, and results in a lateral force that has the same direction whether the tyre is rolling forwards or backwards. This is also a property of camber thrust, so conicity is also known as pseudo-camber. 80% or more of conicity forces are caused by the belt being off-centre, the force sensitivity being typically 30 N/mm of belt offset. For a given design of tyre, the conicity forces occur randomly. The production tolerance is typically equivalent to 100 N lateral force.

A real tyre has both types of force together, hence exhibiting different forces according to the direction of rolling. By testing in both directions these can be resolved into conicity and ply-steer, conicity being the mean force, and ply-steer being half of the difference.

The cornering stiffness itself also varies from tyre to tyre, even amongst those nominally of the same type from a given manufacturer, with a typical standard deviation of 3%, and 6% for aligning torque stiffness. This is simply a matter of the economics of quality control.

Various manufacturers' designs of a given nominal size and type have a standard deviation of about 15% on cornering stiffness. The random variations, although not always negligible, are therefore much less than the systematic differences between designs. The main controlling factor in the cornering stiffness is the carcase structure; basically whether it is radial-ply or bias-ply, Belted bias-ply tyres have a cornering stiffness typically 5% greater than normal bias-ply; radial-ply average typically 40% stiffer than basic bias-ply. Aligning coefficients are ordered similarly, with bias-belted 15% stiffer than bias-ply, and radial 30% stiffer than bias.

Some tyres used to exhibit a reduced cornering stiffness for very small slip angles. This caused unresponsive straight-line handling, and a dead band in the steering ('wide-centre' feel). It is insignificant on most modern tyres, because of better design and quality control.

In examining the effect of tyre structure, a rather different picture emerges for camber coefficient. Belted-bias and bias are much the same, both having coefficients giving a camber thrust resulting in the total force acting approximately directly up the tyre centre-line, corresponding to a camber coefficient of 1/rad or 0.018/deg. For radial ply it is typically 0.4/rad or 0.008/deg. Since radial-ply tyres, compared with bias-ply, exhibit a high cornering coefficient and low camber coefficient, then the different carcase constructions are strongly characterised by the ratio of camber stiffness to cornering stiffness, which of course also indicates the slip angle needed to overcome the force caused by one degree of camber angle. This ratio is typically 0.15 for bias and 0.05 for radial.

Speed has been found to have some effect on force coefficients. Beyond about 8° slip angle, the frictional effects usually dominate, and high speed gives lower forces. At less than 8°, the main effect of higher speed is to stiffen the carcase, increasing the cornering forces. Plotting the cornering stiffness coefficient against the logarithm of speed, a straight line is obtained. The speed sensitivity k_v is defined as the stiffness increase resulting from a one order of magnitude (10×) speed increase; it has a value of 0.06 to 0.10. Hence

$$C_S = C_{S0}\left[\, 1 + k_v \log_{10}(V/V_0)\,\right]$$

A limited amount of data suggests that speed tends to reduce the camber coefficient, particularly beyond 30 m/s.

Lateral force coefficients also depend upon vertical load. The change of cornering stiffness coefficient with load seems to vary substantially between tyre designs. A simple model of the variation is a linear one, declining with load:

$$C_S = C_{S0}\left[1 + k_l\left(1 - \frac{F_V}{F_{V0}}\right)\right]$$

where subscript 0 is the reference point (standard load), and k_l is the load sensitivity. At zero load C_S goes to $(1 + k_l)C_{S0}$. Typically k_l is 0.6. This gives realistic characteristics provided that the loads are not too large, i.e. not approaching those for which C_S goes to zero.

Another model sometimes used expresses the cornering stiffness as

$$\frac{C_\alpha}{C_{\alpha 0}} = \left(\frac{F_V}{F_{V0}}\right)^f$$

which is equivalent to

$$\frac{C_S}{C_{S0}} = \left(\frac{F_V}{F_{V0}}\right)^{f-1}$$

The exponent f varies widely, especially with F_{V0}, but is typically 0.5.

The above models are convenient, but for computer simulations a more accurate model in most cases has the cornering stiffness coefficient declining exponentially with load:

$$C_S = C_{S1}\, e^{-k_1 (F_V / F_{V1} - 1)}$$

This is particularly good for wide load variation.

It is clear from all the above that the broad specification of a tyre structure is indicative of its properties, but that marked variations can occur because of the details. As a very rough guide, the cornering stiffness coefficient is typically 0.12/deg for bias-ply, 0.16/deg for radial-ply, and the camber stiffness coefficient is 0.018/deg for bias and

0.008/deg for radial. Camber-to-cornering stiffness ratios are typically 0.15 for bias-ply and 0.05 for radial-ply.

2.15 Stiffness – design variables

Even given a specific tyre construction, the cornering and camber stiffnesses are found to depend upon the tyre size, section height, section width, tread width, rim width, tread depth and form, inflation pressure, load, and so on. The following is offered only as an approximate guide to likely effects, since there is considerable variation between designs.

2.15.1 Design load (size)

Cornering stiffness when carrying the design load is more or less proportional to design load (i.e. the cornering coefficient is independent of size). Cornering aligning stiffness, if expressed as N m/deg, increases with size. This is reasonable because the pneumatic trail would be expected to increase in proportion to footprint size, and hence with design load and diameter. At constant load, pressure and rim width, the cornering stiffness varies typically as rim diameter to the power 0.5. The cornering aligning coefficient, normally defined as aligning moment stiffness divided by normal load (N m deg^{-1}/N = m/deg) might logically be better defined by also dividing by a linear dimension such as rim diameter. It varies typically as rim diameter to the power 0.8. Camber stiffness coefficient is roughly proportional to design load (camber coefficient independent of size). Camber aligning stiffness (as N m/deg) is rather widely scattered, but small and usually neglected. The constancy of cornering coefficient with size is of course related to the practical observation that the operation of a vehicle in terms of maximum cornering ability and of tyre slip angles does not have a first-order dependence on size.

2.15.2 Construction

More plies seem to have little effect on cornering force, but seem to reduce aligning torque (number of plies to the power –0.3). A large crown angle for the cords gives reduced cornering stiffness.

2.15.3 Section width

Cornering stiffness coefficient varies typically as section width to the power 0.3 (constant load, pressure, rim width), although if an appropriately reduced inflation pressure is used, about half of the increase is lost. The self-aligning torque is affected increasingly as slip angle increases, but typically varies with section width to the power 1.0 at 6° slip angle.

2.15.4 Section height

An increased section height implies a larger outer diameter, and typically a lower design inflation pressure for a given design load, with a longer footprint and more flexible sidewalls. The expected result would be reduced cornering stiffness.

2.15.5 Tread width and contour

Increased tread width increases the cornering stiffness slightly. A rounded contour generally increases the stiffness, possibly owing to a longer contact patch length.

2.15.6 Rim width

Increasing the rim width upon which a given tyre is mounted also incidentally increases the section width, which should be distinguished from a design change to the section width with a corresponding increase of design rim width. Cornering stiffness is roughly proportional to the rim width to the power 0.5, which therefore offers a useful degree of control – there is of course a limit to the rim width that a given tyre can safely accept.

2.15.7 Inflation pressure

At very low pressure the cornering stiffness is correspondingly small, because it depends mainly on the inflation pressure rather than on inherent carcase stiffness. Moderate increases in inflation pressure above the design value generally raise the cornering stiffness, because of an increase of carcase tension. The stiffness typically peaks at about 20% more than the design value for a pressure 70% more than design. It then reduces, probably because of contact-patch shortening. Increasing pressure usually causes a reduced aligning stiffness because of the shorter contact patch, and tends to increase the camber stiffness. Low pressures are desirable to improve the ride, but there is a corresponding loss of tyre life, increase of self-aligning torque, and increase of squeal. The loss of stiffness generally causes poorer handling and feel. The warmed-up operating pressure is typically

30 kPa higher than the cold-set value. This is a 15% gauge pressure increase, 10% absolute. The more water vapour there is in the tyre, the greater is the temperature sensitivity, so dry air is preferable.

2.15.8 Load

Cornering stiffness is zero at zero load, and initially increases in proportion to load. Loads up to 30% more than the design value may increase the cornering stiffness up to 10%, but it then peaks, and declines for larger loads. High inflation pressures, and wide rims, tend to result in an increase in stiffness sensitivity to load. This can be important because, if the cornering stiffness increases with load, the sensitivity of the vehicle handling to load changes is reduced. The cornering stiffness coefficient is greatest at 'zero' load, declining smoothly and exponentially, as discussed in the previous section. The aligning coefficient shows a continuous increase due to the lengthening contact patch. Camber coefficient usually shows a reduction away from the design load.

2.15.9 Tread depth

The introduction of tread grooves, for drainage in wet conditions, increases the lateral compliance of the tread part of the tyre, although it is too simplistic to consider the tread simply as a compliance added to the carcase. Evidently this depends to some extent upon the tread pattern, and where possible the width of circumferential ribs is at least twice the groove depth, i.e. 16 mm for a typical new groove depth of 8 mm. Wide ribs, although aiding tread stability, cause water clearance difficulties. The cornering stiffness of a bias tyre with the tread worn almost entirely away is typically 15% higher, and sometimes as much as 50% higher, than for the new tyre, the difference being pressure- and load-sensitive. Radial-ply tyres often show a reduced stiffness after wearing. The corner aligning torque coefficient generally increases with wear for all types, and may double over the tyre life.

2.16 Friction forces – design variables

The second principal variable controlling the shape of the cornering force versus slip angle curve is the maximum available force, which occurs at 90° slip angle for the simple model. The maximum cornering force is controlled to a large extent by the tread rubber material and the road surface, but other factors play a part, particularly the surface area of rubber presented to the road in relation to the load, i.e. the mean contact pressure. A harder rubber has been known to result in a higher maximum cornering force, but in a smaller maximum braking force. This cannot be simply a friction effect. Presumably it is related to the different distortions experienced by the tread in these two modes of operation. The maximum central force is influenced by both friction and stiffness characteristics because of the $\cos\alpha$ factor. In practice there is found to be a scale effect, with truck tyres exhibiting a friction coefficient typically 40% lower than that of passenger cars (Figure 2.11.1). This is probably attributable to the higher inflation pressures and to the harder tread materials used, because of service requirements.

Changes of road conditions have a major effect, the most extreme case being surface ice. However, even in dry conditions road characteristics vary through a wide range, and this is emphasised by damp conditions (wet conditions are considered later). The self-aligning torque is also affected by low friction, so the driver is able to detect this through light steering at quite low steering angles. The temperature of the road surface affects maximum friction, but in a way that depends on the road material. In one investigation, on asphalt the maximum cornering force coefficient μ_C was 0.9 at 0°C, declining linearly at 0.0025/°C to 0.8 at 40°C. On concrete, it was also 0.9 at 0°C, but increased to a peak of 1.06 at 26°C and then declined to 1.0 at 40°C, in a smooth curve.

Increase of load shows an increase of limiting force, but not in full proportion. The maximum cornering force coefficient variation may be modelled well by

$$\frac{\mu_C}{\mu_{C0}} = \left(\frac{F_V}{F_{V0}}\right)^p$$

The limited amount of actual tyre maximum force data indicates a value of $p = -0.15$ for the typical tyre maximum cornering force coefficient sensitivity to vertical force for passenger car tyres. Racing tyres seem to be more sensitive at –0.20 to –0.25. This relationship is an important one in the investigation of limit handling, for example in considering the effect of centre of mass position, or load transfer.

An increase of contact area gives a higher limiting force – this has contributed to the trend to very wide racing tyres with some tendency in the same direction for passenger cars. This suggests a maximum force coefficient rising as the 0.15 power of tyre tread width, for constant footprint length. The introduction of drainage channels reduces the contact area, and hence reduces the limiting force in dry conditions. The racing 'slick' tyre has no tread pattern at all, to give maximum contact area. The rubber area in direct contact with the road is normally less than the nominal footprint area, and also the pressure distribution is likely to be highly variable, due to the distorting effects of the forces applied by the road. Wider rims seem to help a little in this respect. There is an optimum inflation pressure that gives the most uniform footprint pressure, and this optimum inflation increases with increasing normal load. Too low an inflation pressure gives high contact pressure at the tread edges because of the carcase stiffness, whereas too high a pressure loads the centre. A tyre pressure optimum for wear in normal straight running is therefore rather low for extreme cornering because of severe load transfer in this condition. High pressures can create difficulties for the tyre in adapting to road roughness. An interesting racing development that has found application on road tyres is the depressed crown, which becomes flat once correctly inflated.

The effect of camber angle on maximum force seems to depend on the tyre crown shape. A very round profile seems to develop maximum lateral force with a large negative camber, the optimum slip angle then being quite small. A flatter crown tyre seems to be best in the 0–4° negative camber range. Possibly in some cases negative camber compensates for slip-angle distortions to help keep a flat crown on the road. The camber tolerance range seems to increase with increasing normal force, presumably because the mean contact pressure is increased, so that the proportional pressure variation is reduced.

The tyre speed can affect maximum force measurements. Firstly, this will occur if the Turner number exceeds about 0.8, because of wave effects. Normally vehicle conditions are comfortably below this, and test conditions lower still. Secondly, a speed influence may be exhibited when sliding occurs, because the friction coefficient is speed-dependent. Hence for small slip angles, where elastic effects dominate, tests can be conducted at relatively low speed, although corrections should be applied (see Section 2.14). For large slip angles this cannot be taken for granted. Fortunately the actual contact patch sliding speed, even at maximum lateral force, is usually quite small. According to the simple tyre model advanced in Section 2.9, the rear of the contact patch will slide, giving a relative speed of $V \sin\alpha$. Hence for a slip angle of 15°, at a vehicle speed of 60 m/s, the sliding speed is about 15 m/s. The effect of speed-sensitive friction on lateral force is therefore generally less than it is on locked-wheel braking. Nevertheless, low-speed tests are not entirely adequate in this respect.

A second major difficulty of all force measurements is that they are rather sensitive to the tread contour, as modified by use and wear. The idea of the performance of an engine changing as the parts wear to fit each other is a familiar one – the tyre too behaves like this. Hence a tyre conditioned by gentle driving, suddenly called upon to provide a large lateral force, will not be matched to the road in its new distorted shape, and will present a highly non-uniform pressure in the contact patch. If the demand for this new force is maintained, the tyre will wear the highly loaded regions, and hence adapt, and in due course would be expected to be able to give a higher force. In high wear-rate tests this effect is sometimes observed, but sometimes the opposite occurs. Presumably the tyre force is best assessed as the one immediately available on a tyre with a wear history typical of the relevant application. Obviously the normal tyre testing machine tends to operate at higher average slip angles than the usual driver – it would be uneconomic to do otherwise. Hence some doubt about the applicability of the results must remain.

Vertical load is important because handling in the high lateral acceleration regime causes large lateral load transfers. Figure 2.16.1 shows some example lateral force results, plotted against slip angle for various loads.

Figure 2.16.1. Tyre cornering force against slip angle.

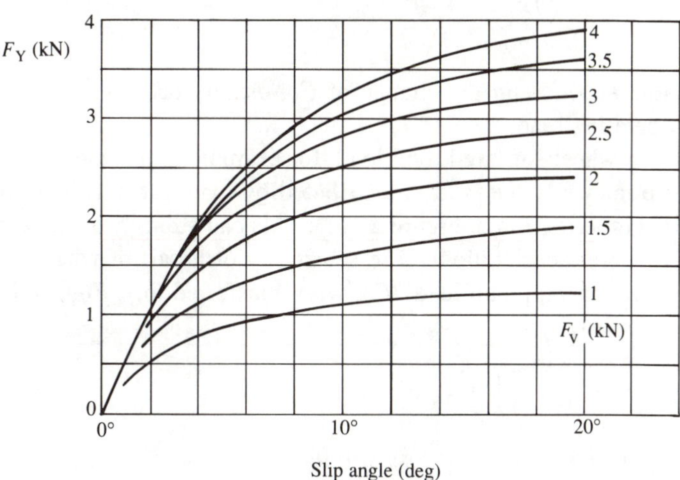

Figure 2.16.2 shows lateral force plotted against load. An increase of load at fixed slip angle does not give a proportionately greater lateral force. The rate of change of lateral force with normal force at a particular slip angle is called the load sensitivity, mathematically defined as:

$$C_{FV} = \left(\frac{\partial F_Y}{\partial F_V}\right)_\alpha$$

Figure 2.16.2. Tyre cornering force against vertical force.

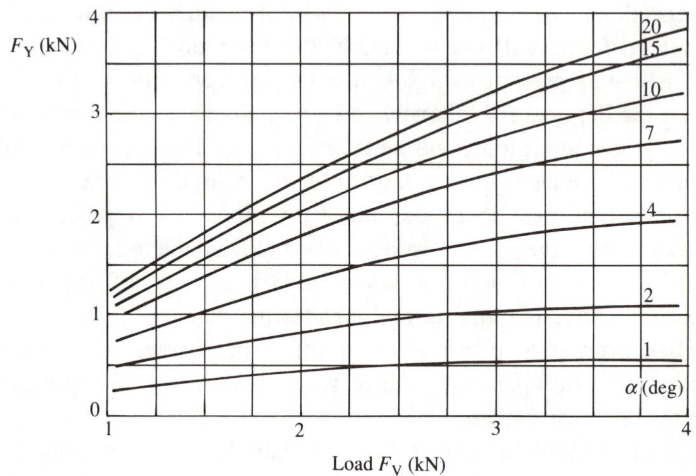

If both ∂F_Y and ∂F_V are normalised, then the result is the normalised load sensitivity:

$$C_V = \frac{\partial F_Y / F_Y}{\partial F_V / F_V} = \frac{\left(\frac{\partial F_Y}{\partial F_V}\right)_\alpha}{F_Y / F_V}$$

The value depends on the change of F_V, usually taken as 0.2 times the mean operating load.

For two wheels of fixed total load, for example at the ends of an axle, the more unevenly the load is distributed by load transfer, then the less the total cornering force, Figure 2.16.3. This can strongly affect handling at large lateral accelerations. The sensitivity of a pair of wheels to load transfer, at constant total load $2F_V$, with transferred load F_{VT}, is the load transfer sensitivity:

$$C_{FT} = \left(\frac{\partial F_{Y2}}{\partial F_{VT}}\right)_{\alpha, F_V}$$

The normalised load transfer sensitivity is

$$C_T = \frac{\left(\frac{\partial F_{Y2}}{\partial F_{VT}}\right)}{F_{Y2} / F_{VT}}$$

where F_{Y2} is the sideforce for two wheels, and F_{VT} is the transferred vertical force. Because this is a second-order effect (i.e., locally, a line curvature) it is normally evaluated for a fairly large normalised load transfer, typically $F_{VT}/F_V = 0.6$, and has a typical value of 0.2. In other

Figure 2.16.3. Effect of load transfer on axle cornering force.

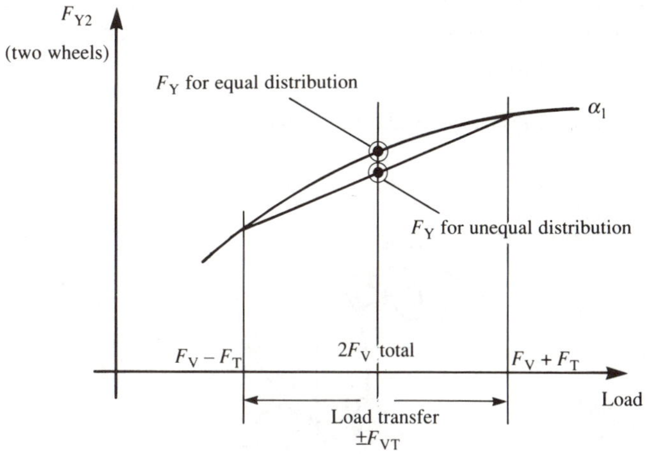

words, 60% load transfer gives 12% reduction of F_{Y2}, the reduction depending on the square of the load transfer.

An adequately realistic model of the variation of F_Y with F_V for one wheel, using the vertical force transfer factor (load transfer factor)

$$e_V = \frac{F_{VT}}{F_{V0}}$$

is

$$\frac{F_Y}{F_{Y0}} = 1 + C_1 e_V + C_2 e_V^2$$

For the two wheels of an axle this gives

$$\frac{F_{Y2}}{F_{Y20}} = 1 + C_2 e_V^2$$

Realistic values of the constants are

$$C_1 = 0.667$$

$$C_2 = C_1 - 1 = -0.333$$

2.17 Longitudinal forces

Whereas lateral forces are achieved by steering the wheels, longitudinal forces are created by applying torques about the wheel spin axis; in the latter case, unless the wheel is not rotating, there must be a corresponding power flow either providing kinetic energy (engine) or dissipating it (brakes). The S.A.E. uses 'longitudinal' to refer to tyre forces along the tyre X'-axis, Figure 2.17.1, and also for the heading vector direction of the vehicle as a whole. Driving will be used for positive forward (X'-axis) tyre forces, braking for negative (backward) forces, and longitudinal for either.

Figure 2.17.1. Forces and moments on a driven wheel.

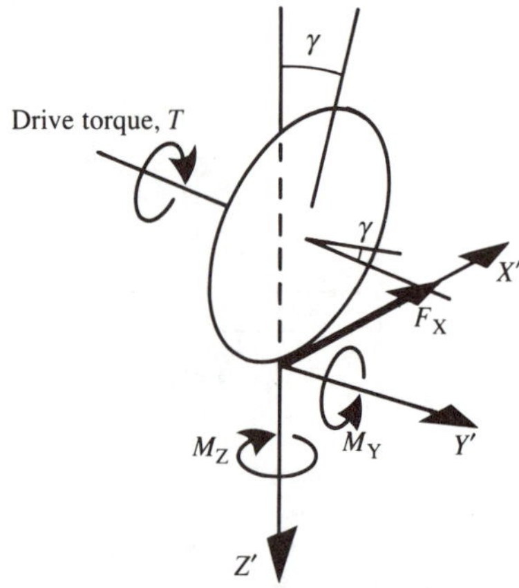

In the S.A.E. system, a positive applied torque T, from the driveshaft or brake, is defined to cause a positive longitudinal force. Figure 2.17.1 shows the corresponding wheel free-body diagram with forces. For moments about the wheel axis:

$$-T + F_X R_l + M_Y \cos\gamma + M_Z \sin\gamma = I\dot{\Omega}$$

For small camber angles, wheel accelerations and rolling resistance, then

$$T = F_X R_l$$

For forward acceleration a torque T is applied through the driveshaft; the brakes may apply a couple to the wheel, but of course the result at the tyre is dependent simply on the effective torque.

If a braking torque is applied to a wheel that is maintained at constant translational speed, then the wheel rotational speed changes slightly. This means that there is an effective difference of speed between the tyre at the contact patch and the road; ΩR_l and V respectively relative to the wheel axis. However, there is not necessarily any actual sliding. The nominal speed difference is $V - \Omega R_l$. This can be non-dimensionalised to give a longitudinal slip S:

$$S = \frac{(V - \Omega R_l)}{V} = 1 - \frac{\Omega R_l}{V}$$

The effective rolling radius is $R_e = V/\Omega$, so the slip can be expressed as

$$S = 1 - \frac{R_l}{R_e}$$

Longitudinal slip is analogous to lateral slip: a small longitudinal slip produces forces by elasticity, and a large slip results in frictional effects coming into play. Again we can introduce a foundation stiffness model, which can be pictured as short cantilever 'spokes' (the 'brush' model), which are stressed in the contact patch.

During road contact, the displacement of the wheel side of the foundation stiffness, Figure 2.17.2, is

$$l = t\Omega R$$

where t is the time that any element is in contact with the road. The displacement at the road side, for zero sliding, is

$$l' = tV$$

hence the total deflection is

$$d_1 = (V - \Omega R)t$$

and the deflection at a distance x along the contact is

$$d = \left(\frac{x}{l}\right)(V - \Omega R)t$$

Noting that $l = \Omega R t$,

$$d = \frac{xS}{(1 - S)}$$

Figure 2.17.3(a) illustrates this deflection for a non-slide condition, i.e. small slip; its similarity to the lateral displacement model is apparent. Once again we can introduce a foundation stiffness, which when multiplied by the displacement area gives the retarding force. It may seem rather anomolous that the wheel can have an angular speed not

Figure 2.17.2. Displacement of brush model.

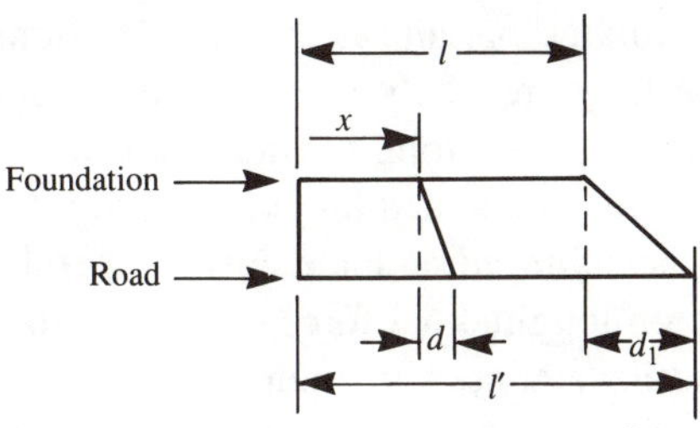

Figure 2.17.3. Longitudinal displacement: (a) small slip, (b) larger slip.

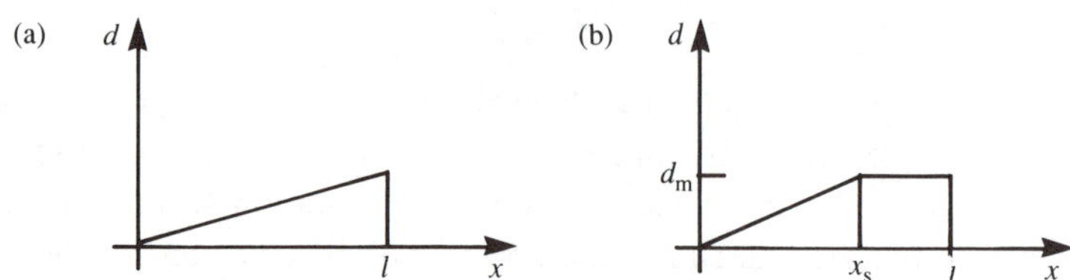

equal to V/R_1 without there being any sliding, but this is possible because the elastic strain that occurs in the contact patch can be recovered when the tread elements leave contact with the road.

When a large slip is called for, then the model predicts that because of friction limitations sliding will begin at the rear of the contact patch, the sliding area growing with slip, Figure 2.17.3(b), as in the case of lateral force generation. The force versus slip curve, according to this model, gives maximum retarding force at a slip of 1.0 (locked wheel). This model can be refined by admitting reduction of friction coefficient with sliding speed, which gives realistic results, with the retarding force reaching a maximum at a slip of about 0.15, and then declining, the rate of decline being greater for high speed.

This incidentally reveals a limitation of the use of slip. The concept of longitudinal slip arises essentially from non-dimensionalisation of the apparent speed differential at the contact patch. If we consider a locked wheel, it is evident that the slip is 1.0 regardless of the wheel translational speed, but since the rubber friction will really depend on the actual speed of sliding, the use of slip clearly has its limitations.

If a tractive force, rather than braking force, is required of the tyre, then the behaviour of the contact patch is similar, with stress and strain both adopting opposite signs from the braking case. Hence sliding will still begin from the rear of the contact patch.

2.18 Combined forces

Combined longitudinal and lateral accelerations are of practical importance because they are likely to occur during accident avoidance manoeuvres. Basically, the longitudinal and lateral forces are a combined function of the lateral and longitudinal slip. Longitudinal forces can generally only be achieved at some cost in lateral force, and lateral force exerts a price on longitudinal force. For $S = 1$ the wheel is locked, so curves of F_X and F_Y against slip angle are basically sine and cosine curves representing the resolution of a roughly constant total force into the $X'Y'$ coordinate directions. The most convenient presentation of such force combinations is probably the plot of lateral force coefficient against longitudinal force coefficient, Figure 2.18.1. This is roughly symmetrical for the radial-ply tyre, although sometimes for bias-ply it is found that braking can initially give a small increase of the cornering force. Evidently, especially in the case of the bias-ply, elastic effects in the carcase play a significant part. The dominant controlling factor is, however, the tyre-to-road friction.

Figure 2.18.1. F_Y/F_V against F_X/F_V, for various slip angles.

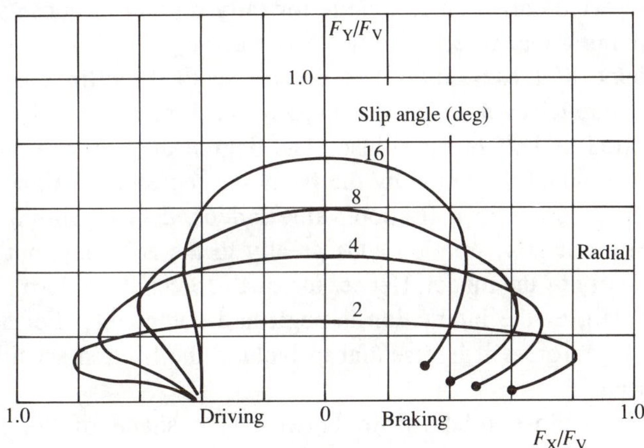

For a foundation stiffness model tyre, operating at a moderate slip angle, the contact patch is laterally stressed, especially at the rear; the lateral displacement from cornering is denoted by d_c. This corresponds to the shear stress to be resisted by friction. If a longitudinal force is also called for, there is a longitudinal (braking) strain d_b. Neglecting interactive strain effects in the carcase, which of course the simple model does not admit, increasing d_b will result in sliding at the rear when

$$d_b^2 + d_c^2 = d_M^2$$

where d_M is the maximum (friction-limited) stress. As written, this corresponds to equal foundation stiffnesses in both directions, but this could easily be generalised if required. The qualitative conclusion is evident: except at low slips in both directions, the available friction force must be shared. Increasing the longitudinal force will reduce the lateral force. To follow this implication through in detail requires consideration of the operating conditions throughout the contact patch. The force variation of Figure 2.18.1 is the practical result.

There is an asymmetry of this figure that is of interest: under driving, F_Y ultimately goes to zero, whereas under braking it does not. A wheel locked in braking will have a total force on it governed kinematically: the force will act in the direction of motion of the ground relative to the tyre. Hence the cornering force does not go to zero, but the central force does. For a wheel spinning due to gross excess power input, the limiting kinematic case of very-high-speed spinning gives the force in the direction of wheel heading rather than wheel motion. In this case the cornering force goes to zero, but there is a residual central force which may be significant at high slip angles. For example, the accepted rear-wheel-drive cornering technique for rally driving on loose surfaces is with spinning rear wheels at very large yaw angle.

The shape of these curves has given rise to the ellipse model, of attractive simplicity, in which the shape for each particular slip angle is approximated as half of an ellipse. The degree of precision available from such a model is limited by the frictional representation, since it is tantamount to adopting a friction value independent of sliding speed. Incorporation of this variation adds greatly to the accuracy, but also to the complexity of the model. Hence, for small forces the ellipse model is often adopted, or the interaction is neglected completely. For accurate results at high forces it is essential to include the friction sensitivity to sliding speed.

There is a close relationship between the shape of the lateral–longitudinal force curve and the friction–speed curve. This is perhaps best seen by considering the limiting F_Y versus F_X curve for small slip angles which lies close to the F_X axis, Figure 2.18.1. Viewed sideways, this corresponds to the shape of the force–speed friction curve for the rubber.

The self-aligning torque when cornering is usually increased by moderate braking, and decreased by driving. This is because the tyre distortion shifts the lateral force line of action, i.e. it changes the pneumatic trail.

The complexity of Figure 2.18.1, applicable to only one speed, with zero camber for one tyre type in a given state of wear on a given wheel and road surface, illustrates the problem of acquiring comprehensive data for the performance of even one tyre design.

2.19 Wet surfaces

During severe braking or cornering in damp conditions, sufficient heat is produced at the footprint for a very thin water film (i.e. dampness, 0.01 μm) to be boiled away. Even under nominally dry conditions, the atmospheric humidity will result in such a film.

Faced with a wet surface, i.e. 1 or 2 mm depth of water rather than just a damp film, the tyre must clear away the water in order to establish contact with the solid road below. Hence the tyre is provided with a tread pattern, i.e. channels, slots and sipes (cuts) to provide passages for water movement. The footprint area is considered to operate in three sections: in the first the bulk water is cleared from beneath the ribs, in the second the residual film is cleared, and in the third the tyre rubber is in good contact with the road surface.

The basic mode of operation is that the water is first squeezed sideways from beneath the tread elements either out of the side of the footprint or into the longitudinal channels, where it then passes through the footprint within the channels. Operation of the channels depends upon the depth of water ahead of the tyre, and the degree of wear of the tyre. The cross-sectional area of the channels in front view, related to the cross-sectional area of the approaching water, within the footprint width, gives a non-dimensional measure of the water drainage problem, i.e. a 'drainage' number. Provided this is less than about 1.0, the approaching water, if shifted sideways into the channels, will not fill them up. In this case there is simply a lateral flow of water. If the approaching water is too deep, and more than fills the channels, then they operate in a new role, acting genuinely as channels rather than reservoirs, allowing the water to flow relative to the tyre through the contact patch and to emerge at the rear. If the wheel is locked, the channels are obliged to operate in this mode anyway. For water approaching the tyres near to the outer edges, it may be preferable to use lateral drainage to the edge of the tyre, especially for a convex tread. A porous road surface can also contribute to footprint clearance for all tyres. The details of optimisation of tread design for drainage vary considerably between manufacturers, and according to application. Evidently a racing tyre, with a footprint aspect ratio (width/length) of three or more, may require a different solution to a conventional road tyre with footprint aspect ratio of one or less, where moving the water out from the side of the footprint can make a proportionately greater contribution.

Once the bulk drainage is essentially complete, it remains to remove the thin residual water film. Both of these operations are facilitated by high contact pressure and the presence of a good road texture. Lateral slots help to reduce resistance to fluid flow into the main channels; sipes (small cuts) aid final wiping, by producing high local contact pressures through local distortion under stress.

Aquaplaning (hydroplaning) can occur in two different ways.

Dynamic aquaplaning occurs when the dynamic pressure of the oncoming water is sufficient to support the tyre. Viscous aquaplaning occurs when the water viscosity prevents the water from being successfully squeezed from beneath the tread elements. The total time that any rubber element is in the footprint is inversely proportional to speed, and is typically 6 ms at a speed of 30 m/s (footprint length 180 mm). Hence water clearance must be achieved in only 2 or 3 ms.

In wet conditions, the influence of vehicle speed and tread depth on maximum braking force is considerable. In one investigation with 2.5 mm of water, at very low speed all tread depths had a maximum braking coefficient μ_B of 1.0, but at 20 m/s it was down to 0.77 for 8 mm tread, 0.58 for 4 mm tread and 0.20 for zero tread. At 40 m/s it was 0.48 for 8 mm, 0.22 for 4 mm and only 0.05 for zero tread. Evidently, at higher speeds much of the tyre is resting on the water.

Because the part of the footprint that contacts the road is the rear part of the usual contact area, this is equivalent to a short footprint tyre with a large mechanical trail (large castor angle). The behaviour is therefore quite unlike a low-friction ice surface. The large total trail means that aligning torques stay high so the driver does not receive the same warning through the steering as on a low-friction surface.

The relative values of track front and rear can affect handling in wet conditions. If the tracks are equal then, when running straight, the rear tyre has the benefit of a partially cleared surface. Once rounding a corner, the front–to–rear alignment changes; at an attitude angle of 6° on a large radius the off-tracking is about 300 mm, moving the rear tyres onto an uncleared surface. With narrower rear track, as on some vehicles, the rear tyres will run on the cleared surface at some particular lateral acceleration depending on the tyre cornering stiffness.

One investigation of how the maximum cornering force coefficient varied with speed for two tread designs of footprint surface solidity 60% and 80% (i.e. pattern void area 40% and 20% respectively) on a low-macrotexture surface, where the road contributes little drainage beneath the footprint, gave for speeds up to 35 m/s:

$$\mu_C = 0.62 - 0.013V \quad \text{(for 80\% solid)}$$

$$\mu_C = 0.50 - 0.005V \quad \text{(for 60\% solid)}$$

At low speed, gross drainage is easily achieved and the higher-solidity tread, offering more rubber area to the ground, achieves a higher maximum coefficient, as for dry conditions. However the low-solidity tread provides better drainage, so it deteriorates more slowly with speed. Hence there is a performance crossover, in this case at 15 m/s.

The linearity suggests that the fraction of the footprint supported by the water is proportional to speed. This is compatible with the notion that drainage rate is more or less a constant, or that, for a given tread and water depth, a fixed time is required for a tread element to make effective contact. The corresponding times calculated for the above equations are 4 ms and 2 ms. The same tyres on a good drainage surface exhibited time factors of about 1.6 ms and 0.8 ms. Evidently the road surface drainage can play as great a part as the tread itself, although a good tread is still helpful even on a good surface.

For a freely-rolling tyre, the time that a point on the tyre is in the footprint is l/V where l is the footprint length. Representing the effective clearance time constant by τ, we can define a viscous hydroplaning number

$$N_V = \frac{\tau}{l/V} = \frac{\tau V}{l}$$

which indicates the fraction of the footprint that is effectively supported off the ground. Full viscous aquaplaning would occur for $N_V = 1.0$.

Figure 2.19.1 shows how the maximum cornering coefficient varied with speed at 6 mm water depth in one investigation. Depending on detailed conditions of tyre and road, the linear relationship, arising from viscous resistance to flow, breaks down as a transition occurs to full dynamic hydroplaning, caused by inertial (dynamic pressure) effects, at 22 m/s in this case. Even in the absence of inertial hydroplaning, full separation would still occur eventually due to viscous effects. Inertial (dynamic) hydroplaning occurs when the tyre is fully supported by the dynamic pressure of oncoming water. Defining a dynamic hydroplaning number

$$N_D = \frac{\frac{1}{2}\rho_W V^2}{p_i}$$

where ρ_W is the water density and p_i is the tyre inflation pressure, then full dynamic flotation occurs at typically $N_D \approx 1.5$.

Different tread rubbers have different frictional–speed dependence in wet conditions. At low speeds the normal friction coefficient effects are dominant. However, at high speed, elastohydrodynamic lubrication effects discrimate against a soft tread surface, so the rather soft butyl rubber deteriorates more rapidly with speed than most. The performance of a blended compound is not merely an average of its constituents, so in fact butyl rubber can still play a useful role in tyre blends that must perform well in wet conditions.

Moderate water depth affects maximum cornering force rather than

Figure 2.19.1. Maximum cornering force coefficient against speed (6 mm water depth).

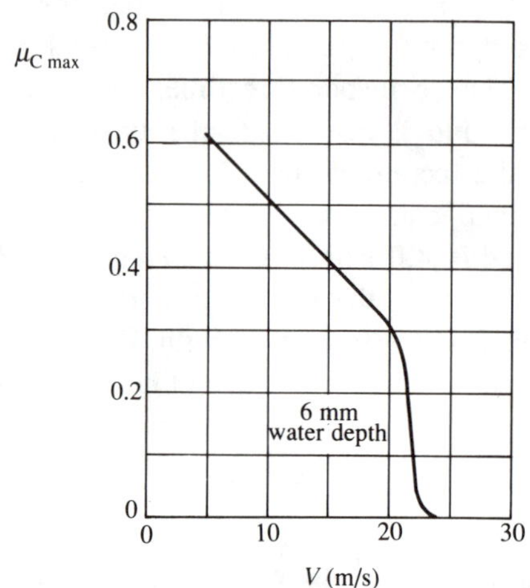

cornering stiffness. In one investigation with 1 mm of water, the very small angle cornering stiffness was independent of speed, but the maximum cornering force coefficient was 0.7 at 10 m/s, 0.5 at 20 m/s and 0.25 at 30 m/s, the optimum slip angles being 7°, 5° and 1.5° respectively.

The slight increase of cornering stiffness with speed for a dry or damp surface becomes a very marked reduction with speed when the water is deep. For example, in one investigation, at 10 m/s there was little effect, but at 20 m/s a 2 mm depth of water reduced the cornering stiffness by 20%, 5 mm by 45% and 10 mm by 70%. At the speed of full hydroplaning, the cornering stiffness goes to zero. Water depth typically affects traction and cornering friction in an equal way.

2.20 Tyre transients

If a step change is made to the slip angle of a rolling tyre, the new steady-state value of cornering force is not developed immediately, but in an exponential manner which depends on the distance rolled, i.e. the lateral force deficit from the final value decays exponentially. The relaxation distance of rolling, L, is the distance required to give $1 - 1/e$ (= 0.632) of the response change.

$$\frac{F}{F_\infty} = 1 - e^{-X/L}$$

L has a value of approximately one tyre radius. It is not very sensitive to load, and tends to decrease with increasing inflation pressure. A useful rule of thumb for assurance of approximate equilibrium conditions is a roll distance of one revolution.

If the slip angle is oscillated sinusoidally, then the amplitude of the force developed depends upon the distance travelled per cycle. For a small distance per cycle, the lateral force amplitude tends to zero; for a long distance it tends to the steady-state value. The force has a phase lag of 90° at high frequency, zero at low frequency. This can be analysed using the relaxation distance concept given above.

During a ramp input of steering angle, passing through zero slip angle, the lateral force will not be zero at the time when the slip angle is zero. For a ramp steer gradient of a (deg/s), speed V and relaxation length L, the spatial steer gradient is a/V deg/m, and the angle lag will be aL/V deg. The force lag will be $C_\alpha aL/V$ newtons. For most practical cases these are small effects, although they may be significant during severe evasion manoeuvres when the steer input may be 500 deg/s or more. To consider it another way, at any speed, neglect of the tyre transient can be expected to introduce a longitudinal path error of size equal to the relaxation length.

Subjective evaluation of handling behaviour suggests that tyre transients may be more important in influencing the perceived vehicle behaviour than would be expected from the vehicle dynamic analysis.

Analysis of the effect of road roughness requires the consideration of transient effects. Obviously the relaxation length is important in smoothing out the effect of road irregularities on cornering forces. On the other hand, a tyre making irregular road contact would relax rather rapidly when off the road, but then recover its proper lateral force generating shape only over the normal relaxation length, resulting in a reduced average cornering force. A larger tyre of lower inflation pressure often exhibits better cornering ability on knobbly surfaces, owing to improved ground contact.

2.21 Problems

Q 2.2.1 Describe the differences between radial-ply and bias-ply tyre constructions.

Q 2.2.2 Summarise the difference in handling characteristics between radial and bias-ply tyre constructions.

Q 2.2.3 Describe the features of a typical tread pattern, and explain their function.

Q 2.3.1 Describe how the visco-elastic properties of rubber may explain the friction of rubber on a lubricated but rough surface.

Q 2.4.1 Define wheel angular position and related terms, and all the wheel forces and moments, giving appropriate diagrams.

Q 2.5.1 Define and describe the various kinds of wheel radius.

Q 2.5.2 Explain the meaning of the 'vertical stiffness' of a tyre. To what extent is this a justifiable model of the tyre vertical characteristic? What parameters affect this stiffness?

Q 2.5.3 Considering a thin-walled toroidal tyre, whose stiffness is the result only of the contact patch area and inflation pressure, find an expression for the vertical force as a function of the deflection, the unloaded radius and section radius. Obtain an algebraic expression for the stiffness for small deflections.

Q 2.5.4 Considering a thin-walled cylindrical (racing) tyre otherwise as for Q 2.5.3, analyse the vertical force. Draw graphs of vertical force and vertical stiffness against deflection.

Q 2.5.5 At speed, the tyre vertical force at a given deflection increases because of the impact of the perimeter against the road at the front of the contact patch. This can be important on cars with underbody venturis because it tends to increase the ground clearance, or at least offsets the reduction of ground clearance with aerodynamic downforce as speed increases. Because it gives a vertical force towards the front of the contact patch where it can increase friction, it also contributes to the negative aligning moment at high slip angles.

(1) Obtain an expression for the impact force for the simple case of a cylindrical (racing) tyre, in terms of the peripheral mass, the unloaded radius, the wheel spin speed and the impact angle θ equal to half of the angle subtended by the contact patch.

(2) Use the intersecting chords theorem to show that the deflection $z \approx l^2/8R_u$.

(3) Obtain an approximate expression for the impact force in terms of the vertical deflection, as a function of vehicle speed (with no wheel slip).

Q 2.5.6 A racing car tyre has peripheral mass of 2.0 kg, a radius of 0.340 m, and footprint length 0.170 m. Calculate the vertical impact force at the front edge of the contact patch at a vehicle speed of 80 m/s, with negligible wheel slip.

Q 2.6.1 Describe the formation of waves in a tyre perimeter at high speed, including a sketch. Give relevant equations, and explain the Turner number.

Q 2.6.2 Modelling the tyre as a crown only, neglecting side walls, obtain an equation relating crown tension per unit width to the radius, peripheral speed, crown density per unit area, and inflation pressure (analyse a small segment of length $R\theta$). Investigate realistic values. Neglecting pressure, what is the Turner number?

Q 2.6.3 Consider an isolated tread element of height h (8 mm), area A, and density ρ (1300 kg/m^3). Obtain an expression for the base tension from centrifugal stress. What is the stress for $R = 0.3$ m at 50 m/s?

Q 2.7.1 Explain the different ways of representing rolling resistance on a wheel free-body diagram.

Q 2.7.2 Explain how the rolling resistance of a tyre arises.

Q 2.7.3 A vehicle has wheels with a mean rolling resistance coefficient of 0.021. On what angle of slope will it just move?

Q 2.8.1 Describe briefly the differences between the foundation stiffness, string and beam models of the tyre.

Q 2.9.1 Define and explain the difference between cornering force, cornering force coefficient, cornering stiffness, cornering stiffness coefficient, maximum cornering force, and maximum cornering force coefficient. Why is cornering power a bad name?

Q 2.9.2 Explain the difference between tyre force, lateral force, cornering force, camber force, tractive force, central force, drag force, and longitudinal force. Draw appropriate figures.

Q 2.9.3 A simple foundation stiffness tyre model has the following parameters: contact patch length 200 mm, vertical force 4.8 kN, friction coefficient 1.2, foundation stiffness 2.8 MPa. Calculate:

(1) The slide-free maximum slip and slip angle.
(2) The cornering stiffness
(3) The cornering stiffness coefficient.
(4) The slide-free maximum force.
(5) The lateral force at 5° slip angle.
(6) The maximum lateral force.

Q 2.9.4 Draw a polar diagram of tyre cornering force against slip angle. Show the central and drag components for a typical cornering force. Show the maximum central force.

Q 2.9.5 Plot a graph of central/drag force ratio including rolling resistance, for $\mu_R = 0$, 0.015 and 0.030, for a cornering stiffness coefficient of 0.16/deg.

Q 2.10.1 Discuss the relative merits of non-dimensionalising tyre lateral forces by dividing by F_V and by dividing by $F_{Y\,max}$.

Q 2.10.2 Explain non-dimensionalisation of tyre forces (1) for the force, (2) for the slip angle. Draw example graphs.

Q 2.11.1 At approximately what sliding speed is typical tyre friction halved from its low-speed value?

Q 2.12.1 Explain the difference between lateral force, cornering force and camber force.

Q 2.12.2 Explain how tyre camber forces arise, giving example values of camber stiffness and coefficients.

Q 2.12.3 A mad inventor has devised a scheme for perpetual motion of his car. It weighs 12 kN, equally distributed. The tyres all have camber stiffness coefficient 0.016/deg and cornering stiffness coefficient 0.16/deg. He plans to set all the wheels at 20° positive camber and 1° toe-in. Give a calculation of the effect of this arrangement, supporting his view, and discuss it in relation to the principle of conservation of energy.

Q 2.12.4 Explain the path curvature stiffness of a tyre and how it may be related to C_γ.

Q 2.13.1 Describe the various possible types of tyre force testing methods, and discuss their relative merits.

Q 2.14.1 Describe and explain typical tyre conditioning effects.

Q 2.14.2 Describe and explain the influence of production tolerances on tyre handling parameters.

Q 2.14.3 At a reference load of 3 kN a tyre has a cornering stiffness coefficient of 0.160/deg. Estimate the cornering stiffness coefficient and cornering stiffness at a vertical load of 3.3 kN.

Q 2.14.4 A tyre has a speed sensitivity of 8%, and a cornering stiffness of 700 N/deg at 5 m/s, found by testing. Estimate the cornering stiffness at 40 m/s.

Q 2.14.5 Give equations for three models of the load sensitivity of tyre cornering stiffness coefficient, and show how the various load sensitivity parameters are related for small load changes.

Q 2.14.6 The cornering characteristics of a tyre are to be modelled by an exponential cornering stiffness coefficient with $C_{\alpha 1} = 820$ N/deg at $F_{V1} = 4000$ N, and $k_1 = 0.74$. Evaluate (1) C_{S1}, and (2) C_{S0} (at $F_V = 0$). Draw graphs of C_S and C_α versus F_V, for F_V taking values from zero to 10 kN.

Q 2.14.7 For the tyre model of the last question, determine analytically (1) F_V for $C_{\alpha max}$, (2) $C_{\alpha max}$.

Q 2.14.8 Describe the exponential decay model of tyre cornering stiffness coefficient variation with vertical force. Give equations, and develop expressions for any significant values of C_S or C_α.

Q 2.14.9 A tyre is found to have a maximum cornering stiffness $C_{\alpha 2}$ of 1195 N/deg at a vertical force F_{V2} of 4780 N. It is to be used at a standard load F_{V1} of 2868 N. Calculate (1) the cornering stiffness coefficient C_{S2}, (2) the load sensitivity, (3) the cornering stiffness coefficient C_{S1}, (4) the cornering stiffness $C_{\alpha 1}$, according to the exponential cornering stiffness coefficient model.

Q 2.14.10 Using the exponential cornering stiffness coefficient model, a tyre has a standard vertical force of 3800 N, a load sensitivity of 0.722, and a standard cornering stiffness of 810 N/deg. Calculate (1) the vertical load for maximum cornering stiffness, (2) the maximum cornering stiffness.

Q 2.14.11 For the exponential cornering stiffness coefficient model, obtain an algebraic expression for the ratio of the peak cornering stiffness to the standard value, in terms of the load sensitivity.

Q 2.14.12 Explain the load dependence of tyre cornering force at small slip angles. Give appropriate equations and example values.

Q 2.16.1 Describe the effect of variation of vertical force on tyre maximum lateral force coefficient.

Q 2.16.2 Describe the effect of lateral load transfer on the total cornering force generated by an axle at a given slip angle.

Q 2.17.1 Describe the generation of longitudinal forces in the contact patch, according to the simple foundation stiffness model.

Q 2.18.1 Describe, and explain by reference to the simple foundation stiffness model, the inter-relationship of lateral and longitudinal force.

Q 2.18.2 "A locked braking wheel will have a cornering force but no central force. A severely spinning driven wheel will have a central force but no cornering force." Discuss, with plan-view wheel force diagrams.

Q 2.19.1 Describe the role of the tread pattern in clearing water from beneath a tyre.

Q 2.19.2 A tyre takes 2 ms to clear the approaching water depth, and has a footprint length of 200 mm. At what speed will half of the footprint be supported by viscous aquaplaning?

Q 2.19.3 Describe and explain the process of aquaplaning.

Q 2.19.4 A vehicle has a wheelbase of 3 m, front track 1.7 m, and rear track 1.5 m. At what attitude angle and approximately what lateral acceleration will the rear outer wheel run on the same line as the outer front?

Q 2.20.1 Describe tyre transients, and their influence on handling on smooth and knobbly surfaces.

Q 2.20.2 A tyre has a distance constant of 0.25 m, and a cornering stiffness of 800 N/deg. When stationary it is set at a slip angle of 1.5°. Estimate the sideforce after it has rolled (1) 0.1 m, (2) 1 m.

Q 2.20.3 Using the rule of thumb that the tyre relaxation distance is one radius, after a tyre has rolled one revolution what is the fractional force deficit?

2.22 Bibliography

The information in this chapter has been gleaned from a large number of papers from the S.A.E., I.Mech.E. and various journals, too numerous to list usefully here. For elaboration of any particular points, the S.A.E. is the best single source of papers. For a general introduction to the tyre, refer to Setright (1972), Shearer (1977), Tompkins (1981), Norbye (1982) and French (1989). None of these, however, consider lateral forces in detail.

Clark (1981) provides a massive compilation of tyre theory and data in one volume, including innumerable references. An interesting reference on the theory and data of rubber friction is Hays & Browne (1974). For a more extensive discussion of tyre properties relevant to vehicle handling, see Moore (1975). A full definition of terms is provided in S.A.E. J670e (1978). For non-dimensionalised experimental tyre data refer to Radt & Milliken (1983). Sakai (1981) provides a detailed discussion of advanced tyre models. Some 280 references, including many foreign language ones, are given in Frank & Hofferberth (1967).

3

Aerodynamics

3.1 Introduction

Analysis of the response of a vehicle to aerodynamic effects may be considered in two parts:

(1) determining the forces acting on the vehicle,

(2) finding the response of the vehicle to those forces.

This chapter considers the determination of the forces. The consequences of the forces are dealt with in Chapters 6 and 7. Drag is a major factor in the fuel economy of a vehicle, although the influence of aerodynamics is in general secondary to that of the tyres as far as handling and stability are concerned.

Atmospheric properties may influence the handling of a vehicle in various ways. First, they affect the aerodynamic forces on the vehicle. This depends primarily on the density of the air, but also in principle on the Reynolds number and the Mach number, and hence also on the viscosity and the speed of sound. Engine power is also affected. The tyre–road interface is affected by dampness or depth of free water depending upon the rate of rainfall, or by snowfall or ice if the temperature is low. Ambient air temperature and solar radiation can affect the tyre–road friction. The vehicle is also affected by existing motions of the atmosphere. The worst cases of wind effects are usually where a vehicle passes from a sheltered region into a strong side wind, i.e. where there is a spatial variation of side wind. Another common problem is the disturbance caused by other vehicles, especially large ones on motorways.

The genesis of aerodynamic forces on the vehicle is described in this chapter, and examples are given for the various force and moment coefficients. Finally, an overview is given of the development of aerodynamics in competition, where it has become a major factor because of the great advantages of downforce in enhancing the forces at the tyres.

3.2 Atmospheric properties

It is desirable to specify a standard set of atmospheric conditions so that different studies are made on a comparable basis. On the other hand, the influence of variations of the real atmosphere from the standard should be recognised. Many standard atmospheres have been proposed, including the International Standard Atmosphere, the I.S.O. Standard Reference Atmosphere, the I.C.A.O. atmosphere, the U.S. Standards (N.A.S.A., U.S.A.F., U.S.N. etc), A.S.M.E., N.B.S., S.A.E. Engine Testing, and so on. The I.S.O. Standard Reference Atmosphere is somewhat unusual in specifying a 65% relative humidity, where all the others specify dry air, i.e. no water vapour whatsoever. Table 3.2.1 gives a representative specification for sea level.

The provision of such accurate standard values, for example a density to five significant figures, is in a sense misleading, because there are considerable variations in the real atmospheric properties arising from variation of pressure, temperature and humidity. The properties are not significantly influenced by changes of the proportions of dry constituents within the altitude range of roads.

Table 3.2.1. *Standard sea-level properties of dry air*

Constituents by mass:	Nitrogen	0.7555	
	Oxygen	0.2313	
	Argon	0.0127	
	CO_2	0.0005	
Temperature	T	15	°C
	T_K	288.15	K
Pressure	p	101325	Pa
Density	ρ	1.2256	kg/m^3
Dynamic viscosity	μ	17.83×10^{-6}	Ns/m^2
Kinematic viscosity	ν	14.55×10^{-6}	m^2/s
Molecular weight	M	28.965	
Specific heats	C_p	1006	J/kg K
	C_v	717	J/kg K
Ratio of specific heats	γ	1.402	
Speed of sound	V_s	340.6	m/s

The typical variation of atmospheric properties with altitude is recognised directly in the aeronautical standards by specifying the temperature variation with height, which when combined with a value for the gravitational field g allows the properties to be calculated. In the troposphere (below 11 km) the standard temperature declines with altitude at 6.5 K/km giving

$$T = 15 - 0.0065z$$

This results in a density relative to sea level of

$$\frac{\rho}{\rho_0} = \left(1 - \frac{z}{z_0}\right)^{4.256}$$

where z is the altitude and z_0 is a constant of value 44 300 m. Thus at 1500 m (about 5000 ft), which is a typical operating height in some parts of the world, the density is reduced by 14%. If this occurs in the tropics then the density may be even lower because of high temperature.

The presence of the small amount of water vapour in the real atmosphere has a considerable effect on weather patterns because of the large latent heat of vaporisation of water. There is a small effect on density, because the molecular weight of water is only 18, which displaces dry air of molecular weight about 29. The density corrections, although small, are nevertheless made in wind-tunnel testing. The absolute humidity is the water content of the air expressed as a density, i.e. as kg/m^3. The water content is normally expressed as the relative humidity, which is the absolute humidity divided by the maximum amount of water that the air can carry in uniform mixture, i.e. without condensation, at that temperature. This maximum varies rapidly with temperature, which is why cooling often leads to condensation, i.e. fog or rain, the relative humidity rising to 100% or more at constant absolute humidity. At 15°C and 100% relative humidity the absolute humidity is 0.0128 kg/m^3. Thus a typical 65% relative humidity corresponds to 0.0083 kg/m^3, or about 0.7% of the total air mass. In the British Isles, humidity is usually 60 to 95%.

Even in temperate climates at low altitude, there are substantial variations in temperature and pressure, and hence in density. Dry air, considered as an ideal gas, obeys the 'perfect' gas equation

$$p = \frac{R_G}{M} \rho T_K$$

where R_G = 8314.3 J/mole K is the universal gas constant, M = 28.965

is the mean molecular weight, ρ is the density and T_K is the absolute (kelvin) temperature. Thus the density is proportional to the pressure, and inversely proportional to the absolute temperature. The ambient pressure varies typically plus or minus 3 kPa from the mean, i.e. by about 3%. The ambient temperature varies by typically 15 K from the mean, 5% of the absolute temperature. Thus at sea level the density can vary by 8% or more from the mean, with more variation with altitude, and with temperature range variations with geographical location. The standard conditions should be seen in this light. These variations are, of course, one of the reasons for the different atmospheric standards in different countries. The extremes of recorded pressure are 87.6 to 108.4 kPa for the world, and 92.6 to 105.5 kPa for the British Isles. The recorded temperature extremes for the British Isles are –27°C to +37°C in the shade.

Appendix A gives an accurate method for calculation of density and other properties. Where a more approximate value of density will suffice, as is normally the case in the dynamics of handling, then it is usual to neglect the water vapour content, and to treat the air as a simple ideal gas, giving, in SI units,

$$\rho = 0.00348 \frac{p}{T_K}$$

Within the ambient temperature range the dynamic viscosity is often taken as having a constant value, for example $\mu = 17.8 \times 10^{-6}$ Ns/m^2, and the speed of sound, if required, is often taken as constant at $V_s =$ 340 m/s.

As a matter of interest, the standard recommended by S.A.E. J670e (Vehicle Handling Terminology) is dry air of density 1.226 kg/m^3 (2378 × 10^{-6} slug/ft^3), pressure 101 kPa (29.92 in Hg) at 15°C (59°F), with a viscosity of 17.9 × 10^{-6} Ns/m^2 (373 × 10^{-9} slug/ft s).

The main influence of rain is to give a layer of water on the road, depending on road camber and the quality of drainage. Drop sizes and terminal speeds for drizzle are 0.05 to 0.5 mm diameter at 0.7 to 2.0 m/s, and for rain 0.5 to 2.5 mm diameter at 3.9 to 9.1 m/s. The intensity of rain, as coverage, is less than 0.5 mm/h for light rain, 0.5 to 4 mm/h for moderate rain, and over 4 mm/h for heavy rain. A coverage of 10 mm/h is 10 kg/m^{2}h, or about 3 g/m^2 s. With a terminal speed of 8 m/s, the rain density is then 0.4 g/m^3, increasing the mean density of the atmosphere by only about 0.03%. Because of the arrangements for drainage, the depth of water on a road is mainly a function of the rainfall rate.

In contrast, because snow is not drained until it melts, the depth of snow is a cumulative effect, and can therefore be a particular problem, not least because in some countries there are floods when it melts. There are various types of snow. On continental land masses far from the sea, dry snow is common, of density 100 kg/m^3. This has small powdery crystals that do not easily bond under pressure, and can be cleared by blowers. For the maritime borders of continents in latitudes 40° to 60°, wet snow is most common, with density up to 300 kg/m^3, for which the crystals are bonded into flakes, which easily stick together. This must be cleared by plough or shovel. In snow or ice conditions, of course, special tyres, sometimes with chains or studs, may be used.

3.3 Wind and turbulence

A vehicle moves in a non-stationary mass of air. Although time-stepping computer simulations can deal with the complex situation of a cornering vehicle in a turbulent windy air, investigations of the effect of air movement are mainly related to the straight running condition. This is because vehicles are most critically sensitive to wind disturbance when running at high speed, for example on a motorway, when a small angular deflection of the path can quickly lead to a collision.

Movements of the atmosphere can best be considered in two parts. More than about 600 m above the surface is the geostrophic wind. Below 600 m is the atmospheric boundary layer, in which air movements are driven by the geostrophic wind but reduced by the friction of the land or sea surface. The geostrophic wind arises from non-uniform heating of the air, which gives the pressure differences that result in the wind. The geostrophic wind in any particular region is a combination of two contributions. The first is the regular global wind pattern; for example, Great Britain is in the 'Westerlies' of the temperate zone, with regular south-westerly winds. In addition there are cyclones and anti-cyclones, familiar from the weather reports, which introduce a low-frequency (i.e. period of about four days) more or less random fluctuation to the wind at any point. The total of these two effects is the geostrophic wind.

The speed in the atmospheric boundary layer varies from zero at the ground to the full geostrophic value at the top. Except for very slow winds, less than about 5 m/s at 10 m height, the boundary layer flow is turbulent, giving random speed fluctuations, the period of which is typically a few seconds, but which can be as slow as five minutes. Thus it is convenient to refer to a sustained wind speed, averaged over some period between ten minutes and an hour, which is free of the turbulent fluctuations. In engineering applications the sustained wind is represented as a function of height by a power model

$$\frac{U}{U_r} = \left(\frac{h}{h_r}\right)^n$$

where U_r is the speed at a reference height h_r. The standard reference height is 10 m, and this is often referred to, perhaps illogically, as the ground level wind. The exponent n varies according to the effective friction of the surface, being typically 0.1 for sea or for land flats, 0.15 for open terrain, 0.25 for suburban conditions, and 0.35 for a city centre. Thus a car, with a characteristic height of about 1 m, is in a strongly shearing wind distribution, the airspeed increasing rapidly with height from zero at ground level.

The likelihood of a given wind speed being exceeded is usually calculated from a Weibull equation. The fraction of the time that the wind speed U will exceed V is given by the long-term probability

$$P_{U>V} = \exp\,[-(V/V_{\mathrm{ch}})^{\beta}]$$

where V_{ch} is the characteristic wind speed, which is typically 3 to 5 m/s for normal land sites at 10 m height, and β is the wind shape factor, typically 2 and normally in the range 1.5 to 2.5. For exposed sites, for example high bridges, the characteristic speed will be high, giving a correspondingly high probability of winds that are sufficient to disallow certain types of vehicle from using the road, because of handling problems or the danger of overturning. For a typical inland site, with a characteristic speed of 4 m/s and shape factor of 2, the probability of exceeding 12 m/s is only 0.012%, and hence wind speeds are usually much less than road vehicle speeds.

The actual speed, measured with fast-response instruments, is found to vary erratically about the sustained speed; i.e. the actual instantaneous speed is the sustained speed U plus a random turbulent component V. These random fluctuations have a period of typically a few seconds. The probability distribution of the instantaneous speed about the sustained speed is found to be well modelled by a Gaussian distribution, which is characterised by the turbulent speed deviation W, Figure 3.3.1. The instantaneous speed is within W of the sustained speed U for 68.3% of the time, within $2W$ for 95.5%, and within $3W$ for 99.7%.

The turbulence of an airflow is usually specified by the turbulent intensity, which is the ratio of the turbulent speed deviation W to the sustained speed V_S:

$$I = \frac{W}{V_S}$$

Figure 3.3.1. Probability distribution of speed for turbulent wind.

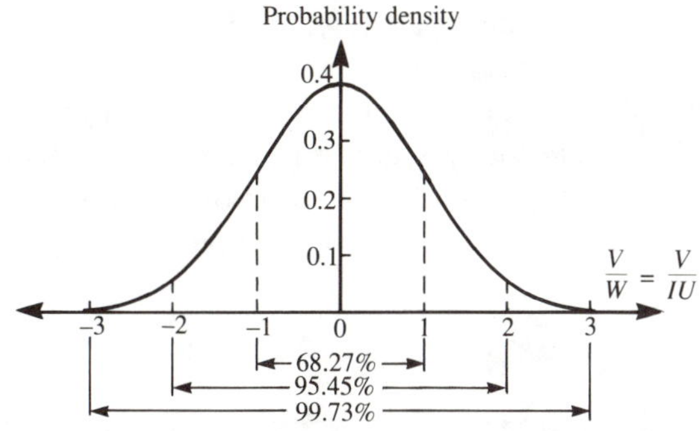

Considering the flow relative to a moving car instead of relative to the ground, the relative sustained speed will be different, but the turbulent deviation W will be the same. We are considering here the properties of the natural atmospheric turbulence, not the turbulence created by the car. Because of this change of relative sustained speed, the turbulent intensity I depends upon the coordinate system in which the flow is observed, for example relative to the car or ground, although the actual turbulent fluctuations are of the same value. The typical turbulent intensity of the wind, measured relative to the ground, is about 20%, varying somewhat with height, and tending to be smaller at sites where the speed–height exponent is smaller. Relative to the moving car, the typical turbulent intensity is usually less because of the car speed; e.g. in a wind of 10 m/s with $I = 0.20$, the turbulent deviation is 2 m/s, so relative to a car moving at 30 m/s directly into the wind the turbulent intensity is only 2/(30 + 10) which is 0.05. Modern wind tunnels tend to have extremely low levels of turbulent intensity, e.g. 0.1% or less. In some cases special facilities are provided to generate turbulence.

The typical state of the atmospheric boundary layer, when the wind speed is high enough to be of significance, is a turbulent one, so the vehicle is in a turbulent, shearing flow. There is also likely to be additional turbulence, caused by vortex shedding or trailing vortices from other vehicles, or from obstructions to the wind at the roadside. These are all variations of speed at a given point, i.e. unsteady flows.

Another effect that can have a major influence on a vehicle is the spatial variation of speed because of steady flow patterns as the wind passes around objects at the roadside. The most familiar example of this is when a vehicle emerges from a sheltered region, such as a row of houses, into a strong sidewind. This kind of flow is difficult to predict analytically, and is usually studied in wind tunnels. The response of vehicles can be investigated by specifying a sidewind distribution for the vehicle to pass through, either in theory, or in a wind tunnel, or in a full-size test. The sidewind distributions are typically either a sharp step change, a ramp-step change, or a sinusoidal-step change.

One classic problem for cars is disturbance when passing, or being passed by, large vehicles at speed, for example on a motorway. In this case the car is subject to an airflow that varies systematically in both space and time.

3.4 Principles

The handling response of a vehicle to aerodynamic influences involves determination of the force acting on the vehicle, and finding the response of the vehicle to that force. The first of these will be introduced in this chapter.

The aerodynamic influences on ordinary vehicles, as far as handling is concerned, are generally secondary to the tyre forces. This is fortunate because accurate determination of the aerodynamic forces in any particular case is rather difficult. This is because the forces are sensitive to detailed changes in body shape. To date, theory is rather limited in its ability to predict the forces from first principles. However, theory is essential in interpreting and using the results of the extensive wind-tunnel tests that are required to assess the aerodynamic properties of a vehicle.

The aerodynamics of a ground vehicle are fundamentally different from those of an aircraft in flight for a variety of reasons, including:

(1) There are severe practicality constraints.
(2) A ground vehicle is a bluff body.
(3) The flow around a ground vehicle separates.
(4) Ground proximity has to be taken into account.
(5) The presence of wheels has to be taken into account.

These factors are of course inter-related. The practicality constraints disallow a long streamlined tail which means that the body is a bluff one, with separating flow, for which the drag is largely the consequence of trailing vortices and shed vortices. In contrast, an aircraft is basically streamlined, with a substantial skin drag, some form drag and trailing vortex drag but relatively little turbulent vortex shedding. The aerodynamic forces on an aircraft with its attached flow are more amenable to theoretical analysis than those on a ground vehicle with its largely separated flow. The presence of the ground is an additional complexity, encountered by aircraft only at take-off and landing. Thus the very extensive knowledge of aircraft aerodynamics has only limited application to ground vehicle aerodynamics, which forms a separate subject in its own right.

The aerodynamic forces are found to depend primarily on the vehicle size, the vehicle shape, the vehicle attitude to the airstream, the air density, and the square of the relative airspeed. The influence of size is such that the forces are proportional to areas, because the other factors combine to give resulting pressures at the vehicle surface. Hence the total forces are usually expressed in terms of force coefficients:

$$F = qCA$$

where

$$q = \tfrac{1}{2}\rho V^2$$

is the dynamic air pressure, F is the force magnitude, C is the force coefficient, and A is the reference area. Different force magnitudes will result from a given coefficient value if different reference areas are used, so it is essential to specify this. In ground vehicle aerodynamics, it is most common to use the total vehicle frontal area, i.e. the area inside the vehicle front profile. Alternatively, the product CA is used, for example the drag area C_DA. Sometimes a reference area of one square metre is used, which gives a drag coefficient numerically equal to the drag area.

The frontal area will be used here except where some other area is declared in some special cases. When modifications to a vehicle are being tested, the frontal area may be changed by the modifications; in such cases the usual policy is to relate all the coefficients to the frontal area of the baseline vehicle; otherwise comparison of the coefficients is not meaningful.

The great utility of force coefficients is that they remain approximately constant with variations of speed and size and of fluid density, provided that the vehicle shape and attitude are the same. This allows wind-tunnel tests to be made on a smaller vehicle, often quarter size, to find the coefficients, which then predict the real vehicle forces. However the coefficients are not truly constant unless the Reynolds numbers (Re) and Mach numbers (Ma) of the flows are the same for the two cases, where

$$\mathrm{Re} = Vx / \nu$$

$$\mathrm{Ma} = V / V_s$$

the parameter x being some vehicle characteristic dimension; ν is the fluid kinematic viscosity (μ/ρ) and V_s is the speed of sound. In practice, unless a full-scale test is performed then the Reynolds numbers and the Mach numbers are not both correct. Also, of course, the force coefficients are used for a range of speeds of the real vehicle and so over a range of Reynolds and Mach numbers.

The problem is usually dealt with in the following way. First, model test Reynolds numbers are generally smaller than real, because the model is small, so if possible the test speed is increased to compensate. This makes the Mach number too high, but in practice this is less important for normal vehicles. Corrections may be applied for Mach number. For a typical wing, the lift force coefficient is influenced by the Mach number Ma, according to Prandtl's correction, by

$$\frac{C'}{C} = \frac{1}{\sqrt{(1-\mathrm{Ma}^2)}}$$

Thus, relative to Ma = 0, there is a 1% correction at Ma = 0.14, which is a speed of 48 m/s (107 mile/hr). Thus Mach corrections are negligible in ordinary road vehicle use, but may be significant in the wind tunnel, or for racing cars, which may reach 100 m/s in some cases. For transonic conditions, as experienced by land speed record vehicles, this method of correction is no longer appropriate. The forces are then highly sensitive to Mach number, and the best that can be done is for wind-tunnel testing to be performed at the correct Mach number, with Reynolds number playing a secondary role, possibly with roughness strips to induce transition in the boundary layer.

The influence of variation of Reynolds number on force coefficients is potentially complex, but fortunately the effect is usually small. Basically the Reynolds number indicates the relative significance of the viscous and inertial forces in the fluid, and indicates the state of the boundary layer, which in turn controls any separation of the flow. Thus where flow separation is well located by sharp edges, Reynolds number will have little effect. Where separation occurs from a curved surface, the separation point, and hence force coefficients, may change significantly with Reynolds number. If the flow is fully attached, for example as on a wing at low incidence, then the Reynolds number usually has a progressive effect because of change in the boundary layer thickness. In practice for a typical car, an increase of flow speed and hence of Reynolds number, in the usual range of operating conditions, can sometimes give a sudden change of flow pattern with a step change of force coefficients of perhaps 10%.

In studying the response of a vehicle to gusts it may be necessary to consider the vehicle to be subject to a non-uniform wind, for example a sidewind that varies along the length of the car. In most cases, however, it is adequate to consider the vehicle to be in a uniform wind.

The ambient wind is measured by the magnitude and direction of the bulk air velocity relative to the ground, i.e. measured in the Earth-fixed axes *XYZ*. Because of the atmospheric boundary layer, this varies with height. Vertical components are normally neglected. Following S.A.E. terminology, the magnitude of the ambient wind speed is v_a. (Speeds are denoted by v (vee) and angles by ν (nu).) The direction, measured clockwise from the *X*-axis direction, is the ambient wind angle ν_a. The relative air velocity may be found by the velocity diagram of Figure 3.4.1, by plotting the vehicle's velocity relative to the ground, v_v at ν from the *X*-axis direction, to give point V, and plotting the air velocity relative to the ground, v_a at ν_a, to give point W. The air velocity relative to the vehicle is then given by the position of W relative to V, with magnitude v_r and direction ν_r. The actual evaluation of the relative velocity is done by considering the *X* and *Y* components and using Pythagoras's Theorem. The speed of the vehicle relative to the air may conveniently be referred to as the airspeed, as distinct from the wind speed or the ground speed of the vehicle.

Figure 3.4.1. Velocity diagram for vehicle and wind.

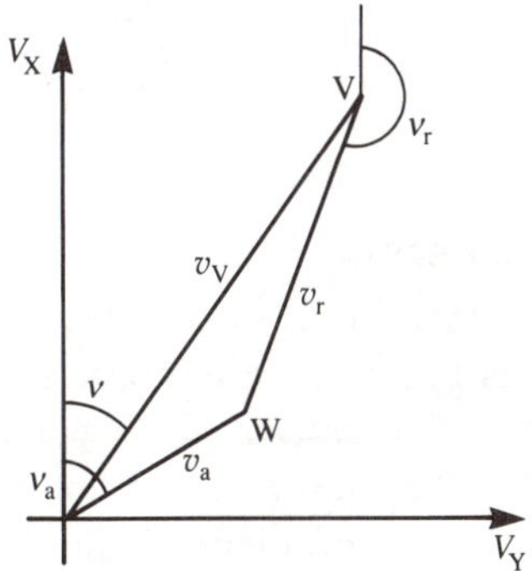

Figure 3.4.2 shows the air impinging on the vehicle. In this figure, the air approaches the vehicle at the angle $\nu_r - 180°$. The path angle ν, vehicle heading angle ψ, and sideslip angle β are as defined in Section 1.3. The aerodynamic sideslip or yaw angle is

$$\begin{aligned}\beta_A &= \nu_r - 180° - \psi \\ &= \nu_r - 180° - \nu + \beta\end{aligned}$$

In still air $\beta_A = \beta$.

Figure 3.4.2. Vehicle and wind angles.

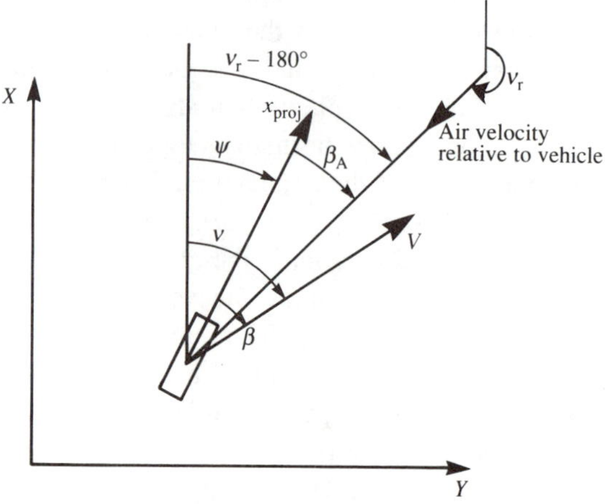

The other angle required to specify the vehicle's aerodynamic attitude is the aerodynamic angle of attack, α_A. Because the air velocity is essentially parallel to the ground, this is simply the angular position of the vehicle's x-axis. This can easily be extended to other cases, for example a non-horizontal wind because of sloping ground.

In principle it is also necessary to specify the roll angle of the vehicle. However, experiments have shown that for normal vehicles the roll angle has only a rather small effect on the aerodynamic forces, so this effect is generally neglected.

3.5 Forces and moments

The air exerts one single total force on the vehicle, but for convenience of analysis this is usually considered to be resolved into rectangular components of force at a given point, plus moments about the axes. Therefore there are three forces and three moments. These are usually expressed as coefficients. The forces are reduced to coefficients by using

$$C_F = \frac{F}{qA}$$

where $q = \frac{1}{2}\rho V^2$ is the dynamic pressure and A is the reference area, as already mentioned. To reduce the moments to coefficients, an extra length dimension is required. Ideally an aerodynamically significant length would be used, such as the overall length, but in practice the wheelbase is usually preferred, giving

$$C_M = \frac{M}{qAl}$$

This means that in comparing coefficients between vehicles, differences of the aerodynamically irrelevant wheelbase may create problems; in practice this is not a serious handicap. It has the considerable virtue of relating pitching and yawing moments to the wheelbase, over which they are reacted by the wheels.

The coordinate system chosen for resolving the aerodynamic force may be the vehicle-fixed axis system *xyz*, giving force coefficients for longitudinal force C_x, sideforce C_y, and normal force C_z, and moment coefficients C_{Mx} for roll, C_{My} for pitch, and C_{Mz} for yaw.

It is often preferred to express the total aerodynamic force in components related to the direction of the airflow. In this case the coefficient in the actual direction of the airflow relative to the vehicle is the drag coefficient. The force perpendicular to the flow is generally called the lift; however there is some consequent ambiguity because the term lift may be applied to the total perpendicular force, the vertical force or the lateral force. The term lift or vertical lift is normally used for the component perpendicular to the ground, and the term lateral lift or side lift is used for the component parallel to the ground. The coefficients will be denoted C_D, C_L, and C_S, and the moment coefficients about the corresponding axes are C_R for roll, C_Y for yaw, and C_P for pitch. C_S is taken positive to the right, i.e. approximately as the *y*-axis, with C_P positive for pitch up. C_Y is positive when it tends to increase the yaw angle. With these coefficients, β_A is usually taken for a direction such that $dC_S/d\beta_A$ is positive.

The majority of aerodynamic investigations are concerned with drag, lift and pitch in straight running, in which case the directions of these axes essentially coincide with the vehicle-fixed axes *xyz*, although $C_D = -C_x$ and $C_L = -C_z$, i.e. C_D is positive to the rear, and C_L is positive upwards.

The use of front and rear lift, instead of lift and pitch, is convenient because this shows directly the influence on the tyre vertical forces, and also because the lift is usually measured in practice by the reactions at the wheels. If the stated front and rear lifts are to be applicable to the real steady-state operating condition then the drag must be taken to act at ground level, because the drag is overcome by a wheel thrust at ground level.

The selection of the vehicle centre of mass as the axis centre would be somewhat arbitrary in an aerodynamic sense, since the position of the centre of mass is not aerodynamically significant. Also, the centre of mass will be changed in the real vehicle by loading conditions. Hence, in vehicle aerodynamics, the standard axes are taken at the centre of the wheelbase at ground level; Figures 3.5.1(a) and (b) show the partial vehicle free-body diagram in side view, with two alternative representations of the standard aerodynamic forces. In particular, in Figure 3.5.1(b) note that, for example, L_f is not a wheel reaction, it is an effective aerodynamic force exerted on the vehicle, and when it is positive it will reduce the force exerted by the ground on the wheel.

Because of the use of the wheelbase as the reference length for non-dimensionalising the moments, there are particularly simple relationships between the coefficients:

Figure 3.5.1. Standard aerodynamic axes for road vehicles, side view (only aerodynamic forces shown).

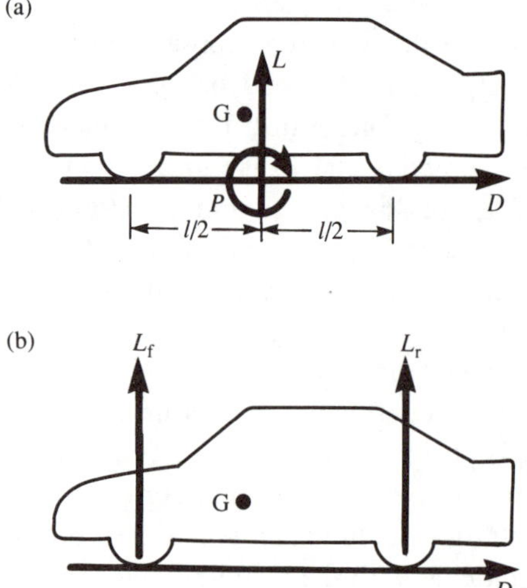

Figure 3.5.2. Standard aerodynamic axes: (a) plan view, (b) rear view (only aerodynamic forces shown).

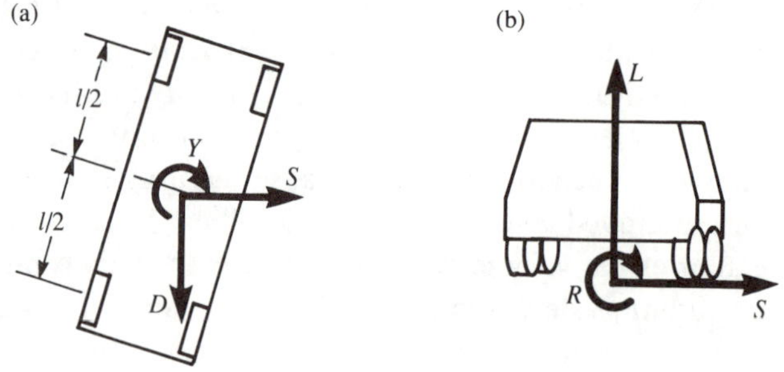

$$C_{Lf} = \tfrac{1}{2}C_L + C_P$$

$$C_{Lr} = \tfrac{1}{2}C_L - C_P$$

$$C_L = C_{Lf} + C_{Lr}$$

$$C_P = \tfrac{1}{2}(C_{Lf} - C_{Lr})$$

The sideforce and yaw moment, which occur when the vehicle is yawed, Figure 3.5.2(a), can be handled in a similar way. The standard axis origin is on the centre-line, still at the mid-point of the wheelbase, and of course the drag is deemed to act at the origin. Alternatively the sideforce may be treated as forces at the front and rear axles. This gives similar relationships as for lift:

$$C_{Sf} = \tfrac{1}{2}C_S + C_Y$$

$$C_{Sr} = \tfrac{1}{2}C_S - C_Y$$

$$C_S = C_{Sf} + C_{Sr}$$

$$C_Y = \tfrac{1}{2}(C_{Sf} - C_{Sr})$$

The sideforce plus yawing moment can be combined to give the sideforce a new line of action, no longer at the axis centre. The distance of the line of action of the single total sideforce behind the axis centre (the static margin from the wheelbase mid-point), h_A, is given in terms of the wheelbase by

$$\frac{h_A}{l} = -\frac{C_Y}{C_S}$$

We must distinguish here between the position of the line of action of this side lift and the position of the next increment of side lift, analogously with the centre of pressure and aerodynamic centre of a wing. Rearward movement of the side lift with increasing yaw angle implies that the increments are further back than the sideforce. In a steady-state handling analysis, it is the position of the side lift that is involved in determining the trim state, i.e. the required tyre slip angles, and hence the steering wheel position, but it is the position of the next force increment that determines the stability of that trim state. Thus it is not correct to say that a side lift in front of the centre of mass necessarily tends to destabilise a steady state. It really depends on the position of the aerodynamic force increment in relation to the position of the tyre force increment, see Chapter 6. The position of the force increments behind the axis centre is given in terms of the wheelbase by

$$\frac{h'_A}{l} = -\frac{dC_Y}{dC_S}$$

$$= -\frac{dC_Y/d\beta_A}{dC_S/d\beta_A}$$

$$= -\frac{C'_Y}{C'_S}$$

In aeronautical engineering, the distance of the force increment behind the centre-of-mass position is called the static margin.

In rear view, Figure 3.5.2(b), the origin of coordinates is at ground level at the centre of the track. The side lift acts at ground level. Because of the attitude angle in cornering, nose inwards at higher lateral accelerations, the aerodynamic roll moment generally opposes normal body roll.

3.6 Coefficient values

The forces, and even force coefficients, vary considerably between vehicles, even of similar design. This section will give some examples. For a given vehicle all forces and moments depend upon the three rotations, i.e. the aerodynamic pitch, yaw and roll. The effect of roll angle on the aerodynamic forces is generally small and usually neglected. The height position of the vehicle, i.e. the ground clearance, is significant. This complex inter-relationship can be simplified somewhat in practice because a vehicle is roughly prismatic, i.e. vertical longitudinal sections are roughly the same, and also the vehicle has lateral symmetry. If we consider the vehicle to be truly prismatic, and considering that pressure forces dominate over skin friction shear forces, then drag and lift forces, and the pitching moment, must be produced by the upper and lower, and front and rear, faces of the body. In handling analysis, these forces affect the tyre vertical reactions, and also the tractive force requirements. On the other hand, lateral forces, and the yawing and rolling moments, are produced primarily by pressures on the vehicle sides. These forces directly affect the lateral dynamics.

First, consider the flow over the top of the vehicle. The vehicle body acts somewhat like a wing section, giving flows of increased speed and low pressure, giving a lift force and a pitching moment. The actual flow around a vehicle is very much a three-dimensional phenomenon, so the drag, lift and pitching moment cannot be calculated satisfactorily from the centre-line pressure distribution. By adding a spoiler at the rear of the boot (U.S. trunk), a positive pressure increment can be produced over the boot and rear window, and possibly even on the roof. The positive pressure increment on the rear window tends to reduce drag and usually more than offsets the drag on the spoiler itself for spoiler heights up to about 20 mm. The pressure increment also reduces lift at the rear, i.e. it gives a negative lift increment and a pitch-up increment. This effect can be made quite large with a large spoiler, beyond the drag optimum. The pressures are sensitive to the shape, for example to the screen inclinations, and satisfactory results can often be achieved without extraneous devices.

The actual flow pattern is characterised by flow separation at the rear. The pattern depends largely on the angle of the rear window because in the range 20° to 40° to the horizontal there are strong trailing vortices, so the drag has a peak at 30° rear-screen angle. Typically, less than 20° is called fastback, 20° to 40° is hatchback, and more than 40° is squareback. The so-called K-tail, sometimes Kamm tail, in which the rear is sharply truncated with only limited tapering of the body, originated in 1935. Although the flat rear has a base drag because of its low pressure, the avoidance of the strong trailing vortices means that it is often better than an apparently more streamlined shape of the same length.

At the front of the vehicle the trend is toward low bonnet (U.S. hood) fronts, and raked windscreens, with a rather small angular deflection between the bonnet and windscreen. This wedge-shaped front may help to reduce front lift. A low stagnation point at the front means that the air flows over rather than under the vehicle, which is generally favourable.

On commercial vehicles, which are basically box-shaped, even quite modest radiusing of the front edges can have a considerably beneficial effect – for example for a typical box the C_D is better than halved if the leading edges are given a radius of 5% of the box height.

Underneath the vehicle, the influence can be considered in two parts: first the mean pressure, and then the effect of flow. The higher the mean pressure underneath, the greater the lift. The mean pressure is a compromise between the high stagnation pressure at the front of the vehicle and the low base pressure at the rear because of flow separation. Thus a minimal pressure will be achieved underneath with a large flow resistance at the front edge of the underside and a small one at the rear. This will be achieved with a front airdam and with the underbody open to the low pressure at the rear. Side skirts can help if the average pressure is below atmospheric. The practicalities of ground clearance, and especially kerb clearance, limit the effectiveness of such devices on passenger vehicles, although they are applied in racing. The mean underpressure can also be influenced by suction at the wheel side apertures.

The flow under a vehicle can be used to great advantage in racing (Section 3.7). On normal vehicles the underbody flow seems principally to increase the drag because the underside of most vehicles is extremely irregular. With suspension and exhaust components to be accommodated, this is clearly a problem, but improvements have been made. The front airdam is helpful in minimising the flow beneath the vehicle. Also it tends to give a pitch-down increment, so it can be balanced against a rear spoiler.

The aerodynamics of cars have long been a consequence of packaging and aesthetic requirements, rather than a decisive influence on design. In general, drag coefficients have been lower in Europe than in the U.S.A., possibly because of the higher price of fuel. The fuel price rises since 1970 have triggered considerable attention to drag, and hence also to other aerodynamic aspects, and there have certainly been dramatic improvements. Some of these have been without significant styling repercussions; others do influence styling, but many would say favourably.

In the 1920s for typical cars C_D was about 0.8. Improved metal forming in the 1940s gave better shapes and C_D reduced to typically 0.65. In the 1970s it was typically 0.55 in the U.S.A, somewhat under 0.5 in Europe. In 1990, some production vehicles are below 0.30, with research vehicles below 0.25. Various studies suggest that the aerodynamic drag is typically distributed 65/15/5/5/10 in form drag, interference drag, internal (engine and cabin cooling) drag, skin drag, and lift-associated drag, but this will vary substantially between vehicles. Skin roughness has negligible effect, although mismatch of panel edges may be significant.

Table 3.6.1 gives typical drag figures for a range of vehicles, showing the frontal area A, the drag coefficient C_D and the drag area $C_\mathrm{D}A$. The frontal area of a typical car is 0.81 of its width times its height, and is about 1.8 m^2 (European medium or U.S. compact).

The figures above refer to the vehicle in standard condition, at zero incidence and yaw, and at standard ground clearance. For a typical car

$$\frac{\mathrm{d}C_\mathrm{D}}{\mathrm{d}h} \approx 0.5/\mathrm{m}$$

$$\frac{\mathrm{d}C_\mathrm{D}}{\mathrm{d}\alpha} \approx 0.015/\mathrm{deg}$$

This latter figure is largely the consequence of increased airflow

Table 3.6.1. *Typical drag coefficients and drag areas*

Vehicle	Frontal area A (m^2)	Drag coefficient C_D	Drag area A_D (m^2)
Motorcycle	0.7	0.90	0.63
Kart	0.35	0.80	0.28
Good car	1.80	0.30	0.54
Poor car	1.80	0.50	0.90
Larger car	2.3	0.46	1.06
Small commercial vehicle	5	0.50	2.50
Light truck	7	0.73	5.10
Coach	7	0.66	4.60
Heavy truck	9	0.78	7.00
Heavy truck + trailer	9	0.90	8.10
Light aircraft	5	0.12	0.60
Cyclist – touring	0.50	1.00	0.50
– racing	0.33	0.90	0.30

Table 3.6.2. *Typical aerodynamic coefficients (car)*

Vertical lift	C_L	0.30
Drag	C_D	0.35
Pitch	C_P	–0.05
Side lift	$C'_S = dC_S/d\beta$	0.040/deg
Yaw	$C'_Y = dC_Y/d\beta$	0.005/deg
Roll	$C'_R = dC_R/d\beta$	0.005/deg

beneath the car. Thus pitch moment is of interest in maintaining correct attitude to control drag. The drag itself has some influence on handling because it must be overcome by a tractive force, which affects the tyre lateral force characteristics. Lift and drag coefficients may vary with speed because of changing height or pitch angle. In particular, this can lead to a rapid increase of front lift on fast cars.

Representative main aerodynamic coefficients for a car are summarised in Table 3.6.2. The total lift coefficient of a typical car is about 0.3, ranging widely from zero to 0.6 or more. The lift coefficient is quite sensitive to height and incidence, with typically

$$\frac{dC_L}{dh} \approx -0.2/\text{m} \qquad \frac{dC_L}{d\alpha} \approx 0.06/\text{deg}$$

The pitching moment coefficient varies considerably between vehicles; it should preferably be small. It is typically –0.05. It varies with incidence, typically with

$$\frac{dC_P}{d\alpha} \approx 0.03/\text{deg}$$

The lift force and pitching moment, or front and rear lift, can have a substantial effect on high-speed handling, because by changing the tyre vertical forces the cornering stiffness of the tyres is altered. Front lift leads to unpleasantly light steering with low limit lateral forces, and rear lift leads to directional instability, discussed in more detail in Chapter 6. Passenger car designers aim for zero lift and pitch or a small amount of balanced downforce. Positive lift can be dangerous at high speeds.

It remains now to consider the influence of pressures on the vehicle's sides, i.e. to look at the lateral lift, the yawing moment and the rolling moment. Although there may be significant pressures on the sides in straight running, they balance out. A net resultant force only occurs when there is an aerodynamic yaw angle β_A. In normal conditions the vehicle speed is substantially greater than the wind speed, so it is yaw angles up to about 30° that are of most interest. The following comments are restricted to that range, and should not be extrapolated beyond 30°.

The principal effect of yaw is to give a lateral lift and a yawing moment. Various theories of lateral lift have been proposed, none entirely satisfactory because of the complex flow pattern. In practice a typical experimental value of the lateral lift coefficient is 0.04/deg, with individual examples ranging from 0.02 to 0.05/deg. This is usually fairly constant up to 30° yaw.

The yawing moment is the consequence of the lateral lift and its line of action. The yaw moment coefficient C_Y usually increases fairly linearly up to 15° yaw angle at a rate of about 0.005/deg, i.e. $dC_Y/d\beta_A =$ 0.005/deg, but then peaks at a value of typically 0.10 at an angle of 25°, and then declines. This reflects the fact that the line of action of the lateral lift starts well forward, but moves back towards the axis centre, tending to reduce the moment. The yaw moment is normally positive, and hence tends to increase the yaw angle. Better aerodynamic directional stability is afforded by vehicles with a smaller yaw coefficient, i.e. greater side area at the rear, such as estate cars, or those with a high boot profile, and with a low bonnet line.

An aerodynamic yaw angle also causes a roll moment coefficient. Typically this increases smoothly with β_A up to 30°, at 0.005/deg. This is because the side lift really acts above the ground level, actually at a height $(C_R/C_S)l$ which is suprisingly low at about the wheel radius, presumably because the flow can easily pass around the upper edges.

The yaw angle also causes an increase of drag, typically quadratic in nature, i.e.

$$\Delta C_D = k\beta_A^2$$

where in one case $k = 1.6 \times 10^{-4}$ deg^{-2}. Yaw also causes a quadratic increase of lift coefficient up to about 20°, beyond which it levels off and then declines beyond 30°. In this case typically $k \approx 1.2 \times 10^{-3}$ deg^{-2}. The pitch coefficient also tends to increase, typically in this case with $k \approx 1.2 \times 10^{-4}$ deg^{-2}.

The angle of the steered wheels will have an effect on the side lift and yaw, although this is probably fairly small for the usual semi-enclosed wheels. Measurements on one Formula 1 racing car, with exposed wheels at steer angle δ, gave

$$\frac{dC_S}{d\delta} = 0.008/\text{deg}$$

$$\frac{dC_Y}{d\delta} = 0.024/\text{deg}$$

Using the above coefficients as estimates or, preferably of course, using vehicle-specific wind tunnel data, and the following equations, the drag force, lift force, side lift force, pitch moment, yaw moment and roll moment may be found:

$$D = C_D qA$$

$$L = C_L qA$$

$$P = C_P qAl$$

$$S = C_S qA \approx C'_S \beta_A qA$$

$$Y = C_Y qA\,l \approx C'_Y \beta_A qAl$$

$$R = C_R qA\,l \approx C'_R \beta_A qAl$$

In principle the angular speed of the vehicle in yaw, pitch and roll may also influence the forces and moments. This is important for aircraft, where the fin and tail provide the pitch and yaw damping, and the outer wings provide the roll damping. Such effects seem to be rather small for cars, where roll and pitch damping are primarily provided by the dampers, and yaw damping by the tyres (see Chapters 6 and 7).

3.7 Competition vehicles

The aerodynamics of competition vehicles, like their other features, is more extreme than for normal passenger or commercial vehicles. For land speed record vehicles, drag has always been important, and lift/downforce has been considered since the 1920s. The role of aerodynamics in racing-car design has increased dramatically since the mid-1960s, and is now of great importance. The history of competition vehicle aerodynamics will be treated briefly here because of its technical interest. Also, road vehicle design has been influenced to a limited extent by competition experience, for example by underbody shape and ground effects.

Early racing and land speed record vehicles appear crude by modern standards. In many cases aerodynamics was however influential in the layout because frontal area, governed by the large engines mounted high on rigid axles, was generally minimised within the perceived options. Streamlined body forms, although with many excrescences including the driver, were naturally applied to land speed record vehicles in the very early days, for example Chasseloup–Laubat and Jenatzy in 1899, both with battery electric vehicles at about 25 m/s (56 mile/hr). Much improved streamlining was achieved by the Baker Electric Torpedo in 1902 and the Stanley Rocket (steam) in 1906. In view of the limited aerodynamic knowledge of the day, these were quite respectable efforts, and in marked contrast to most of their competition.

In 1927 the first rocket-powered speed contender ran. This was the Opel Rak 1, of uncompetitive performance but of historical interest because it featured small wings mounted on the body behind the front wheels. The Rak 2 of 1928 had larger wings of about 1 m^2 each. Presumably these were intended to produce a downforce. Judging from photographs, however, the design was very odd, using a fairly thin well-cambered section fitted as if to give lift, but then set at a negative incidence that would have resulted in little vertical force at all. Nevertheless, this vehicle was the beginning of an awareness of the vertical aerodynamic component on ground vehicles.

Prevost (1928) first described proposals for a venturi-bodied land speed record car, Figure 3.7.1, with the stated intention of preventing lift. In 1929, the Irving-designed Golden Arrow set a land speed record of 103 m/s (231 m.p.h). This used a venturi-shaped underbody, Figure 3.7.2, which according to wind-tunnel tests would give 2440 N downforce at the design speed of 112 m/s (250 m.p.h.) (Irving 1930). In Prevost's proposal the venturi had a large entry depth, but in Irving's

Figure 3.7.1. Prevost's proposal for a venturi underbody (1928).

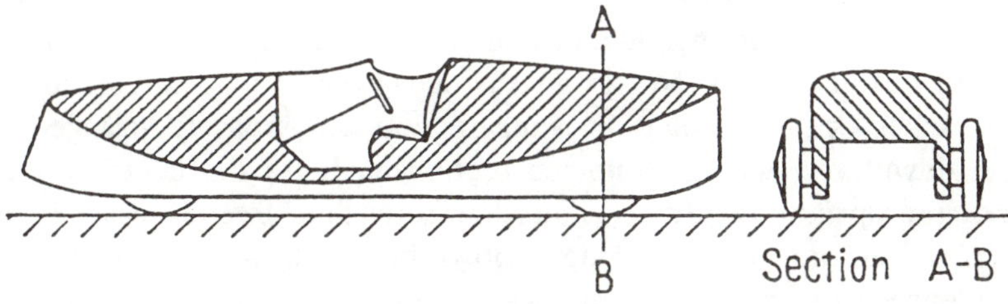

Figure 3.7.2. Basic body shape of Irving's Golden Arrow (1929)

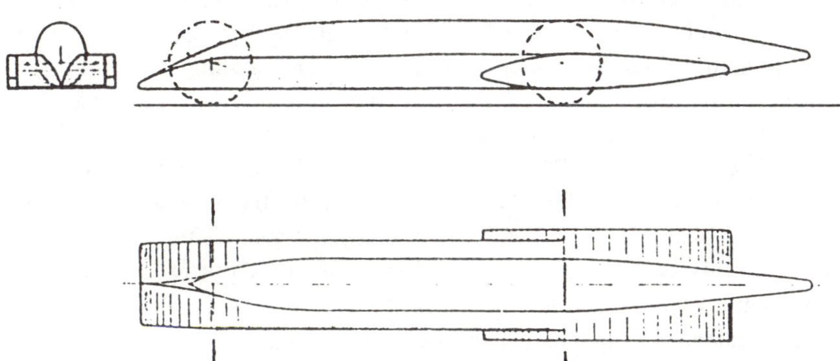

design the venturi had acquired its modern form, with the entry barely any deeper than the throat, necessary because a deep entry gives front lift.

The massive Eyston–Andreau Thunderbolt of 1938 had the eight wheels enclosed within the body using a narrow front track to allow this, and achieved 154 m/s (345 mile/hr) on 3500 kW. In contrast, in 1939 the smaller and highly streamlined Railton Mobil Special achieved 166 m/s (370 mile/hr) on only 930 kW, with the emphasis on low drag including ice cooling to reduce the internal flow losses. It was not unusual in that period of wheel-driven speed vehicles, in contrast to the later thrust reaction types, to have traction problems, and this led to the use of heavy vehicles and in some cases to multiple drive axles. For example, Thunderbolt required a thrust of 23 kN with a weight of 70 kN, for a traction requirement of 0.33 even if all eight wheels had been driven, and a much higher value with only the rear four driven, as used. It should also be borne in mind that record runs were mainly made on natural surfaces, such as sand or salt, with less friction available than on a prepared hard surface.

The possible application of aerodynamic downforce to improve traction was recognised in the 1939 Daimler Benz T80 speed record contender, which was built but never ran. This featured tapered stub wings of low aspect ratio and area about 0.8 m^2 each, positioned on the body sides in front of the rear wheels. However, in marked contrast to the earlier Opel, this time the wing section was properly arranged to give downforce, with a cambered section having a flat upper surface and appropriate incidence, and therefore this should perhaps be admitted as the first true winged downforce vehicle. An adjustable-incidence rear aerofoil on struts was also proposed, but not used. The drag coefficient was 0.18 on 1.7 m^2 frontal area.

It seems that the potential for downforce to improve racing vehicles (as opposed to speed record vehicles) remained unrecognised, with aerodynamic attention limited to drag and engine cooling, and provision of a cool buffet-free cockpit. The next interesting development was the use of an hydraulically-operated air brake on the Mercedes 300SLR sports racer, tested in 1952 and used at Le Mans in 1955. This was intended to relieve the brakes on fast circuits, but it was found that cornering also improved if the brake was kept open. This was because the air-brake, a rear-hinged boot lid that rose to about 60°, acted as a half-metre-high spoiler that gave downforce.

At Indianapolis in this period there was a vogue for full bodywork, which dramatically reduced drag compared with the usual fully exposed wheels, but the extra mass and the adverse aerodynamics in lift and yaw, not yet adequately understood, were sufficient to halt the trend. Compared with most European circuits which have a wide variety of corners and a wide speed range, at Indianapolis speeds are relatively uniform and high, for example a minimum more than 75% of the maximum (about 90 m/s) so aerodynamic effects are particularly strong. However, even in Europe the Mercedes W196 of 1954 used full streamlining to good effect on some circuits.

In 1956 a Porsche sports racer appeared at the Nurburgring with a downforce aerofoil. It was used in practice, but disallowed from racing by the scrutineers, rejected as being dangerous. Over the next ten years up to the mid-1960s there seems to have been no development of the concept. Perhaps in the early 1960s designers were too busy coming to terms with the rear-engine revolution, which itself improved aerodynamics by reducing drag because of the lower driver position.

In 1960 spoilers appeared at the rear of Porsche and Ferrari sports racers, giving some rear downthrust, better high-speed stability and slightly better cornering. In 1965 Herd for McLaren in England and Hall in the U.S.A. made preliminary tests on downforce wings, with encouraging results. In 1966, Hall started to use a downforce wing on his Can-Am series Chaparral, placed high on struts above the rear wheels. The original intention was apparently to improve high-speed stability. In 1967 the wing was of roughly symmetrical section, but had pitch control from a driver's pedal, and was used to give large downforce in corners, but smaller downforce and drag on the straight. In Formula 1, in 1967, limited aerodynamic devices were tried, especially small front 'bib' or 'ear' spoilers on the nose, but these were often removed because of lack of pitch balance. The existence of body lift and the variation of lift and pitch moment with pitch angle were proving a problem, and the Lotus answer was the wedge body that also did well at Indianapolis in 1968. This then acquired front stub wings and an upswept engine-cover rear spoiler.

Then, in mid-1968, the wing idea was widely adopted in Europe, and high mounted wings appeared on many cars, some acting on the body but most acting directly on the hub carriers, to avoid loading the suspension. Balance was achieved by a variety of front devices from bib spoilers to stub wings. In 1969 front wings were also mounted high and on the hub carriers, and many wings had pitch control. However the severe vibration and aerodynamic load reversals caused some dramatic accidents. After a period of uncertainty, new rules limiting wing design were introduced throughout most of motor racing, including Formula 1 and Indianapolis. Essentially these limited the size and height of wings, and required the wings to act on the sprung mass, not directly on the carriers. Wings were still not universally used; an alternative is to try to use the whole of the body's upper surface as a wedge. The wedge body shape usually produces good front downforce, so it is well suited to combination with a large rear wing, with front wings used just for trimming. A wedge body with hydraulically controlled incidence has also been used. A shallow spoiler was used across the front of bodies to reduce body lift or improve downforce. Generally, 1970 saw improved understanding of wings and their use, and the slatted rear wing appeared.

In sports car racing around 1970 the vehicle front profile underwent a notable change from rounded to vertical, i.e. the air-dam front arrived to reduce underbody pressure. Sometimes a 'splitter' plate was added – a thin horizontal lip protruding forward at the bottom. The splitter helped the downforce, but tended to increase pitch sensitivity (sensitivity of the downforce distribution to pitch angle). The sides also became flatter and closer to the ground.

Concave longitudinal section front and rear decks were found to be worthwhile to minimise sensitivity to pitch angle, although increasing drag by 10%. As an example of the drag influence of bodywork, wind tunnel tests to attempt to purge the drag of one Porsche open sports racer reduced C_D from 0.70 to 0.49, and to 0.36 with a long tail. The better short-tail version gave its best track performance with the addition of front and rear spoilers that increased C_D to 0.58. Twin rear fins were found to reduce drag, as was a wing mounted between them, because this prevented flow separation from the body. Front-edge fences can considerably increase front downforce without increasing drag. Full-length edge fences were briefly popular. Before movable aerodynamic devices were banned, flaps at the rear operated by the suspension were used to minimise pitch changes. Separate left and right flaps can in principle reduce roll, but in practice this did not prove helpful.

More revolutionary, in 1970 Hall produced the fan car for sports racing. This had flexible sealing skirts all round, with twin fans that extracted the air and blew it backwards, reducing the underpressure. One advantage of this system over wings is that it functions at full effectiveness at low speed, so low-speed cornering and low-speed traction are greatly improved. With 6 m^2 of enclosed area, only 1 kPa pressure reduction would give 3 kN downforce, about half the vehicle weight. The principle was very successful, and was promptly banned. The same idea appeared briefly in Formula 1 in 1978, with the same results.

1971 saw a marked improvement in the application of wings. Previously the rear wing was mounted rather low over the engine, with very restricted flow to its underside, so that it really acted as a large spoiler. On the 1971 McLaren for Indianapolis, which was basically a wedge with wings and side radiators, the rear bodywork was improved to give a good airflow below the wing, greatly increasing the downforce. The downforce was equal to about half the car's weight at maximum speed, about 3 kN at 90 m/s or a total lift area C_LA of about 0.6 m^2. In Formula 1 tall air boxes also appeared, to deliver air to the engine, although there was disagreement about whether they made airflow to the rear wing better or worse. In 1972 there were further dramatic cornering speed increases, with the rear wing going further to the rear. Total downforce increased to about 6 kN at 90 m/s, a lift area of 1.2 m^2. The actual wing areas were about 0.6 m^2 at the rear and 0.25 m^2 at the front, these being limited by the rules. One advantage of the rearward wing is that the suction on the wing undersurface is not compromised by suction on the top surface of the body beneath, because the wing is over the road instead. Also, because the rear wing is of low aspect ratio and in turbulent air, its maximum lift is limited, so moving it back whilst retaining overall aerodynamic balance allows a larger downforce at the front giving an increased total downforce.

The period 1973 to 1977 saw refinement of the aerodynamics with increasing lift coefficients from improved aerofoil contours, tip plates, adjustable flaps, slots, trailing-edge lips, and so on.

The next revolution in racing was to use the underbody to produce downforce, i.e. ground-effect venturis as used on speed record cars in the 1920s, introduced to racing by Lotus in 1977. The essential concept of the car was a very wide chord wing on each side of the body. The conceptual improvement was large end-plates with brushes that rubbed against the ground to provide an air seal. The wing span was severely limited by the body width regulations, but despite the very low aspect ratio and the poor seal of the brushes, considerable downforce was generated at a good lift/drag ratio. The cooling radiators were buried in the wings. The wing and ground considered in longitudinal section act as a venturi. The wing to ground clearance (the venturi throat) is about 100 mm. The large exit draws a maximum of air through the throat where the correspondingly high airspeed results in a low pressure on the wing undersurface giving an effective downforce. A later development was to have a long parallel throat to maximise the area where the pressure is very low. This effect is assisted by the extractor effect of the rear wing which helps to draw air through the venturi. The Lotus 79, for 1978, developed the idea further, the whole layout being governed by optimisation of the ground effects; this included minimising the width of the true body to maximise the width of the venturi tunnels. To do this the driver was moved forward, and the fuel was placed in a single cell behind the driver. This car firmly established the ground-effect concept.

In 1979 ground-effect cars also appeared at Indianapolis, and wherever it was allowed, ground effect became essential. Optimisation of the ground-effects tunnels affected many other features; for example, where previously the flat opposed-cylinder engine layout was considered ideal because of the low centre of mass, the V8 became preferable because it is narrower and matches the front profile of the driver with narrow hips and wide shoulders, giving more room for the tunnels.

Front wings were often omitted because the downforce from underbody tunnels was already well forward, tending to give limit oversteer at high speed. To help to correct this, front wings were sometimes used to give an upward force. It is important to position the downforce correctly; fine balance is usually achieved by adjusting the rear wing flap. Front suspension was revised, moving the springs and dampers into the body to improve airflow to the tunnels.

The downforce did create problems in demanding extremely stiff suspension springing to minimise height and attitude changes, even with sliding or flexible sealing skirts. In 1981, in Formula 1, a 60 mm ground clearance was required, but this rule was circumvented by hydraulic suspension lifters. Flat bottoms between the axles and greater clearance were required for 1983, which reduced downforce and allowed suspension action to be restored. The body can still produce some downforce, but less than the wings.

1981 was also the year of the abortive Lotus 88, an imaginative concept that fell foul of the rules. This had two chassis, the inner one being the main chassis with conventional suspension, and the outer one being an aerodynamic shell with much stiffer connection to the wheels allowing good control of the ground clearance. It was disallowed on the basis that the outer shell was a movable aerodynamic device.

At their peak, ground effects were such that downforces of about 22 kN were being achieved in Formula 1 at 75 m/s, corresponding to a lift area of 6.4 m^2. Of this, about 20% came from the wings and 80% from the body. On a vehicle mass of 660 kg and with a tyre friction coefficient of over 1.0, this allows lateral and braking accelerations of about 40 m/s^2 ($4g$). An interesting aspect of downforce is that for a given force the performance becomes more sensitive to the total mass. This is because for a large downforce the tyre vertical forces are almost independent of weight, so the maximum lateral acceleration becomes inversely proportional to mass. Also, the position of the centre of mass becomes critical. The drag area was about 1.4 m^2, giving a lift/drag ratio of 4.6. The current restrictive rules, specifying maximum wing sizes and calling for flat-bottomed areas under the body, considerably reduce the downforce values, although some favourable underbody effect is achieved by swept-up diffuser areas at the rear. The engine exhaust is often fed to the diffusers, possibly 'blowing' an improved airflow beneath, and also helping the engine by reduced exhaust back pressure, although often compromising the length of the exhaust pipe secondaries for optimum pressure pulse tuning.

Over recent years there has been detailed optimisation rather than revolution, and aerodynamics is now just a large part of the picture, rather than completely dominating it.

Wind-tunnel tests of ground-effect vehicles have sometimes shown considerable discrepancies from track behaviour. This is caused by the boundary layer on the floor of the wind tunnel. Nowadays it is considered highly desirable to have a moving-belt floor in the wind tunnel, the speed of which matches the airspeed. Corrections can be applied for static floors, but this is not completely satisfactory.

Downforce has been used in other competition motoring events, including rallying, drag racing and kart racing. Small stub wings have even been used on motorcycles, to discourage lifting of the front wheel in acceleration, but wings are now banned. In any case, because a motorcycle leans considerably in a corner, downforce wings would be of limited effectiveness in cornering. Racing sidecar motorcycles have more extensive bodywork, so some downforce can be produced.

3.8 Problems

Q 3.2.1 What is the standard atmospheric density at 500 m above sea level?

Q 3.2.2 Temperature, pressure and relative humidity are measured as 22.3°C, 99.4 kPa, and 54%. Calculate the density according to the approximate method of Section 3.2.

Q 3.3.1 In open terrain the 10 m wind is 8.4 m/s. Estimate the wind speed at 1 m height.

Q 3.3.2 An airflow has a mean speed of 14 m/s and turbulent intensity of 24%. What speed range will the instantaneous speed be within for 95.5% of the time?

Q 3.3.3 In a wind of 8 m/s and turbulent intensity 28%, a car drives cross-wind at 27 m/s. What turbulent intensity does the car experience?

Q 3.3.4 The wind at a high bridge may be modelled by a Weibull distribution with characteristic speed 7 m/s and shape factor 1.9. What is the probability for winds exceeding 10 m/s? If sustained winds exceeding 20 m/s make the bridge unsafe for use, what proportion of the time is it likely to be closed?

Q 3.4.1 Discuss the differences between aircraft aerodynamics and road vehicle aerodynamics.

Q 3.4.2 Discuss the problems of obtaining accurate force prediction for a real car by testing a small model in a wind tunnel.

Q 3.4.3 Describe how force coefficients are likely to change with Reynolds number for various types of body.

Q 3.4.4 A vehicle travels at 26 m/s at a heading angle of 40° with a sideslip angle of 4° in a wind of 16 m/s at 160°. Find the air velocity relative to the vehicle. What is the aerodynamic yaw angle?

Q 3.5.1 A vehicle has mass 1200 kg, centre of mass 1.350 m back on a 3.000 m wheelbase, and $C_L = 0.412$, $C_P = 0.083$, $C_S = 0.160$, $C_Y = 0.020$, and $C_D = 0.392$, based on standard vehicle aerodynamic axes. The air density is 1.200 kg/m^3. The frontal area is 2.35 m^2 and the speed 29 m/s. What are the actual forces and moments?

Q 3.5.2 For the vehicle of the last question, find the C_{Lf} and C_{Lr}.

Q 3.5.3 The vehicle of Q 3.5.1 has a lateral acceleration of 4 m/s^2 at a speed of 45 m/s. What are the front and rear axle cornering force coefficients?

Q 3.6.1 Summarise typical values of car aerodynamic coefficients.

Q 3.6.2 A typical car is given a steady-state cornering test at 8 m/s^2 on a 50 m radius, at which it has an attitude angle of 8°. Make an estimate of the aerodynamic forces and moments, and discuss their significance.

Q 3.6.3 Explain the concepts of centre of pressure and aerodynamic centre in the context of lateral aerodynamic force and yaw moment on a vehicle, including any relevant equations.

Q 3.7.1 The Daimler Benz T80 land speed record contender had a mass of about 2000 kg and a probable design speed of 200 m/s, with a drag area of 0.306 m^2. Discuss the possible value of the wings (total area 1.6 m^2) in enhancing traction, giving calculations.

Q 3.7.2 Describe the features of the bodywork of a typical modern sports racer, and discuss the reasons for it.

Q 3.7.3 Discuss the evolution of Formula 1 or Indianapolis racing car design in the light of aerodynamic effects and rule changes.

Q 3.8.1 See Q 5.18.3 – 5.18.6 at the end of Chapter 5.

3.9 Bibliography

A general introduction to automobile aerodynamics is provided by Howard (1986). Ludvigsen (1970) reviewed historical developments in vehicle aerodynamics. A good overview of vehicle aerodynamics, with innumerable references, can be found in Dorgham & Businaro (1983). An excellent condensation and systematisation of the extensive experimental data on drag and lift is given in Hoerner (1965) and Hoerner & Borst (1975).

From the many books on competition motor vehicles, four that can be recommended (not just for aerodynamics) are: Frere (1973), Nye (1978), Huntingdon (1981) and Posthumus & Tremayne (1985).

For those interested in the application of wings to competition vehicles, the theory of wings may be found in fluid dynamic texts, for example Hoerner (1965). A basic introduction is provided by Massey (1983). Detail is given in more specialised texts, such as Houghton & Brock (1970). For theory and extensive wind-tunnel results of wing sections, see Abbott & von Doenhoff (1959). Rae & Pope (1984) is a useful reference on testing techniques. For technical analysis of some aspects of competition vehicles, see Pershing (1968 and 1974).

4

Suspension components

4.1 Introduction

Chapter 2 showed that the lateral force exerted by the road on a tyre depends upon many factors; principal amongst these are the slip angle, the camber angle, the vertical tyre deflection and the longitudinal slip (or the vertical force and the longitudinal force). The angular positions and the vertical force depend upon the suspension system, which locates the wheel relative to the vehicle body. Thus the suspension system design plays an important role in the cornering and handling characteristics, and requires detailed consideration. In analysing the suspension, our main interests are in the geometry in order to find the wheel camber and steer angles, and in the distribution of vertical force between the four wheels. This chapter looks mainly at the properties of individual suspension components. Chapter 5 then looks at the complete suspension, and at suspension analysis in the context of handling, including ideas such as the roll-centre and roll-axis.

In a practical suspension system, the wheel is connected to the body through various links; these permit an approximately vertical motion of the wheel relative to the body, controlled by the spring and damper. The steering system controls a different type of wheel motion – rotation about an approximately vertical axis. For a normal road vehicle, the suspension links include rubber bushes. These reduce the transmission of noise, vibration and harshness into the passenger compartment. The compliance in the links, bushes, steering system and chassis has the result that the slip angle and camber angle of the wheel are to some extent dependent on the forces acting. However, in order to obtain a clear picture of the operation and behaviour of suspensions, it is easiest to begin by analysing them as though all links were perfectly rigid, and to add the effects of compliance later. Thus we should consider:

(1) The geometry of idealised suspensions, investigating possible arrangements of the links and the consequent implications for the wheel motion relative to the chassis;

(2) The compliance of suspension components, in particular those that are specifically provided to control the wheel motions, such as springs and anti-roll bars, but also the compliance of links and rubber bushes;

(3) Friction, both that deliberately introduced by dampers, and residual friction in the joints;

(4) Inertia of the components in the types of motion relevant to suspension action.

The intention of this chapter is not to provide a comprehensive systematisation of all possible suspensions; rather it is to reveal some order among the wide range of types, and to outline the features that influence handling.

4.2 Mobility analysis

The following few pages attempt to place some systematic order on suspension systems, by using a degrees-of-freedom (mobility) analysis. Mobility analysis can be very helpful in the design of precise mechanisms, but there are some difficulties in applying it to vehicle suspensions. Most suspensions include many compliant rubber bushes, so that changes of link lengths are possible, contrary to the geometrically ideal rigid bodies of mobility analysis. Even worse, sometimes the link itself is highly compliant, such as a leaf spring supporting an axle, or a trailing twist axle. Such cases must be dealt with by finding geometrically equivalent rigid linkages. Further, some suspensions are redundant, i.e. they contain 'too many' links, because in the idealised analysis one or more of the links merely confirms rather than controls a particular motion. Finally, a suspension that in an idealised analysis appears to be a structure, rather than a mechanism, may function adequately because of compliance in the bushes or in the links themselves.

Notwithstanding the above limitations, most suspensions are amenable to analysis, and a degrees-of-freedom analysis does provide a helpful framework for categorising the wide range of real suspension designs. Even if it does not point directly to an optimum design, it provides an appreciation of certain limitations of the simple mechanisms that are met in practice.

Every real suspension works in three spatial dimensions. A full kinematic analysis of the required generality to cope with the range of practical suspensions is highly complex. Therefore, the presentation here is based on an appeal to relatively simple notions in common engineering usage. Computer software packages now make possible advanced three-dimensional analysis. Nevertheless, it is helpful to begin by considering the behaviour of two-dimensional suspension-like mechanisms, especially since the interpretation and discussion of actual handling behaviour is based largely on two-dimensional motion concepts such as camber and steer. Some principles of kinematics are reviewed here because they provide the key to important later results.

A two-dimensional mechanism is one with all motions parallel to a given plane. The dimension perpendicular to that plane is neglected. This could be said to correspond to a two-dimensional drawing, such as a front elevation. To specify the position of any particular object then requires three coordinates in a specified coordinate system, and a freely mobile object is therefore said to have three degrees of freedom when in two-dimensional motion. For example, the position of a link could be specified by two coordinates (x,y) for a given point on the link, plus the angle of the link (θ). If such an object moves without change of the angle θ, it is said to be translating; if it moves with θ changing but the location point remaining at (x,y), then it is said to be rotating about the point (x,y).

Even in complex motions, at any instant it is possible to identify a point that is the centre of rotation at that instant. This point is called the instantaneous centre, or centro. Imagining the moving object extended as a sheet in all directions, the centro is then that point of the sheet that is not changing its (x,y) coordinates at that instant, i.e. it is the point that is stationary in the selected coordinate system at that instant.

One important property of rotation is that all velocities are perpendicular to the radius from the rotation centre, and have a magnitude given by $V = \omega r$ where ω is the angular speed at that moment. This property of perpendicularity is very useful in identifying centros (instantaneous centres). If the velocity directions are known at any two points of an object, then the radii can be drawn, and their intersection must be the centro. These known velocities must be in a specified coordinate system or, equivalently, they must be relative to some other specified object, for example the chassis.

When there are several objects all moving, as in the case of a chassis plus several suspension links and a wheel carrier, then we can direct our attention to any pair of these objects, and that pair will have a particular centro for the motion at that instant, i.e. for one object relative to the other. This is in general different from the centro for any other pair of the objects. However, the centro of A's motion relative to B is the same as that of B's motion relative to A. The centro is a property of the objects as a pair. In some cases it is normal practice for one of the objects to be implied; for example to speak of the centro of the wheel normally means the centro of the wheel–body combination, and hence of the centro of the wheel motion relative to the body. The centro concept is a powerful tool in suspension analysis.

One special case occurs when the relative motion of the objects is pure translation. Then the constructed radii intended to reveal the centro do not intersect. They are said to meet at infinity (at either side), and this is where the centro is located. This does not cause any practical analytic problems because lines constructed from the centro are simply parallel to the locating radii.

In many practical cases of connected components, the centro location is obvious without formal construction of radii. This is the case, for example, for a radius rod where the rod pivot itself defines the centro for the rod motion relative to the base member.

A property that we shall require in the investigation of roll-centres is illustrated in Figure 4.2.1; the centros for the three possible pairs of a group of three objects lie on a straight line. This is known as the Kennedy–Arronhold theorem. In this example AB is pivoted on BC, which is pivoted on CDE. Therefore B is the centro for AB with BC, and C is the centro for BC with CDE. Considering motion relative to link AB, evidently C can only move perpendicularly to the radial line BC. Considering the motion of CDE relative to C, the point E must be moving perpendicularly to the radial line CE. The velocity of E relative to B, $\mathbf{V}_{E/B}$, is the vector sum of $\mathbf{V}_{E/C}$ and $\mathbf{V}_{C/B}$, which are both perpendicular to BCE. Thus $\mathbf{V}_{E/B}$ is perpendicular to BCE. Thus the centro of AB with CDE must be somewhere on a line through BC, i.e. the three centros must be on a straight line..

In two-dimensions, the connection of a pair of objects can be classified as either a sliding joint or a rotating joint. In practice a sliding joint is usually straight. Relative to a fixed member, a slider has one

Figure 4.2.1. Kennedy–Arronhold theorem of co-linear centros.

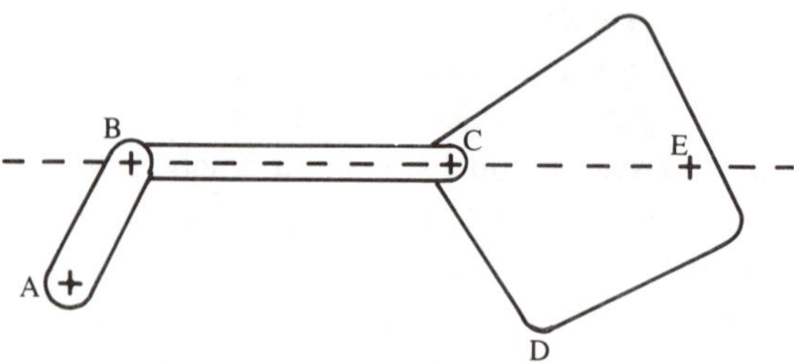

degree of freedom, i.e. its relative position can be specified by a single parameter. The pivoting link also has one degree of freedom (1-d.o.f.), but in rotation, the relative position being specified by a single angle. These elements can be joined serially to allow two degrees of freedom relative to the base member, Figure 4.2.2, where the trunnion (c or d) corresponds closely to the function of a typical suspension strut. The links with two degrees of freedom (2-d.o.f.) require two parameters to specify their positions relative to the hatched members; e.g. in (b) two angles, in (d) one angle plus the extension.

Figure 4.2.2. Serial connection for 2-d.o.f.

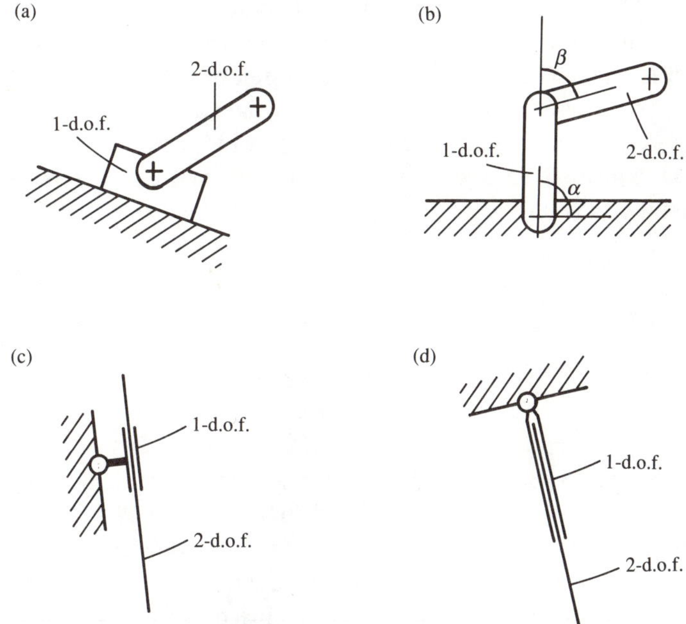

4.3 Straight-line mechanisms

It is often desirable to provide a straight motion path for some point of a suspension member, e.g. for lateral location of an axle, and there are several practical ways to achieve this. This section presents several such methods. Analysis of straight-line mechanisms also demonstrates some wider truths that will be relevant later on. In a real suspension, the straight line need only be approximate; precision of the path may be traded-off against other factors, especially cost.

Geometrically, we wish to constrain a given point on body A to move in a straight line relative to body B. Thus there must remain two degrees of freedom for the relative motion. For example, an axle must rise and fall (heave, 1-d.o.f.), and have freedom to roll (1-d.o.f.), but must be located laterally. For convenience we can assume that the angular rolling freedom is achieved because body B is provided with a pivot hole at the appropriate point to accept a pivot pin; the mechanism to be designed must therefore cause the pin to have a straight-line motion relative to body A.

The most direct solution is to provide a simple slider, Figure 4.3.1. Here the sliding block B has one degree of freedom relative to stationary member A attached to the body. C is the point with straight-line motion. The axle has its second degree of freedom by rotating about C. The locating channel A may alternatively be fixed to the axle and C to the chassis. This is not entirely equivalent, because when the channel is attached to the rolled body, motion of the axle perpendicular to the road requires some lateral motion of the axle or body. In practice the slider method of Figure 4.3.1 is not good for passenger cars because of noise transmission and wear, but it is compact and has been used successfully in racing.

Figure 4.3.1. Slider.

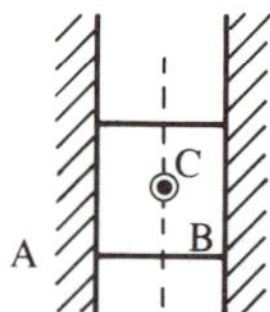

The second type of straight-line linkage is the simple radius rod, Figure 4.3.2, known as the Panhard rod (track rod in U.S.A.) when used for lateral axle location. Of course, in a formal geometric sense, this is anything but a straight-line mechanism, but it is commonly used because its advantages often outweigh the inaccuracy of its straight-line approximation. In seeking to provide a straight line in a given direction it is subject to two kinds of errors. Firstly, the path is non-linear, having a radius equal to the rod length. Thus a total travel h has a lateral path error $l(1 - \cos\theta)$ where $\sin\theta = h/2l$. This error is approximately $h^2/8l$, showing that the error grows with the square of the travel and inversely with rod length. For a Panhard rod with a practical length of 1 m, a suspension range of 0.2 m gives a rod angle of up to 6°, with an error of 5 mm. The second kind of error arises when the rod is not perpendicular to the desired path, this being equivalent to a misaligned channel in the slider-type mechanism. The influence of body roll is not as clear as in the case of the slider mechanism. Both roll and heave of the chassis will alter the height of the pivot, thus introducing errors of the second kind. Also, the consequences of roll are not symmetrical since the pivot, normally being on one side of the chassis, will rise or fall according to direction of roll. The vertical motion of the pivot of a Panhard rod in roll is eliminated if the chassis pivot is on the vehicle centre-line, but this is achieved at the expense of shortening the rod and increasing the curvature errors, and is therefore not normally chosen.

Figure 4.3.2. Panhard rod (radius rod).

A third type of straight-line mechanism is Watt's linkage, Figure 4.3.3. This may be perceived as a logical development of the radius rod, introducing compensating errors. As the link BCD rises or falls, the two equal links AB and DE rotate in opposite senses; C at the mid-point of BD adopts a mean position, resulting in a remarkably good straight line. The errors depend upon the length of the short vertical link, as does the total limiting range of vertical motion. For suspension purposes there is negligible error for about half of the total vertical travel, or over a total range of about one quarter of the total mechanism width, neatly encompassing practical requirements of axle lateral location if the full available width between the wheels is employed.

Figure 4.3.3. Watt's linkage.

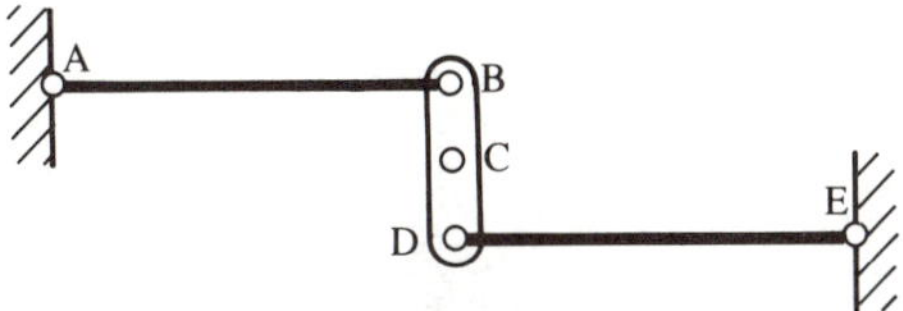

Watt's linkage can be generalised somewhat; it is not essential for the arms AB and DE to be of equal length provided that C is located at the appropriate point along BD to optimally compensate the errors. Simply stated, C must be nearer to the longer arm, where:

$$\frac{BC}{CD} = \frac{DE}{AB}$$

Alternatively, it is possible to have both arms on the same side of the vertical link, Figure 4.3.4. To retain the error compensation, C must now be on the opposite end from the short link, with the above proportions still applied. This layout often appears in the side elevation of rear-axle locations, and also in the front elevation of independent suspensions, although in this context its relationship to the Watt's linkage usually passes unremarked. Where ground clearance is not important, it can be used for solid-axle lateral location to give a low roll-centre, and has been used in this way as a modification in racing.

Figure 4.3.4. Modified Watt linkage.

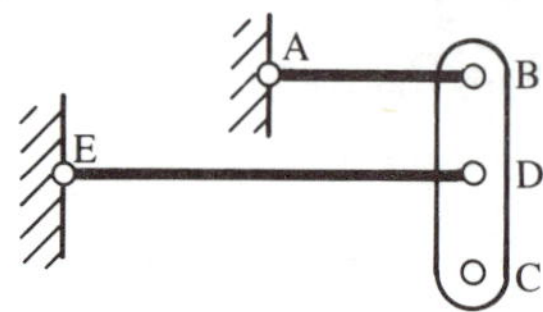

A fifth kind is the Roberts straight-line mechanism, Figure 4.3.5. Once again, this can be considered in terms of compensating errors. As C moves up or down, the inclinations of AB and CB to the horizontal are kept approximately equal. For good results, AB, BC, CD and DE should be equal in length and at equal angles to the horizontal in the middle position. Then C is exactly on AE, and also passes directly through A and E. It has very little error over the whole range A to E provided that the initial link inclination θ is kept below about 25°. Thus links of about 300 mm length at 20° will secure a very good motion over an adequate practical suspension range. Its main problem is the need to provide a low fixture at E; nevertheless it has found application in racing.

Figure 4.3.5. Roberts linkage.

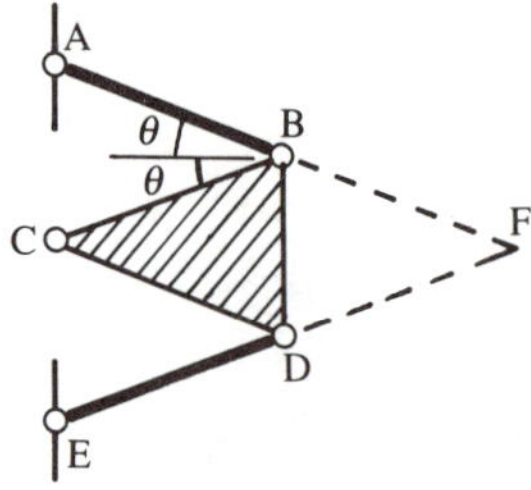

This mechanism provides a useful cautionary tale. The instantaneous centre F for the triangular plate is easily found by the usual construction of Figure 4.3.5, projecting AB and ED, because B must move perpendicularly to AB so the centro is on AB projected. Thus it is sometimes remarked that this mechanism is equivalent to a Panhard rod of length CF. This is incorrect; the path for a radius rod CF is a very poor competitor to the straight line obtained with the true mechanism. This difference arises because the instantaneous centre itself moves as the mechanism moves; indeed the reason for the good straight line is that the centro F keeps pace with C as the latter moves. This keeps CF closely perpendicular to AE so that at every instant the motion of C, perpendicular to CF, has little or no component perpendicular to the desired path AE, which is not the case for the supposedly equivalent radius rod. The truly equivalent radius rod for a given motion must therefore be derived with some caution; this will be relevant later when discussing widely used concepts such as the 'equivalent swing arm' for independent suspensions.

Figure 4.3.6. Tchebichef linkage.

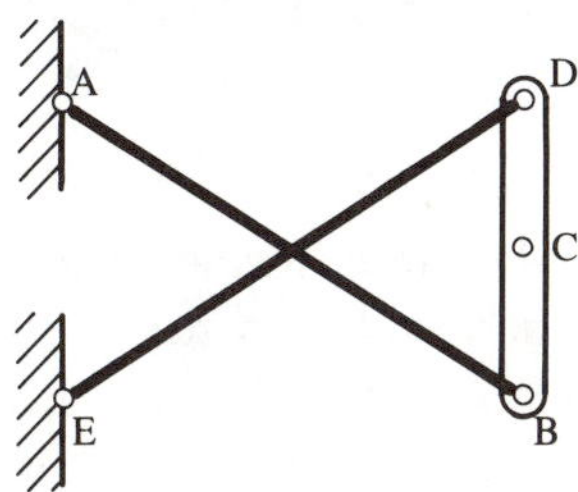

The Tchebichef straight-line mechanism, Figure 4.3.6, has a much more limited useful range; C takes the mean lateral position of B and D, which have compensating motions. The Evans mechanism, Figure 4.3.7, in which the inclinations of CD and DE compensate, looks reasonably practicable. The Aston Martin linkage of Figure 4.3.8, which has been used successfully despite its complexity, is well suited to the space that is available adjacent to an axle, and shows how symmetry can be used to obtain a true straight line. The plate FCJ is preferably fixed to the body. Links GK and HK ensure that BFG and DJH, which pivot on FCJ at F and J, must remain symmetrically positioned on FCJ. In a simplified version, G and H are directly linked, necessary compliance being provided by a rubber bush, or a simple slot.

Figure 4.3.7. Evans linkage.

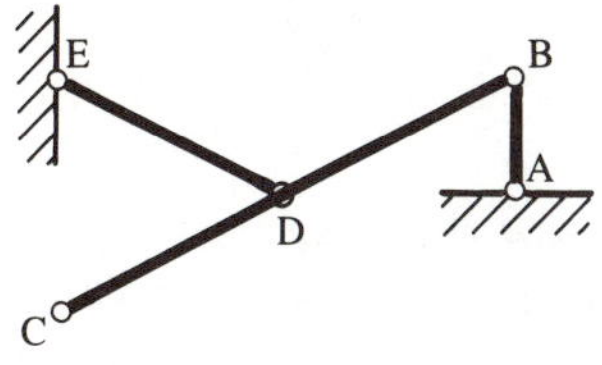

Figure 4.3.8. Aston Martin linkage.

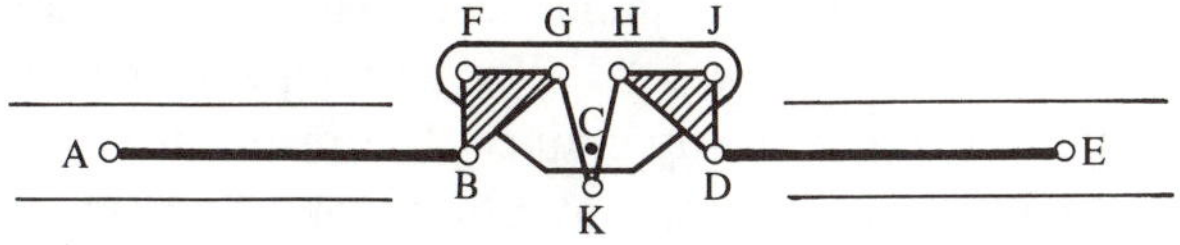

Figure 4.3.9 shows a close relative used on Mallock racing cars. Here BFG and HJD are still pivoted on the body, but symmetry is abandoned allowing the simple connecting link GH. Also, by inclining the links AB and DE, a lower roll-centre is obtained.

Figure 4.3.9. Mallock linkage.

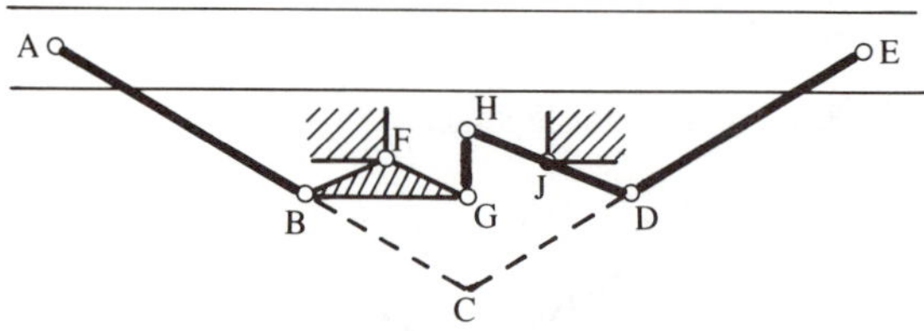

Figure 4.3.10 shows the Adex mechanism, which was used in the 1920s, and was particularly suited to the chassis construction of that period. Here, DE and EB act to keep B and D symmetrical about the centre-line, and BC and CD therefore keep C in the centre.

Figure 4.3.10. Adex linkage.

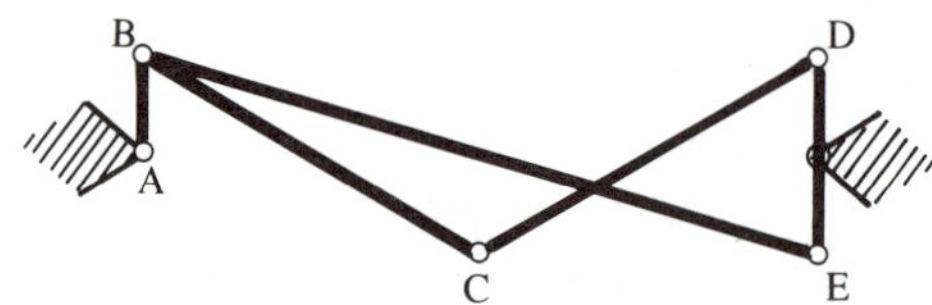

A last example: taking advantage of the third dimension admits a further solution, the Alfa-Romeo T-bar, Figure 4.3.11, in this case shown in plan view instead of rear view; the locus is a true straight line in rear view, but does have curvature in the side view.

Figure 4.3.11. Alfa-Romeo T-bar (plan view).

4.4 Two-dimensional analysis

This section presents a simple analysis of two-dimensonal suspension-like mechanisms. In this case the 'wheel' must have one degree of freedom relative to the chassis structure, for vertical motion. This suggests two particularly simple suspensions: firstly a slider, and secondly a single pivot, Figure 4.4.1. The first type corresponds to the pillar suspension, now practically extinct. The instantaneous centre for the wheel and body is, in this case, at infinity on a line perpendicular to

Figure 4.4.1. 2-D suspension.

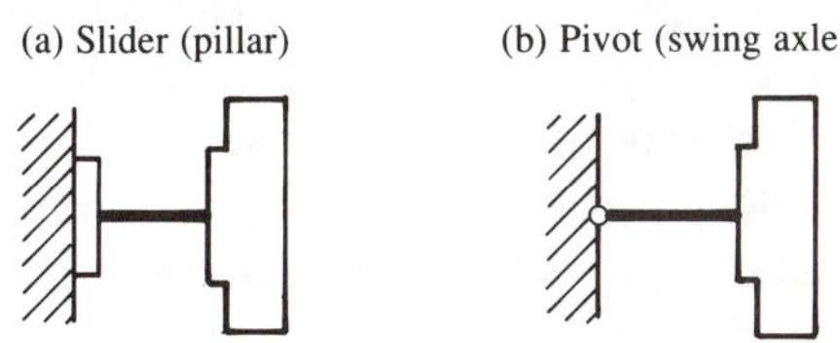

the slider direction. The second type corresponds to the swing axle, for which the centro is at the pivot.

If we allow the wheel an extra degree of freedom by adding a pivot at the outer end of the swing-axle pivot arm, then an additional radius rod may be added, Figure 4.4.2, corresponding to the three-dimensional two-wishbone (two-A-arm) suspension. The centro of the wheel relative to the body is in this case at the intersection of the radius arm lines, at infinity if they are parallel. Alternatively, a trunnion may be added, Figure 4.4.3; this gives the equivalent of the strut type suspension. In general the strut slider does not align with the lower outer pivot. It usually does align with the trunnion top pivot, although it need not do so. To find the centro of the wheel relative to the body, the correct radius line for the trunnion is the line perpendicular to the slider line AB and passing through the top pivot point A; the centro of the wheel motion relative to the body is where this line meets CD extended. This should be distinguished from the steering axis, which is line AC that goes from the trunnion point A down to the radius arm outer ball joint C.

Figure 4.4.2. 2-D suspension using two rods.

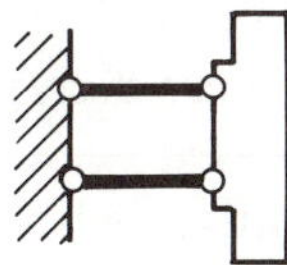

Figure 4.4.3. 2-D suspension using one rod plus one trunnion.

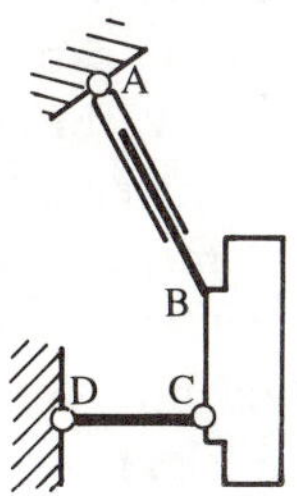

This range of four suspension types realistically exhausts the elementary geometric types within two dimensions. These suspensions can be divided into two types: those in which the location is through a single member (pillar and swing axle), and those in which there are separate upper and lower locations (the double link, and the trunnion with link). The last two types, although somewhat more complex, do permit a much wider choice of location of instantaneous centre. Of course, more complex systems can be devised, since there is no limit to the number of links in a mechanism having one degree of freedom.

4.5 Independent systems

In the full three dimensions, an unconstrained object has six degrees of freedom: three of translation and three of rotation. Relative to the chassis, a wheel has two degrees of freedom for a given steering position. One of these is rotation corresponding to normal rolling. The other is for the vertical action of the suspension. The wheel carrier has only one degree of freedom when the steering is fixed. Since an entirely free object has six degrees of freedom, it is necessary to provide constraints to remove five degrees of freedom from the carrier, including the constraint of the steering.

Figure 4.5.1 outlines some constituents of the appropriate constraints. It is convenient to consider these to be built-up in the following way. A spherical (ball) joint (a) allows three degrees of freedom. A radius rod (b), ball-jointed at each end, allows five degrees of freedom, i.e. constrains one degree. The only constraint of such a radius rod on the two objects that it connects, is that it prevents relative motion directly along the rod. Two such radius rods will constrain two degrees of freedom. These two rods can be connected together to form a wishbone or A-arm. How the names derived from the shape is visible in Figure 4.5.2, but they are now used interchangeably, with the term 'A-arm' favoured in U.S.A., whilst 'wishbone' is used in U.K. The wishbone constraint results in one point on member D being constrained to move in an arc about an axis through the two spherical joints on A, leaving D four degrees of freedom. The same result is obtained by a pivot arm that is ball-jointed to the carrier, Figure 4.5.2(c). The term wishbone will be used generically for wishbones and A-arms, and pivoted arms with a ball-joint at the end. On the other hand, if member D is simply pivoted directly on member A, equivalent to locking the three degrees of freedom of the spherical joint on D, then only one degree of freedom remains. Evidently this gives a complete suspension unit, i.e. the wheel carrier has a 'rigid arm' (as opposed to a wishbone)

Figure 4.5.1. Various suspension elements.

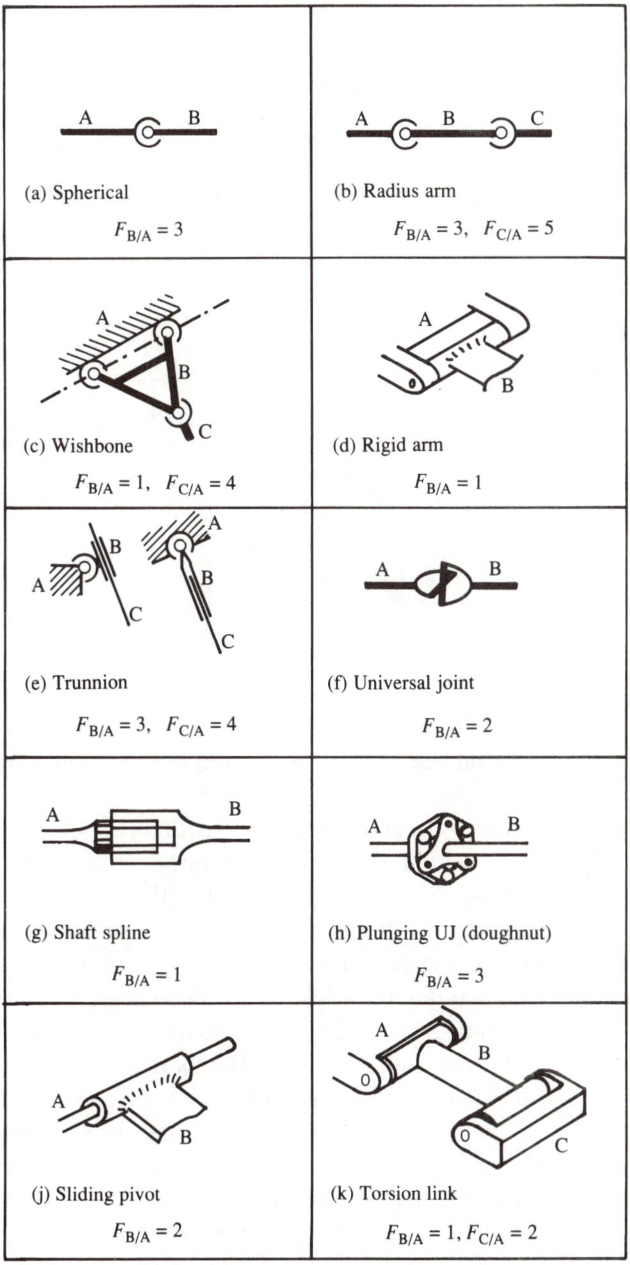

(a) Spherical $F_{B/A} = 3$	(b) Radius arm $F_{B/A} = 3, \quad F_{C/A} = 5$
(c) Wishbone $F_{B/A} = 1, \quad F_{C/A} = 4$	(d) Rigid arm $F_{B/A} = 1$
(e) Trunnion $F_{B/A} = 3, \quad F_{C/A} = 4$	(f) Universal joint $F_{B/A} = 2$
(g) Shaft spline $F_{B/A} = 1$	(h) Plunging UJ (doughnut) $F_{B/A} = 3$
(j) Sliding pivot $F_{B/A} = 2$	(k) Torsion link $F_{B/A} = 1, F_{C/A} = 2$

Figure 4.5.2. Plan views: (a) wishbone, (b) A-arm, (c) pivot arm.

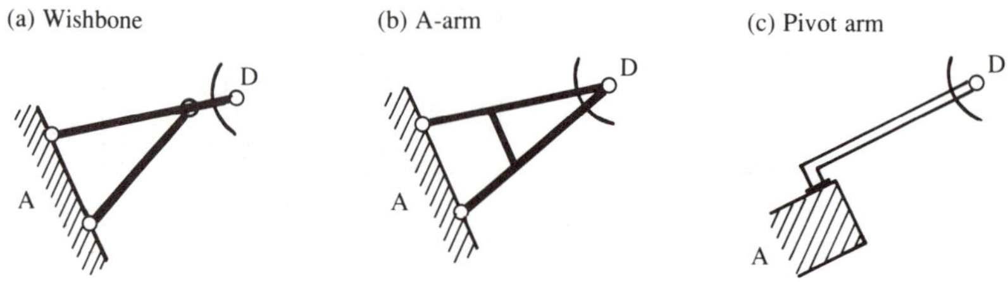

Figure 4.5.3. Rigid trailing arm.

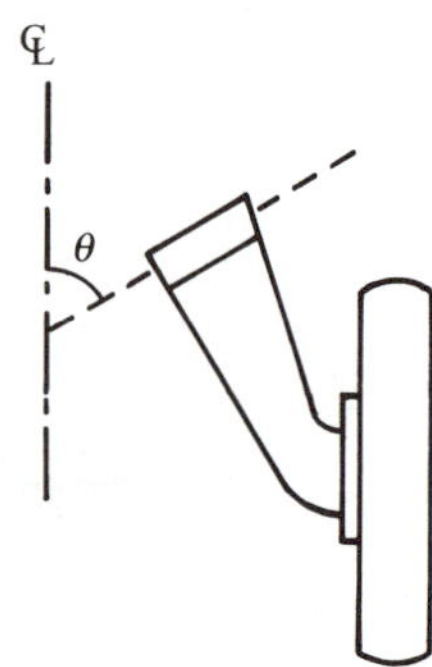

that is pivoted only on the chassis, for example a trailing arm, Figure 4.5.3.

The wishbone and rigid arm are shown in Figures 4.5.1(c) and (d) respectively. The trunnion connection in three dimensions, Figure 4.5.1(e), provides two degrees of constraint – the inner slider (at its notional extended point next to the trunnion pivot) cannot move in either of the two directions perpendicular to the slider alignment. Thus the trunnion constrains two degrees of freedom. The universal joint, Figure 4.5.1(f), constrains plunge and rotation. The shaft spline, Figure 4.5.1(g), allows plunge but constrains rotation. The plunging universal joint (doughnut), Figure 4.5.1(h), constrains rotation and radial motion in two directions. The sliding pivot, Figure 4.5.1(j), is free to slide and rotate. The torsion link, Figure 4.5.1(k), a torsionally rigid link between two pivots, removes four degrees of freedom.

The above classification makes it possible to list the practical range of simple independent suspension types. The practical combinations to constrain five degrees of freedom are, showing the constraints in brackets:

(1) Rigid arm (5)

(2) Two wishbones + 1 rod ($2 \times 2 + 1$)

(3) One wishbone + 1 trunnion + 1 rod ($2 + 2 + 1$)

(4) One wishbone + 3 rods ($2 + 3 \times 1$)

(5) One trunnion + 3 rods ($2 + 3 \times 1$)

(6) One torsion link + 1 rod ($4 + 1$)

(7) Five rods (5)

It is of course necessary that all alignments are appropriately made so that no members are geometrically redundant. Also, the list is not exhaustive. The twin-trunnion plus rod, for example, is certainly possible, even if unlikely in practice. More complex combinations are, of course, always possible.

The directly pivoted carrier, i.e. the rigid arm of Figure 4.5.3, does not admit the possibility of steering, but is widely used at the rear for both front- and rear-wheel drive. It is mainly characterised by the pivot angle θ to the vehicle centre-line. If θ is 90°, it is a plain trailing arm. If θ is zero, it is a transverse rigid arm, sometimes loosely called a swing axle. In the particular case that the axle, with no plunge freedom, forms part of the arm, then it is a true swing axle. Intermediate values of θ, mostly in the range 15° to 30°, give the semi-trailing arm.

The twin wishbone plus rod suspension is widely used for the front, with the rod controlling the steering. The wishbone pivot axes are usually roughly parallel to the car's centre-line. However, pivot axes perpendicular to the centre-line give the twin trailing arm (e.g. VW Beetle). There have even been cases of perpendicularity of the two axes in plan view (e.g. Rover 2000). Twin wishbones at the rear is unusual on road vehicles. The all-round wishbone system is the classic and universal method adopted for purpose-designed racing cars in recent years.

The combination of trunnion and wishbone is widely used at the front, and is continuing to gain in popularity. This includes the classic Macpherson strut, in which the front member of the effective wishbone is the lever of the anti-roll bar.

The use of a trunnion at the rear is relatively unusual, again tending to be restricted to vehicles aspiring to particular handling qualities. A trunnion plus three rods has been used in some cases. One interesting special case is the Chapman strut, where one of the rods is provided by the fixed-length driveshaft.

Where a wishbone is used at the rear, the wishbone plus toe-control rod can become difficult to distinguish from three rods, Figure 4.5.4.

Figure 4.5.4. Rear wishbone plus rod, plan views.

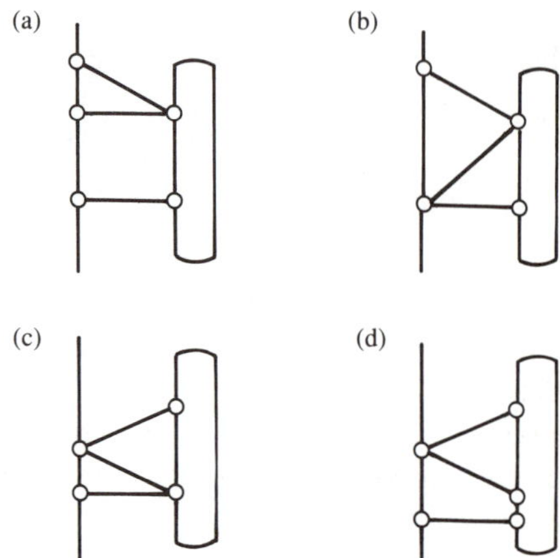

Types (c) and (d) are often referred to as reversed wishbones, since the pivot axis of the 'wishbone' is on the wheel carrier rather than the chassis.

The torsion link plus rod has been used at the rear, with the torsion link below and the fixed length driveshaft acting as the extra rod. In practice the torsion link has inadequate stiffness against tractive forces, so it is supplemented by an additional link forward from the hub.

The five-link independent suspension has only recently come to fruition. It offers the possibility of achieving certain desirable load–distortion properties that will be discussed later. In general, its geometric behaviour requires computer analysis. The other systems can be considered as special cases where simple relationships amongst the links permit a comprehensible analysis. Although the moving two-dimensional body always has an instantaneous centre (in a given coordinate system) even if this is at infinity, the motion of a three-dimensional body does not necessarily reduce to rotation about an instantaneous axis – there is in general an axial velocity left in addition to the rotation. In a number of practical suspensions, however, the linkage system is such that there is an instantaneous axis of rotation for the wheel carrier. Generally, this is so if all motion is parallel to a single plane. This is always true for the rigid arm. It is also true for double wishbones provided that the pivot axes are parallel (and that the toe-control rod is in sympathy). Steering permitting, it will be approximately true for the trunnion and wishbone, because the trunnion will behave as the notional infinite wishbone with its axis parallel to the bottom wishbone axis.

4.6 Dependent systems

A dependent (i.e. non-independent) system is one in which the location of one wheel depends in a direct way on its partner on the other side. The dependence must be essentially geometric – an anti-roll bar on an independent system is not considered to make it dependent.

In the normal dependent system, i.e. the axle in its narrower sense, the wheel carriers are rigidly connected together so that the whole can be considered as one unit. Axles are broadly divided into three kinds:

(1) The dead axle – used at the rear with front drive, having undriven wheels,

(2) The live axle – has driven wheels and carries the differential,

(3) The de Dion axle – has driven wheels but does not carry the differential.

Live and de Dion axles have more stringent location requirements, particularly for rigidity to avoid large displacements or large-amplitude axle vibrations when tractive forces are applied.

The axle must have freedom to heave and roll relative to the body, hence requiring two degrees of freedom, i.e. a four degrees-of-freedom constraint. This can therefore be achieved by four radius rods, Figure 4.6.1(a). It is common practice to introduce an extra link, to provide two links above the axle, Figure 4.6.1(b). Although this is geometrically redundant, it affects the stiffness and hence the response to forces. The Panhard rod lateral location of Figures 4.6.1 (a), (b) and (c) can, of course, be replaced in any of these examples by an alternative such as a lateral Watt's linkage. The upper longitudinal links can be turned round to also give Watt's type linkages, Figure 4.6.1(c). Alternatively the two top links can be angled, to provide lateral location, thus eliminating the need for the Panhard rod, and reducing the number of links back to 4, Figure 4.6.1(d).

This upper link pair is sometimes formed into a single wishbone, or a T-bar. The lower links may also be inclined inwards, Figure 4.6.1(e). There are several systems using a single front ball-joint (three degrees of constraint), plus a lateral location. These may be wide-based at the rear, Figure 4.6.1(f), or the so-called 'torque tube', Figure 4.6.1(g), a singularly unfortunate name since it does not resist significant torque except on those rare cases where it is mounted rigidly to the engine unit.

As these examples show, the lateral location of an axle may be achieved as part of the function of a convergent link pair (or pivot arm), or by a dedicated system such as a Panhard rod. The sliding block and

Figure 4.6.1. Axle location systems.

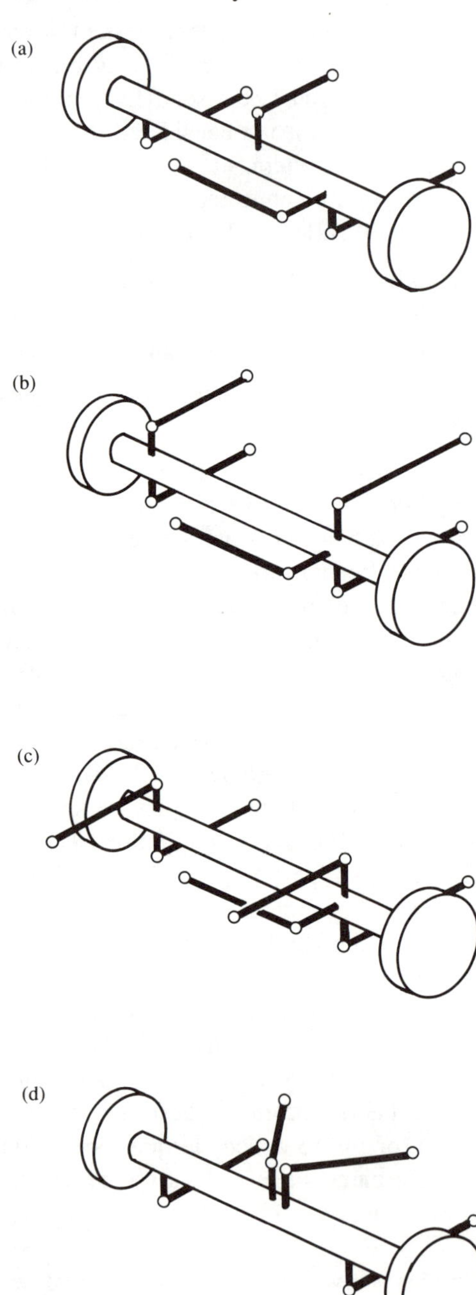

Figure 4.6.1. Axle location systems (cont.)

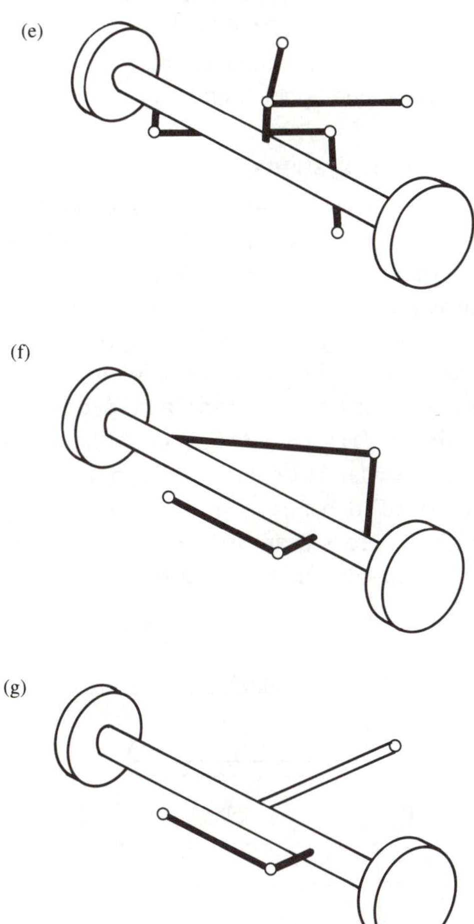

channel gives an accurate straight-line path, but there are large lateral forces, so it has substantial friction and consequently also wears, although a roller-bearing system might resolve this. The depth of channel, which must be significantly greater than the total axle movement, would be inconvenient in many cases. The Panhard rod is simple, and easily adjustable for suspension tuning, although it gives a relatively poor straight line. It is a likely choice at the rear for a front-wheel-drive vehicle, where it has been demonstrated to be highly successful (Alfa Sud, Saab). However, for rear-wheel drive it does not seem to be so satisfactory, and then one is more likely to find good handling where there is a Watt's linkage. The Alfa-Romeo Giulia series used a T-bar with success, although this was later abandoned in favour of the Watt's linkage, possibly to achieve a lower roll centre. The Roberts mechanism offers good straight line accuracy and is compact laterally, but rather deep. It has found use on some competition cars but does not seem to have found application in quantity production.

4.7 Compliant link systems

Although not a new idea, semi-dependent systems have recently come to prominence, particularly at the rear of small front-drive vehicles. Figure 4.7.1 shows a trailing twist axle in which an essential variable is the position of the transverse connecting member. This transverse member is rigid in bending, hence locating the wheels in plan view, but it has torsional compliance, allowing roll and acting as an anti-roll bar. Independent (trailing arm) and dependent (beam axle) represent two particular extreme cases of a continuous range of trailing twist axles. The camber angle of the wheels in single-wheel bump (or when the body is rolled) depends upon the geometry and rigidity of all the members. With the cross beam behind the wheel centres it is even possible to camber both wheels into the corner.

Figure 4.7.1. Trailing twist axle (plan view).

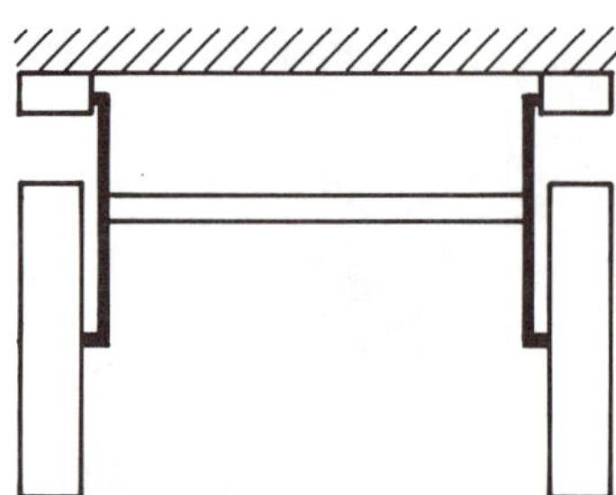

Leaf springs are capable of providing location in both of the directions perpendicular to the direction of maximum compliance. The use of leaf springs has been reducing over the years on passenger vehicles, although the Hotchkiss drive (live rear axle plus two leaf springs) has been persistent, and is still popular on commercial vehicles. New materials such as glass-reinforced plastics for the leaf springs may give it a new lease of life. Two example applications will be considered here; firstly a cantilever ('quarter elliptic'), and secondly a typical rear-axle location by a beam spring ('half elliptic') on each side. The term 'elliptic' stems from stage-coach days when curved back-to-back beam springs were indeed roughly elliptical. Terms such as semi-elliptic and quarter-elliptic are really misleading because in practice there may be little or no curvature visible.

The cantilever type is exemplified by the front transverse multileaf beam clamped rigidly at the centre, since this acts as two independent cantilevers. It was also adopted by Bugatti for rear-axle location, as a leading arm, presumably for reduced unsprung mass. Since the leaf spring is designed, approximately at least, for uniform surface stress, the radius of curvature under load is roughly constant along its length, whilst the length itself is hardly changed at all. The geometrically equivalent radius arm length is approximately $0.75l$. The cantilever form finds little application today; it has been superseded by separate geometric location, with a coil spring.

The leaf spring has survived longer at the rear as a complete beam, Figure 4.7.2(a), in this case often initially curved, and with a shackle (short radius arm) at the rear to permit effective length changes. Equivalent linkages can be constructed usually based again on a constant radius model, typically as Figure 4.7.2(b), where the centre link remains horizontal. Evidently the equivalent linkage is highly dependent on the particular application. If the leaf is very rigid in front of the axle (a common modification in saloon car racing), then the equivalent linkage is effectively a simple trailing arm with a spring applied at the rear.

Figure 4.7.2. Full leaf spring and equivalent linkage.

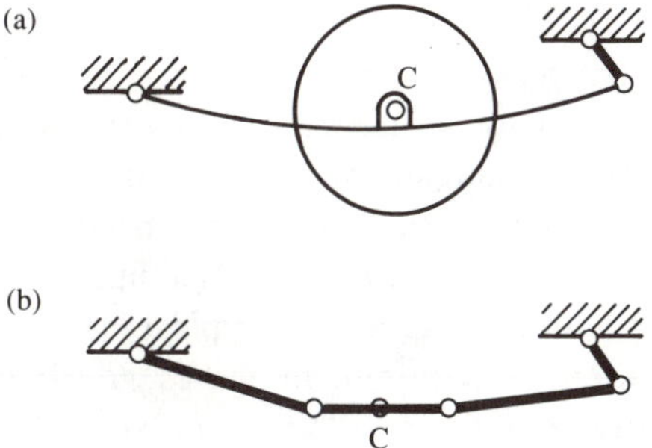

4.8 Spring types

Springs are components designed to have a relatively low stiffness compared with normal rigid members, thus making it possible to exert a force that varies in a controlled way with the length of the member. Springs are generally classified according to the material used and the way that the forces and corresponding stresses occur. As far as vehicle handling is concerned, it is the apparent properties of the 'spring unit' that matter, i.e. force, stiffness, inertia, friction etc., as felt at the wheel, rather than the precise way in which those results are achieved. Therefore the choice of a springing medium, such as nitrogen, rubber or steel, does not in principle pre-empt the achievement of any particular handling qualities. The wide range of systems in practical use today shows that no overwhelming advantage has been demonstrated for any particular type of spring.

The most obvious first decision is that of the state of the springing medium: solid, liquid or gas. A gas is of course highly compliant, and therefore little mass is required for a given energy storage. Even when the necessary associated containers are included, it is competitive. Typically, nitrogen is used at about 2.5 MPa. Of course, most vehicles are already fitted with four air springs: the tyres. Amongst liquids, oil is relatively compressible and is used in aircraft oleo legs, but is not favoured for normal ground transport. Water, and water–alcohol mixtures, have a relatively small compressibility, and are therefore not used as the actual spring medium, although they are commonly used as the working hydraulic fluid for force transfer to a rubber spring, e.g. Hydrolastic, or to a gas spring, e.g. Hydragas.

Amongst solids, rubber is pre-eminent in offering a high natural compliance. Indeed, new synthetic materials with high compliance are usually described as rubber, so this is perhaps better seen as a definition of the term ‘rubber’. In order to achieve the extreme fatigue life necessary, rubber is best used in a combination of shear and compression.

For metal, the stiffness in direct compression and tension is very great, so bending or torsion is used. The beam spring (i.e. in bending – the leaf spring) has a rectangular-section bending stiffness of $ab^3E/12$ where b is the dimension perpendicular to the bending axis, and E is the material normal stress modulus (Young’s modulus) – about 205 GPa for steel. The bending-moment diagram of a typical leaf spring, Figure 4.8.1, shows that the strength requirement varies considerably along the beam; since it is desirable to use a thin beam (small b rather than small a) to give the required compliance, the best shape is a diamond, i.e. one that has a width in proportion to the bending moment, but of constant section depth. This also has excellent rigidity against sideforces, i.e. bending about the other axis. Because of the likely inconvenience of the great width at the centre, it is usual to ‘slice up’ the diamond shape and to place the slices on top of one another. The bending stiffness in the normal load direction is the same. It is reduced for transverse loads, but still higher by a factor $(a/b)^2$. The bending stress distribution means that the material near to the centre-line of each leaf is hardly stressed; the leaf is, in practice, too thin to take advantage of the I-section shape prevalent amongst deeper structural beams. All the above comments

Figure 4.8.1. Bending-moment diagram of beam (leaf) spring.

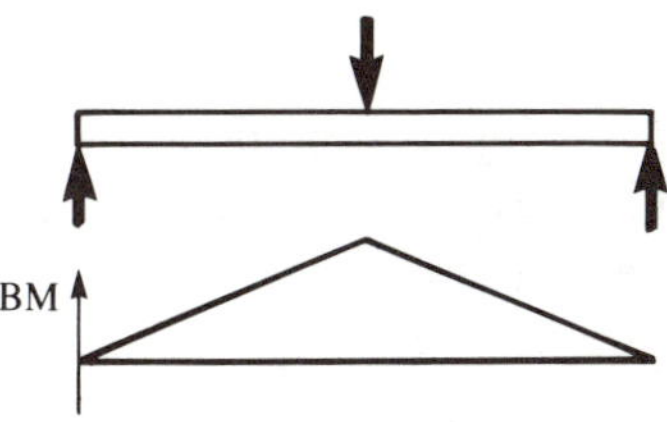

also apply to the cantilever spring, which can be considered to be one half of a beam spring.

A more efficient utilisation of material is achieved by a torsion bar, because with the circular section the bulk of the material is nearer the edge, and is operated nearer to the highest stress. A hollow section may be used, eliminating the relatively understressed core. Controlled overloading and yielding can modify the working stress distribution usefully for solid bars. This reduces the surface stress and hence improves the fatigue life. The angular displacement under a load torque is

$$\theta = \frac{TL}{JG}$$

where G is the shear modulus (80 GPa for steel) and J is the polar second moment of area,

$$J = \tfrac{1}{2}\pi r_o^4 - \tfrac{1}{2}\pi r_i^4$$

for a concentric tube. For a solid rod, r_i is zero, and the torsional stiffness varies as the fourth power of diameter.

The coil spring is essentially a coiled torsion bar, as may be seen by observing the loading on a cross-section of the spring wire. As a complete unit it is generally loaded in compression for suspension applications. For a load of this kind, the deflection is given approximately by

$$x = \frac{8FD^3N}{Gd^4}$$

where F is the force magnitude, D is the coil diameter between centres, N is the number of working turns, G is the shear modulus and d is the wire diameter. Thus the stiffness depends on the fourth power of the wire diameter, and varies inversely with the cube of the spring diameter and inversely with the number of working coils. The coil spring generally has rather little stiffness laterally and in bending, although these may have some influence on the effective stiffness for suspension geometries that give large angular changes at the spring seats. The performance of the spring is sometimes varied by designing it with varying coil spacing, varying coil radius or varying wire diameter, so that the spring progressively closes, reducing the number of working turns and increasing the stiffness.

The modern high-quality coil spring is produced on complex automatic machinery. The leaf spring, in contrast, is relatively easily manufactured without special facilities, and this, coupled with the fact that it also provides fore and aft, lateral and pitch angle axle location very cheaply, was the reason for its very widespread use in the early days of motoring. Although the steel leaf spring is generally regarded as rather heavy, and inferior to a link and coil system, the use of glass-reinforced plastic leaf springs offers possible weight savings, and has resulted in a new interest in this old design. Transverse glass-reinforced plastic (GRP) beam springs with link location have recently been used on a production sports car.

4.9 Spring linkage geometry

The spring can be brought to bear on any moving part of the suspension. For the wishbone and trunnion suspension, it is natural to place a coil spring around the trunnion unit, as for example on the Macpherson strut, although the spring is sometimes placed elsewhere. For the double wishbone, common at the front, the spring usually operates on the lower arm since this is more compact overall. When this space is occupied by a driveshaft, the spring can act on top of the upper arm. For many years purpose-designed racing cars usually had the spring unit attached diagonally, acting at the outer end of the lower wishbone. The era of ground-effect aerodynamics called for a reduction in drag and turbulence from the front, so rocker arms, pullrods and pushrods, combined with internal springs, became more popular. For a rigid arm, such as a typical rear trailing arm, the spring unit may operate down on the arm, or is sometimes placed horizontally, acting on a tongue.

In general the spring compression or extension is different from the wheel displacement, and their ratio is not constant. The force at the spring is generally greater than that at the wheel, but because the required range of motion is correspondingly reduced, the energy storage and hence the material requirements of the spring are not dependent on the ratio. The motion ratio is conventionally expressed as the motion of the wheel divided by that at the spring or damper as appropriate, and is generally between 1 and 2. To achieve a given effective suspension stiffness at the wheel, the spring stiffness required depends on the motion ratio squared.

Neglecting detailed arrangements which may have some secondary influence, for example through rubber bush distortion, the important parameters for handling analysis are the force at the wheel, and the stiffness at the wheel, which is usually called the wheel rate. It is sometimes considered desirable for the wheel rate to increase as the wheel rises, i.e. for the force to increase non-linearly, so that when heavily loaded the vehicle has increased wheel rates. This may be achieved by the characteristics of the spring unit. Gas and rubber springs have this kind of behaviour due to the material itself. Steel coil springs can do so if the coils are unevenly spaced, or if the wire diameter is tapered, so that they close up progressively. Leaf springs do so if initially curved so as to be separate. Alternatively, it may be achieved through a variable motion ratio of the linkage, by having an increasing moment arm for the spring as the wheel deflects. The latter method is more amenable to adjustment and control, and can also give progressive damping.

The use of torsion bars for basic springing requires that space be found for the long bars. This length can be reduced by using a compound rod, i.e. a rod in a tube, fixed together at one end. Alternatively several thin rods in parallel can be used. The rod may be mounted anywhere, and connected to the geometric links by a 'drop-link' – a short radius rod. In practice there are two typical installations. Firstly, a bottom wishbone (front suspension) will be splined directly to the torsion bar which lies parallel to the vehicle centre-line. Secondly, the rod is transverse, and splined to a trailing arm (rear) or a wishbone with transverse axis (front).

Whatever the means of providing springing, to find the effective force at the wheel generally requires some analysis. For coil spring or equivalent units, the spring compression must be found from a series of position diagrams. The effective rate at any particular position may alternatively be found by constructing a velocity diagram. This process can be computerised, of course.

There may be some secondary contributions to stiffness. The distortion of rubber bushes is one, adding perhaps 10% to the wheel rate. This can be calculated for some cases, such as simple cylindrical bushes in rotation. Another source of force is the pressurised gas-filled damper. The typical pressure of 2.5 MPa, acting on the piston-rod cross-sectional area, may exert a force of about 200 N, which will raise the suspension a few millimetres. This pressure is, however, not highly dependent on position, so the effect on suspension stiffness is small.

Inevitably there are limitations on the possible movement of any suspension. If no explicit provision has been made, then eventually there will be metal-to-metal contact between suspension arms and body, or the wheel will strike the body, or the spring will become coil-bound. In view of the harshness of such occurrences to both passengers and vehicle, it is normal to provide bump and droop stops which soften the final blow. These are usually in the form of a moulded rubber block, which is squeezed between a suspension arm and the body at an appropriate point.

It is sometimes not possible to provide such a stop, for example for an axle in droop. In such a case the dampers are sometimes allowed to define the limit, or straps may be used.

4.10 Roll and pitch springing

The provision of stiffness has so far been examined in the context of a single wheel. However the position of the body in relation to the road, as far as suspension location is concerned, should also be considered. It is essentially one of heave, roll and pitch.

Body roll is geometrically equivalent to the raising of one wheel and lowering of the other one, whilst heave is equivalent to equal motion of the two wheels in the same direction. Thus, for conventional suspension, springing at each wheel will contribute to both body heave and body roll stiffnesses. However, the resulting values may not be as required; it may be desired to have more roll stiffness without increasing the heave stiffness, or sometimes vice versa. This can be achieved by appropriate spring design and coupling.

A zero roll-stiffness suspension may be produced by arranging for roll to have no effect on the springs. Figure 4.10.1 shows one example; the spring is connected only to the two parts of the axle, not to the body. The transverse leaf spring can be given this characteristic if provided with a central pivot. The illustration shows a swing-axle type suspension, but equivalent systems for more advanced independent systems or for true axles are easily devised. The use of a swing-axle illustration is in fact appropriate, because they do generally benefit from the use of only a small roll stiffness, to reduce load transfer which causes problems for this type of axle. Another possibility is the Z-bar, Figure 4.10.2. Provided that the two lever arms are equal, and the suspensions are symmetrical, there is no roll stiffness. The positioning of the body pivots for the Z-bar is not critical but they should be well spaced and symmetrically disposed about the vehicle centre-line. Its operation is best imagined by considering the Z-bar rotating – one wheel rises, the other falls; this is equivalent to body roll. Body heave causes opposed motions of the end levers, putting the centre section into torsion. Hence the Z-bar gives heave stiffness with no roll stiffness. In practice a zero roll stiffness system is likely to be combined with conventional springing, to give the desired overall combination of properties.

It is also possible to devise elements that provide stiffness in roll only, with no heave stiffness. In practice, one particular type is used more-or-less exclusively – the anti-roll bar, Figure 4.10.3. It is a direct corollary to the Z-bar; although shaped as a U it is rarely called a U-bar. The end levers must be of equal length with symmetrical suspension, or a heave stiffness results. Again, in practice the anti-roll bar is combined with conventional springs to give the desired overall properties.

A combination of Z-bar and anti-roll bar can provide any required combination of heave and roll stiffness. In principle either one of them, using unequal lever arms, could provide a required combination, although this is not done. In practice, the usual requirement is for the roll stiffness to exceed that obtained from the springing selected for heave stiffness, so the use of anti-roll bars is common, combined with conventional springs. The opposite case, a requirement for particularly low roll stiffness, seems to be restricted to the virtually defunct swing-

Figure 4.10.1. Zero roll-stiffness suspension (rear view).

Figure 4.10.2. Z-bar of no roll stiffness (plan view).

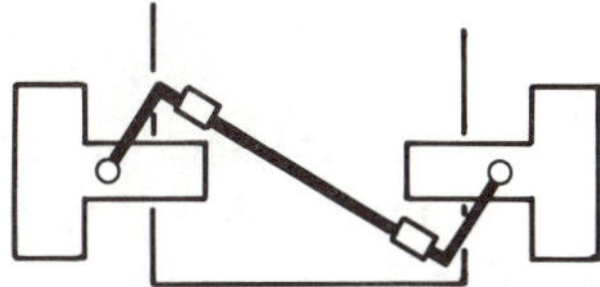

Figure 4.10.3. Anti-roll bar with pure roll stiffness (plan view).

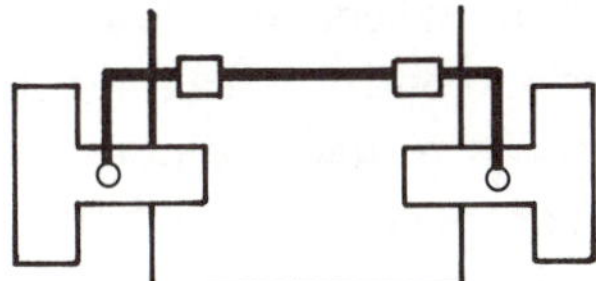

axle suspension, and that is where the Z-bar finds its primary evolutionary niche.

Because of their torsion and stiffness, longitudinal leaf springs have an anti-roll effect which can contribute typically 40% to the total body roll stiffness. If there is extra lateral location, for example a Panhard rod, then the lateral bending of the leaf springs can give considerable additional roll stiffness.

As will be discussed in more detail later, the distribution of roll stiffness between front and rear axles is used for adjustment of handling qualities. To control the limit handling balance, it is quite common on racing cars to have the roll stiffness at one axle adjustable by the driver without stopping. This is typically done by using an anti-roll bar with one lever arm of thin wide section, that can be rotated to change its effective bending stiffness.

In contrast to roll, in pitch the basic springing tends to give rather a high stiffness for comfort, especially for small vehicles. Thus some fore-and-aft interconnection may be used, in the furtherance of ride quality if not handling. The front-to-rear Z-bar has been used to provide basic heave springing, coupled with torsion-bar support for pitch springing, plus front and rear anti-roll bars. The transverse anti-roll bars have no effect on pitch or heave at all. Front-to-rear interconnection to reduce pitch stiffness is perhaps more easily achieved with hydraulic suspension systems.

Front-to-rear connection to increase pitch stiffness, for example by longitudinal U-bars, is not used on passenger vehicles. It has however been suggested for racing vehicles to reduce the pitching and associated wheel camber changes that are detrimental to braking with modern wide tyres, and even more importantly to avoid shift of aerodynamic downforce, which is very pitch-sensitive on many racing cars.

4.11 Damper types

The damper is known as the shock absorber in the U.S.A., and also elsewhere, although the implication that shocks are absorbed is misleading. Their purpose is to dissipate energy in the vertical motion of body or wheels, such motion having arisen from control inputs, or from disturbance by rough roads or winds. As an agglomeration of masses and springs, the car with its wheels constitutes a vibrating system that needs dampers to optimise control behaviour, by preventing response overshoots, and to minimise the influence of some unavoidable resonances. The mathematical theory of vibrating systems largely uses the concept of a linear damper – with force proportional to extension speed – mainly because it gives equations for which the solutions are well understood and documented, and usually tolerably realistic. There is no obligation on a damper to exhibit such a characteristic; nevertheless the typical modern hydraulic damper does so approximately. This is because the vehicle and damper manufacturers consider this to be desirable for good physical behaviour, not for the convenience of the theorist. This characteristic is achieved only by some effort from the manufacturer.

Damper types, which are explained later, can be initially classified as friction (solid elements) or hydraulic (fluid elements), the latter being fitted exclusively in recent times. The friction type came originally as sliding discs operated by two arms, and later as a wrapped belt, the 'snubber'. The hydraulic varieties are lever-arm and telescopic. The lever-arm type uses a lever to operate a vane, now extinct, or a pair of pistons. Telescopics, now most common, are either double-tube or gas-pressurised single-tube.

The history of damper development is virtually as old as that of the car itself. Before 1900, Paris cycle engineer Truffault invented a friction disc system, of bronze and oiled hide, pressed together by conical disc springs and operated by two arms, with a floating body. Between 1900 and 1903, he developed a version for cars, at the instigation of Hartford in the U.S.A., who began quantity production in 1904. In 1901 Horock patented a telescopic hydraulic unit, laying the foundations of the modern type, whilst in 1902 Mors actually built a vehicle which used simple hydraulic pot dampers. In 1905, Renault patented an opposed-piston hydraulic type, and also patented improvements to Horock's telescopic type, establishing substantially the design used today. Renault used the piston type on his 1906 Grand Prix cars, but not on his production cars. Meanwhile Houdaille started to develop his double-arm vane-type. Caille proposed the single-lever parallel-piston variety in 1907. In 1909 a single-acting Houdaille vane type was fitted as original equipment, but this was an isolated success for the hydraulic type, the friction-disc type remaining dominant. In 1915 Foster invented the belt 'snubber' which had great commercial success in the U.S.A. The beginning of a turnaround to telescopics was the introduction by Lancia of the double-acting hydraulic unit incorporated in the front independent pillar suspension of the Aurelia in 1924. 1930 saw the issue of Armstrong's telescopic type patent, and Monroe began manufacture of telescopics in 1934. In racing, at Indianapolis the hydraulic vane type arrived in the late 1920s, and was considered a great step forward; the adjustable-piston hydraulic appeared in the early 1930s, but the telescopic was not used there until 1950. In 1947 Koning introduced the adjustable telescopic: the one remaining major advance was the gas-pressurised single-tube telescopic, invented by de Carbon in the same period and manufactured from 1950.

Each damper type has some advantage, although the hydraulic, particularly telescopic, now reigns supreme. The simple friction-disc was light and cheap, and easily adjusted, even remotely while driving. However its characteristic is that of typical Coulomb friction, reducing with speed. It was sensitive to water or oil contamination, and it is not easy to arrange for controlled asymmetry of force. Perhaps it is fanciful to suggest that materials might be developed with a contact friction rising in a suitable way with sliding speed, although this is not *a priori* impossible. The Coulomb friction gives poor ride and handling by present standards, and there is little incentive to abandon fluid dampers, so the friction-disc damper seems to be defunct.

The belt snubber was a strap wrapped around a pair of spring-loaded wooden blocks, and fixed to one of them. Release of tension allowed the block to expand to take up slack. Pulling the strap gave a tension presumably governed by the usual exponential function for wrapping members as applied to drive belts, ropes on bollards and the like. Thus it was single-acting, on rebound only, although still not speed-dependent. The S.A.E. defines the term snubber to mean any type of damper using Coulomb friction. It seems unlikely that snubbers will return for basic wheel damping.

Of the hydraulics, the vane type is seriously disadvantaged by the long seal length at the vane edges, giving wear problems, and seems to offer no adequate compensating advantages. This leaves those types based on motion of one or more pistons in cylinders. There are various possible configurations, since all that is fundamentally needed is to move the fluid through a restriction; they fall into two distinct categories: the lever type with two pistons, and the telescopic with one piston. The lever types are opposed piston, parallel piston, and parallel piston operated by involute cams. In practice, to provide different bump and rebound characteristics, two passages and two restriction control valves are generally provided. If the fluid, after its pressure drop, passes directly into the other cylinder, which is the normal arrangement, the same fluid is continuously re-used, so good heat transfer around the passages is important to keep temperatures under control. This problem is eased if the fluid is exhausted to the reservoir above the pistons, giving a general circulation, but demanding high-flow replenishment circuits. The small fluid-circuit volume means that lever dampers are sensitive to gas in the circuit, but fortunately the body is mounted to the vehicle sprung mass, and the fluid circuit is normally isolated below the pistons, so aeration is minimised.

The telescopic types, Figure 4.11.1, are most easily understood in reverse order of invention. The arrows shown indicate the fluid flow direction when the valve is restricting the flow in a controlled way. The emulsified type, Figure 4.11.1(a), is simply a piston and rod in a fixed-volume cylinder full of an emulsion of oil and gas (nitrogen). The gas separates out when the car is parked, but re-emulsification is achieved quickly, and this does not seem to be a problem. The piston contains two valves, one that controls the pressure differential across the piston during bump, and the other in rebound. These are calibrated appropriately for the emulsion, not pure oil. During bump (compression) an increasing volume of the piston rod is contained in the cylinder; this causes a pressure rise, and is the reason why it is necessary to have some gas present. This pressure exerts a force on the piston-rod area, trying to expand the damper. Alternatively, the anti-emulsion single-tube damper has the gas separated at one end by a floating piston. Otherwise the function is similar. For these single-tube types there is little problem of reduced effectiveness in sustained use, even on quite rough roads. Hence they are popular in rallying and off-road racing.

Figure 4.11.1. Telescopic dampers

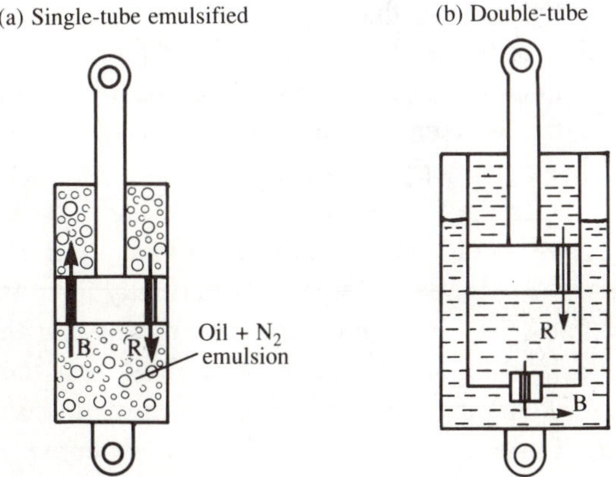

The remote reservoir anti-emulsion type reduces the need for a gas separator piston, especially if the reservoir is chassis-mounted and connected by flexible tubing. This may help cooling in difficult conditions. Arranging the reservoir as an annulus around the main working cylinder gives the double-tube type, Figure 4.11.1(b). This type is inherently poorer at cooling, although it is improved by arranging some circulation from the top of the inner tube out into the reservoir during rebound. All other types can be mounted upside-down with the lighter piston-rod end on the wheel.

The telescopic damper is not entirely free of problems. The double-tube type must be mounted with the reservoir end on the suspension; this is subject to severe agitation, and although baffles or an annular piston are helpful, aeration is generally a problem in severe conditions. This has been alleviated by anti-foaming agents and solved by the single-tube type. However the latter type is more prone to damage by stone impact, because the active cylinder is exposed. A potential problem with the pressurised type is loss of pressure, which may occur over a period of twelve months or so. Oil sealing has only been really successful with the arrival of chromed piston rods of excellent finish running in synthetic rubber seals. Temperature effects have been alleviated by high-viscosity-index oils.

4.12 Damper characteristics

Hydraulic dampers have substantial advantages in that the damping force can be made a function of speed and direction, and even position. Positional change of damping effect has not been used on cars so far, although it is used on motorcycles to act as a bump stop for the front forks. This can be achieved by placing the bump valve in the side of the working cylinder so that the piston passes it, then forcing fluid through a smaller orifice at the bottom. One advantage is absence of the bounce that occurs with a rubber stop. Some positional modification of the damping rate, as seen at the wheel, may be achieved by the connecting linkage geometry, as for springs.

Some advantages may be gained by pressurising the entire damper, typically to 2.5 MPa cold. This is essential for the anti-emulsion single tube to operate the secondary piston; it is normally done to single tubes, but not to double tubes. Such pressurisation means that the flow return valves previously designed to have very low resistance can now contribute to the damping without causing the low pressures (below 100 kPa absolute) liable to cause cavitation. As a consequence of the pressure acting on the piston-rod area, there is a force of typically 300 N attempting to expand the damper, even near full extension, and this increases at about 500 N/m as it shortens. This is at normal temperatures, and may double in severe use.

In order to achieve the desired basic damping characteristic, i.e. the relationship between force and speed, it is necessary to employ a valve system – simple orifices are inadequate. When a fluid passes through a hole or tube there are two contributions to the pressure drop. One is the viscous drag at the walls, which for the usual turbulent flow is roughly proportional to the flow-rate squared. The second is dissipated kinetic energy, at the entry and exit, which depends on the density rather than viscosity, and is again proportional to the flow-rate squared for a given hole. Use of the second effect in preference to the first would reduce sensitivity to viscosity and hence to temperature. In either case, illustrated in Figure 4.12.1, line A shows the variation of force with velocity for a given hole. The addition of a larger hole B in parallel gives line A + B. By applying to B a valve which opens progressively, a transitional characteristic is obtained. Proper choice of the basic jet sizes A and B and the progressive qualities of the valve enable quite a good straight line, or other desired shape characteristic, to be achieved. These are referred to as stages: stage 1 is valve closed, stage 2 is valve partially open, stage 3 is valve fully open. The valve begins to open at typically 0.1 m/s damper speed, and is completely open at several metres per second.

Figure 4.12.1. Damper characteristic.

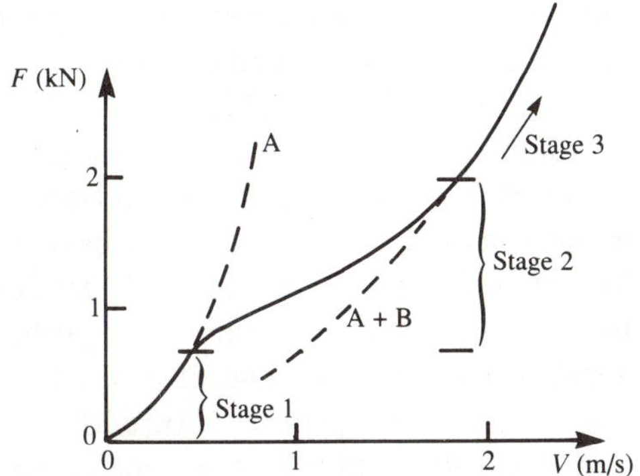

The typical car damper is designed to exert only about half the force in bump as in rebound. A large rebound coefficient helps to stop the wheel dropping into pot-holes. Very early dampers acted only in bump, to discourage bottoming of the suspension (hence the name shock absorber). However it is now considered that the best all-round behaviour is achieved if, for a given total damping, 60–70% is on the rebound stroke. This is expressed as, for example, a 30/70 bump to rebound (compression to extension) ratio, the compression figure always being quoted first.

Adopting a proportional model, the damper characteristic can be expressed simply by the slope of the force–speed curve, i.e. the damping coefficient c, typically 2 kN/m s^{-1} or 2 kNs/m. For a lever-arm damper this is evidently dependent on the arm length. It is, of course, the effective force at the wheel that matters. It is therefore necessary to establish the motion ratio between wheel displacement and damper length, or arm position. This motion ratio R may vary with wheel position – it is typically about 1.3 to 1.5 for a separate damper, and near to 1.0 for one incorporated in a strut. For a given wheel speed the damper speed and force is thereby reduced, and the damper force is further levered down as seen at the wheel, so the damping effect is inversely proportional to R^2, as was found for spring rates. The energy dissipating service required does not depend upon R, so it is a matter of providing a larger-diameter shorter-stroke unit for a bigger R.

The damping forces seen at the wheel can be expressed in various ways. For example, dividing the total damping coefficient at the wheels in bump by the vehicle weight gives a characteristic speed, of order 10 m/s, representing a notional steady sink rate at which the car would settle on its suspension when resisted by its dampers without spring support. However, the optimum damping depends not just on the vehicle mass, but on the spring stiffness too, so it is better to refer to the classic vibration concept of damping ratio. This is a poor approximation to apply when damping in the two motion directions is different, but it is useful in a qualitative way. Figure 4.12.2 shows the behaviour of a system for various damping ratios, ζ, when released from a deflected position. A damping ratio of 1.0 just prevents overshoot. A more lightly damped vehicle, such as the average car, will have some overshoot, e.g. $\zeta = 0.4$. The average damping ratio for body motion is typically in the range 0.2 to 0.6. Manufacturer's opinions regarding the optimum damping ratio seem to vary over a range of at least 2:1, which is not surprising in view of the subjective nature of ride assessment. Competition vehicles are likely to have twice the damping of a normal passenger vehicle.

Operation of the damper results in a temperature rise of the fluid, the kinetic energy of motion of the body being dissipated in thermal energy in the fluid. The dampers must not become too hot during the rated sustained operation. Brief intensification of heating is not immediately dealt with by heat transfer to the air, but by a temporary temperature rise. Thus the fluid mass and cooling arrangements depend on the vehicle and service requirement. In off-road racing, probably the worst case, water cooling to a radiator, with electric water pump and air fan, has been used, but multiple dampers to reduce individual heating is the widely adopted solution. For ordinary road use a telescopic piston area of typically 1 to 2 mm^2/kg of wheel load is provided, referred to the

Figure 4.12.2. Free response with various damping ratios.

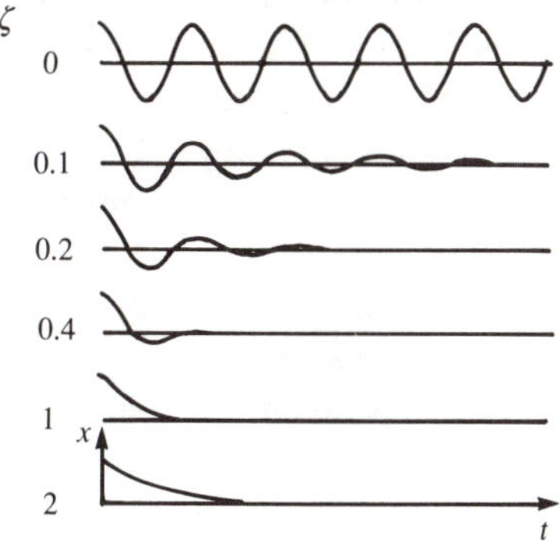

wheel, with a fluid volume of 300 to 500 mm³/kg. The temperature rise at a single stroke depends on the pressure drop, and hence speed, and thermal capacity (about 2.5 J/gK) and will be typically in a range up to 0.5°C for the active fluid only. Thus high-frequency short stroking can cause problems if the fluid is not well cooled, or circulated.

Different makes of dampers do not have a large difference in effect in ordinary driving, although optimisation is important for competition. The main differences are in life, adjustability, resistance to aeration, and consistency one to another. Like a carburettor, the damper depends upon fine control of fluid in small orifices, and production tolerances can result in substantial variation between dampers of the same specification from the same manufacturer, and even greater variations between dampers of the same specification from different manufacturers. The best dampers have close tolerances and are individually calibrated, which is expensive.

4.13 Parasitic friction

Apart from the deliberately introduced damper viscous friction, there are other sources of damping inherent in the system. One is the aerodynamic damping force on the body; this appears to be insignificant for wheeled vehicles. Another is the residual friction in suspension pivots and sliders, driveshaft splines and the like. These generally behave in a Coulomb friction manner. This means that for small changes of vertical wheel force, the suspension remains locked, giving a poor ride quality, e.g. boulevard jerk, a mode where the body and wheels vibrate vertically as one on the tyre stiffness. It is also detrimental to handling because it interferes with operation of the suspension and therefore alters the vertical force on the tyres. Thus it is generally desirable to minimise such friction.

The total parasitic friction on one wheel is typically equivalent to about ±5 mm of suspension movement, for example ±200 N on a one-tonne vehicle, and possibly a good deal worse for a poor design or if there is inadequate lubrication of joints. The friction value of 200 N could result in a diagonal load distribution asymmetry of 400 N, a diagonal bias of about 2%.

The source of this friction is the various sliding joints, including dampers and suspension arms, rubber bushes and especially strut sliders. The friction force depends on the normal forces involved at the sliding face. For example, the main spring may operate on the suspension arm only one third along; the inner pivot must then support twice the load at the outer ball joint. Loads in the joints are affected by any change of wheel force, especially cornering forces. Again, the leverages involved often give large forces at the joints. An interesting case is the strut and pivot arm suspension. This must transmit, through the guide and piston of the strut damper, a force whose true line of action is through the top bulkhead joint. As a consequence it is desirable to make the strut length as great as possible, to maximise the distance between guide and piston. The guide rod itself must be much larger than for an equivalent damper, to spread the load at the guide, and consequently the potential leakage path is also greater. Struts with angled mainsprings have been used to align the spring force with the intersection of the tyre and bottom link force; this can eliminate the strut sideforce in straight running, but at the cost of worse forces in cornering.

The springs themselves can be a source of friction. The coil spring and the torsion bar splined directly to the arm are ideal in respect of friction. The internal friction of metal is entirely negligible. The multileaf spring, however, is usually replete with friction, because of sliding of the leaves. In the early days this was considered useful free damping, although it tends to be rather variable, especially when wet; nowadays it is sometimes minimised by the incorporation of plastic spacers. Lubrication does not seem to be very helpful, since it only reduces the dynamic friction and not the static breakout value.

Rubber springs exhibit the usual rubber hysteresis loop. This is not viscous – it is hardly speed-dependent for speeds of interest – but it is not entirely of a Coulomb friction nature either. Therefore, although a rubber-suspended car may exhibit a large friction dead band in suspension height, the ride will not necessarily be as bad as this would suggest.

In order to minimise sliding friction at suspension joints, it is now normal to use a cylindrical rubber bush that distorts in shear as the suspension operates, so most suspensions have an element of rubber type friction and stiffness. A well-lubricated bearing might be better for handling, but rubber bushes are a practical compromise with low cost, low maintenance and good vibration isolation, but generally poorer for handling because of flexibility. In production cars modified for competition, bushes are replaced with harder rubber for rallying, or spherical joints for racing.

The exact influence of all these effective friction sources is not easy to establish, even if the total result as a suspension height dead band is measured directly, because of uncertainty in the influence of vibration from the engine and road roughness. However, since the development of effective hydraulic dampers, there is little to be said in favour of Coulomb friction, and there has been a continuous campaign to reduce it.

A further difficulty arises in the case of the driven wheels, particularly for independent suspension. The suspension geometry generally requires that there is some ability in the driveshafts to extend and contract, usually facilitated by a sliding spline joint. When drive torque is transmitted, or when braking if the brakes are inboard, the splines are subject to considerable friction because of the large torque acting on the splines at a small radius. The consequent binding of the suspension may particularly affect traction on poorer surfaces. There are a number of possible solutions. The effective friction seen at the wheel depends on the motion ratio, so spline sliding should be minimised; this is an unwelcome additional requirement on the geometry. A wire-reinforced flexible 'doughnut' can be used for limited plunge. An attractive solution due to Roesch in 1933 is to make the driveshaft of fixed length, and to use it as one of the suspension links. The first to actually implement this idea was Chapman on the 1956 Lotus Elite rear suspension, where it was used as part of the bottom arm, with a strut above. Various others have used it since; whilst it solves the friction problem, the driveshaft is not in the most convenient position for a suspension link. A more general solution to this problem is now available – the ball-spline. This uses steel balls to minimise the friction, and is completely effective, although expensive.

4.14 Inertia

The total mass of the vehicle may be considered to be divided into sprung mass and unsprung mass. These terms refer to the component motion relative to the road. Basically the sprung mass is the body, and the unsprung mass is the wheels. This distinction is made because it may be necessary to consider the unsprung mass separately, for example for rough roads, or when calculating the load transfer distribution in cornering.

When the road is not smooth, for good handling it is generally desirable to have a small unsprung mass. For a given spring stiffness, a small unsprung mass will follow the contours of the road better, with a more uniform vertical force, and hence less lateral force variation and a greater maximum mean cornering force. The choice of suspension spring stiffness is mainly governed by the vehicle body mass, so a useful parameter is the ratio of sprung to unsprung mass. This is usually considered separately for the two ends of the vehicle.

Actually the unsprung mass is not strictly a constant for a given vehicle, but depends on the purpose of the analysis. For example, in single-wheel bump an axle does not exhibit an inertia equal to half of the value for double-wheel bump. The reason for this may be seen by considering a single uniform suspension link. Its weight will be distributed equally between the two ends. However its angular inertia about one end is $ml^2/3$, so when one end is pivoted and the other end is moved, the apparent inertia there is only $m/3$. An extra point mass only half-way along contributes an inertia at the end of only 1/4 of its mass, i.e. the effective mass contribution depends on the square of the motion ratio. Apart from the case of a live axle, the contribution to the unsprung mass is dominated by the wheel unit, so these effects can usually be neglected, and the links considered to be distributed half as sprung and half as unsprung with little error. Conventional leaf springs are about 2/3 unsprung.

Table 4.14.1 shows representative sprung and unsprung masses for an axle, expressed as a percentage of the total end-mass. The unsprung mass runs from 13 to 26% of the total mass, and the sprung to unsprung mass ratio ranges from under 3 to nearly 7. Compared with the minimal set-up of wishbones, coils and inboard brakes, a de Dion axle adds 2%, a differential adds 2%, outboard brakes add 5%, and longitudinal steel leaf springs add 4%. A transverse GRP cantilever leaf spring is as good as coil springs. The quite good showing of the swing axle here has possibly contributed to its use, despite its other shortcomings.

The unsprung mass should include an appropriate contribution from all the links, the moving part of the damper, the spring, etc., factored according to the motion ratio. However as these figures show, these are usually a small proportion of the total, so precise assessment is not important.

The excellent ratio of independent systems may appear to account

Table 4.14.1. *Example unsprung masses for heave of an axle*

Type	Unsprung	Sprung	$\frac{\text{Sprung}}{\text{Unsprung}}$	$\frac{\text{Unsprung}}{\text{Sprung}}$
	m_U/m_e	m_S/m_e	m_S/m_U	m_U/m_S
	(%)	(%)	(Ratio)	(%)
Wishbones, coils, inboard brakes	13	87	6.7	14.9
De Dion, coils, inboard brakes	15	85	5.7	17.6
Wishbones, coils	18	82	4.6	22.0
Swing axle, coils	18	82	4.6	22.0
De Dion, coils	20	80	4.0	25.0
Solid axle, links, coils	22	78	3.5	28.2
Solid axle, leaf springs	26	74	2.8	35.1

for the widespread use of such systems on the front of cars. This is not entirely so. The success of front independent systems is due to the elimination or reduction of various steering vibration problems, better compatibility with engine space requirements, and better steering geometry which is very difficult to achieve with soft springs and large movements on solid axles.

4.15 Gyroscopic effects

A rotating wheel also acts as a gyroscope, and any attempt to turn the wheel out of its existing plane of rotation results in a precession effect. For example, if the wheel in bump is cambered a little by the suspension, then it will attempt to rotate about the steering axis. The other wheel will resist, but there will be some effect on the steering. This also arises when the wheels camber in roll. The precession torque is given by

$$T = I\Omega\omega$$

where Ω is the wheel angular speed because of forward motion, and ω is the precession angular speed. I is the second moment of mass about the spin axis, and is typically 0.6 kg m^2 for a passenger car wheel.

4.16 Problems

Q 4.2.1 Explain the Kennedy–Arronhold theorem.

Q 4.3.1 Compare the merits of various straight-line mechanisms in the role of axle lateral location.

Q 4.5.1 List and explain the combinations of links that form the basic independent suspensions.

Q 4.6.1 Describe the basic forms of link combinations for axle location (assume lateral location by a Panhard rod if lateral location is separately provided).

Q 4.7.1 Explain the concept of an equivalent rigid-link system.

Q 4.8.1 Explain the relative merits of the various types of spring and spring materials.

Q 4.9.1 A spring and damper both have a motion ratio of 1.5. The damper coefficient in bump is 0.8 kN/m s^{-1} and the spring stiffness is 6.2 kN/m. Find the effective values at the wheel.

Q 4.10.1 Explain the various ways in which extra roll stiffness arises as a side effect, beyond that corresponding to the normal spring stiffness.

Q 4.10.2 Explain how roll stiffness may be added.

Q 4.10.3 Describe the various ways in which a very low roll-stiffness suspension may be achieved. Why might this be done?

Q 4.10.4 Describe ways in which a small pitch stiffness may be achieved.

Q 4.11.1 Describe the various damper types, and their advantages and disadvantages.

Q 4.12.1 Explain how valving may be used to obtain an approximately linear force–speed relationship for a damper.

Q 4.12.2 A typical damper exerts a force of 3 kN at an extension speed of 2 m/s. What is the damping coefficient? Approximately what force would you expect to be required to compress the damper at 1 m/s?

Q 4.13.1 Summarise the methods used to minimise the influence of Coulomb friction in suspension members and driveshafts. Does Coulomb friction have any advantages?

Q 4.13.2 A two-wheel-drive vehicle of mass 1400 kg and wheel radius 0.3 m accelerates at low speed at 8 m/s^2. The driveshafts have splines of effective radius 15 mm, and a coefficient of friction of 1.2. Find the plunge force required to move the splines axially.

Q 4.14.1 Summarise the advantages and disadvantages of keeping the unsprung mass to a small value.

Q 4.14.2 Discuss what can be done to keep the unsprung mass as small as possible.

Q 4.15.1 A wheel with second moment of mass 0.7 kg m^2 about its spin axis, of radius 0.30 m travelling at 50 m/s, is forced to rotate in camber at 20 deg/s as the wheel passes over a bump. What precessional torque must be applied by the steering if the wheel is held rigidly?

Q 4.15.2 Gyroscopic moments are considered in Chapter 5, questions Q 5.10.12 to Q 5.10.15. If you understand lateral load transfer, do those questions now.

5

Suspension characteristics

5.1 Introduction

This chapter considers characteristics of the suspension as a complete system. In particular this includes consideration of the effect of chassis roll, the roll-centre and roll-axis, load transfer and the distribution of vertical forces, and the influence of pitching in response to acceleration and braking. This is followed by a look at the geometry of steering systems, at the steering effects of wheel bump and chassis roll, and at the steering effects of wheel forces because of system compliance.

The term 'axle' is used here in a wide sense, to include independent suspension, in which case it means the combination of the wheels, hub carriers and links for the two sides.

5.2 Bump and heave

Bump is upward displacement of a wheel relative to the car body, sometimes applied more broadly to mean up or down displacement. It is also known as compression or jounce. The opposite, a lowering of the wheel, is called rebound, extension, or droop. Bump is so called because it occurs when a single wheel passes over a bump. When a pair of wheels rises symmetrically this is called double-wheel bump, and is geometrically equivalent to a negative heave of the body, heave being a vertical upward motion of the body. Bump and heave influence the wheel camber and steer angles relative to the body and road, and also influence the spring and damper forces and hence the tyre vertical force. These all influence the tyre lateral force. Bump velocities also affect the slip angle because of the scrub velocity component. Body roll in cornering gives a combination of bump and droop on opposite wheels, relative to the body. When the car is in combined braking and cornering, the longitudinal and lateral load transfers result in a different combination of heave and bump on each wheel.

For a solid axle with springs of stiffness K_s at spacing S on track T, in double-wheel bump each wheel requires a force equal to the spring force, i.e. in this case the wheel rate equals the spring stiffness:

$$K_w = K_s$$

However, if only one wheel is raised, then by moments

$$K_w = \tfrac{1}{2}K_s\left(1 + \frac{S^2}{T^2}\right)$$

In practice this expression is good for coil springs, but for leaf springs the effective rate is greater because of their torsional stiffness. A representative value of S/T is 0.7. Where coil springs act on links, it is the spacing of the link connection on the axle and the effective stiffness that the spring creates at the link that are relevant.

Figure 5.2.1. Double-wishbone suspension in rear view.

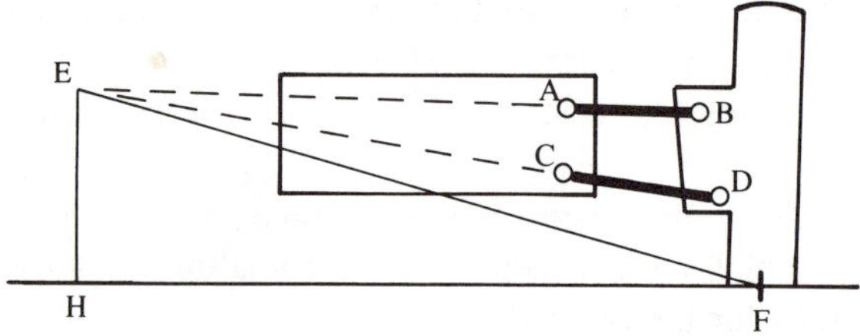

As an example of independent suspension, consider the double-wishbone system of Figure 5.2.1. The complete range of motion of the wheel can be investigated by position diagrams, or preferably by computer analysis. This will also give the spring length, and hence force. If the forces in the links are found, the bush distortions and torques can be included. Such analysis will reveal the variation in camber angle, and also the variation in track, i.e. the lateral scrub of the centre of tyre contact F. Scrub is significant when the wheel is moving in bump, because there is then a scrub speed; combined with the existing longitudinal speed this scrub speed gives a change of tyre slip angle. From the spring and damper force may be found the vertical force at the wheel. The variation of vertical force with vertical position is called the wheel rate. The ride stiffness (or rate) is the wheel rate in series with the tyre vertical stiffness, and may be found from the equation

$$\frac{1}{K_r} = \frac{1}{K_w} + \frac{1}{K_t}$$

In Figure 5.2.1, the intersection of AB and CD extended gives point E, the instantaneous centre (centro) for the wheel relative to the body (S.A.E. swing centre). Thus the direction of motion of F relative to the body is perpendicular to EF. The scrub rate is the rate of change of lateral F position in bump. This is the tangent of the angle between the direction of motion of F and the vertical, which equals the angle EFH. Therefore the scrub rate equals EH/HF. Minus twice this value is the rate of track change in heave. The S.A.E. swing arm radius is defined as the horizontal distance from E to F, i.e. the length HF, taken positive if E is on the centre-line side of the wheel. This is a rather misleading term because the motion of F does not have a radius of curvature equal to HF, or even EF, because the centro E also moves. This was discussed in Section 4.3. In practice E is typically chosen to be approximately at the opposite wheel. However in some cases the arms are parallel, giving an infinite swing arm, or E may even be on the other side of the wheel, known as a negative swing arm.

The double-wishbone system is particularly amenable to design adjustments to obtain properties that the designer considers important. Bollée had an unequal wishbone suspension as early as 1878. In the modern phase of wishbone usage, they began with short parallel equal arms. Longer arms were then adopted because these reduce all unfavourable effects. The top arm was made shorter, this leading to negative camber relative to the body in bump, which is favourable when the body is rolled in cornering. Nowadays a non-parallel unequal-arm layout is usually favoured.

The centro E for other suspensions can also be found quite easily, given the idealisation of neglecting bush distortions, and is discussed in more detail later (Fig. 5.5.2). For example, for a Macpherson strut and link, the top line is perpendicular to the slider (not the steering axis) and through the top connection to the body. For pure trailing arms, E is at infinity, with FE parallel to the arm axis in front view. For a semi-trailing arm, projecting the axis of the arm into the transverse vertical plane of the wheel centres gives E. For a true swing axle, it is the axle pivot, usually close to the differential.

Equilibrium analysis of the wheel and links with the spring, bump stops, bushes and static damper force determines the relationship between wheel position and vertical wheel force. (Under dynamic conditions, wheel inertia and damping effects alter the actual vertical force of course.) This is central to handling analysis because it controls the load distribution between the wheels. It is convenient to consider the isolated wheel force with the anti-roll bar disconnected, and to handle the anti-roll bar couple separately. Figure 5.2.2 shows a typical result for wheel force versus bump position, which may be obtained theoretically or experimentally. Point A is the reference standing height and force. The gradient at A is the wheel rate. Moving into bump, if there is an upward curvature then the suspension is rising rate. Point B is the fully-loaded condition. In further bump, in due course the bump stop is engaged at C, where there may be a discontinuity of stiffness depending on the bump stop design. The force then rises rapidly to the limit position D.

Figure 5.2.2. Wheel vertical force against position.

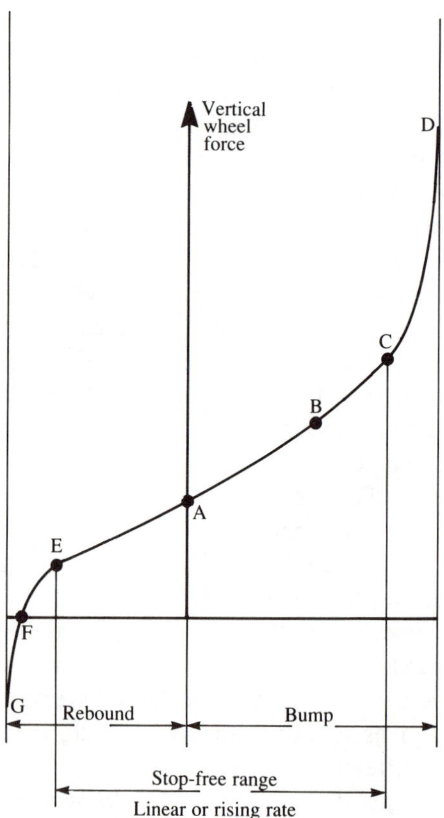

Moving in the droop direction, there is usually a linear or slightly reducing stiffness until the droop stop is encountered at E. Again, here there is likely to be a sharp change in stiffness, especially if the limiter is the damper or an axle strap. Point F is the free droop position of zero vertical wheel force, with the droop stop engaged and acting against the main spring. If the wheel is pulled down it can go on to G. The point of engagement and progressiveness of the bump and droop stops varies considerably between vehicles. In some cases the bump stops are already engaged at normal height, and are used to provide a steadily progressing stiffness. The curve of Figure 5.2.2 is the mean force for both directions. In practice the curve will exhibit some hysteresis, even for essentially static measurements, because of Coulomb friction, or hysteresis for rubber springs.

Figure 5.2.3. Resolved forces for double-wishbone, rear view.

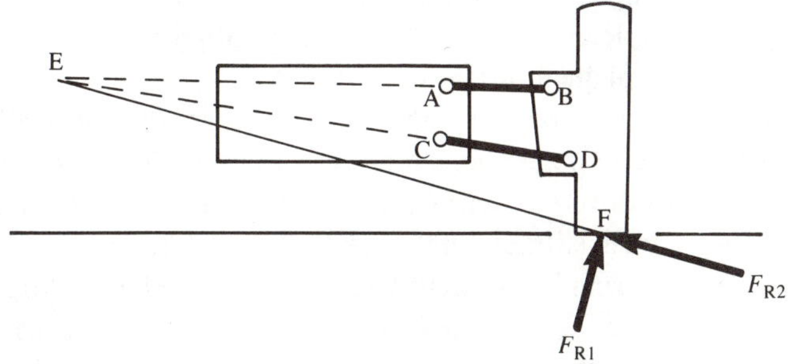

Figure 5.2.3 shows the rear view of the example double-wishbone suspension again. The forces associated with weight, acceleration and load transfer of the unsprung mass have been subtracted from the tyre force, leaving the tyre force exerted on the spring and links. The resulting vertical and lateral tyre force has been resolved into perpendicular components, F_{R1} and F_{R2} (R for right) where F_{R2} acts through E. Consider an idealised (impractical) case, the idealised spring model, where the suspension spring acts on the wheel carrier directly along the line of action of F_{R1}. The force F_{R1} will then be resisted by the spring and have no effect on the links. F_{R2} will be transmitted by the links and have no effect on the spring, because it has no moment about E. Also, considering the link AB, if we neglect its weight and any torques applied by the bushes then it will be in pure tension or compression. Hence the force exerted by the link AB on the body will be along AB. Also the force exerted by CD will be along CD. The total resultant force from the two links must therefore have a line of action through E, because they exert no moment about this point. So, for the idealised spring position the force F_{R2} is transmitted simply by the links, and F_{R1} simply by the spring. F_{R1} and F_{R2} are the net spring force and

the net link force. Consider now a more general case of spring position, for example acting part way along a link; then the force F_{R1} will be partly transmitted to the body by the spring and partly by the links. Such a spring position will give rise to forces in the links, but these will be 'internal' forces that balance out, not affecting the suspension other than through, for example, rubber-bush distortion. However, the important result is this: F_{R2} will still have no effect on the spring because it acts through E, and F_{R1} is still the force to be balanced by the spring.

The equivalent suspension with a spring acting directly along F_{R1} will be called the idealised spring model. The spring characteristic in this model is chosen to give the correct wheel forces. This model has different internal forces but the same external effects. In particular, the link forces are simplified, and reduced to the transmission of F_{R2}, which is of special significance when investigating roll-centres. The idealised spring model is explained further in Section 5.10.

The point E is known as the force centre of the links. It is approximately the same as the centro E in Figure 5.2.1, but strictly not identical because different approximations have been made with regard to link stiffness, bush stiffness, link weight and bush torques.

When the car is running straight then there is little lateral force on the tyre, in which case F_{R2} is quite small. Nevertheless because of the inclination of FE, some of the vehicle's weight is supported through the links rather than on the springs. Our main interest in the F_{R2} force is when it comes into play in a significant way in cornering. However, it is interesting that when a car is lowered from a central jack, a symmetrical condition with each side like Figure 5.2.3 may occur, with large F_2 forces acting inwards on the tyres and holding the body in an unusually high position. These forces are relaxed when the car is allowed to move forward. The fact that F_2 can relieve the springs of some of the car's weight comes into play in cornering, giving the link jacking effect.

As in the case of the centro, the force centre E for other suspensions can be found quite easily, but now given the idealisation of negligible bush torque or friction, discussed later (Figure 5.5.2). For a Macpherson strut, neglecting slider friction the top line is perpendicular to the slider and through the top body connection. For pure trailing arms, E is again at infinity with FE parallel to the arm axis in front view. For a semi-trailing arm, projecting the arm axis into the vertical transverse plane of the wheel centres gives E, because there can be no moment exerted about this axis. For the swing axle it is at the pivot near to the differential, again because there can be no moment about this point.

5.3 Roll

Body roll is the sum of axle roll, from tyre deflection, and suspension roll:

$$\phi = \phi_S + \phi_A$$

Suspension roll is formally defined (S.A.E.) as rotation of the vehicle sprung mass about a fore–aft axis with respect to a transverse line joining a pair of wheel centres. It is positive for clockwise rotation viewed from the rear. This is unambiguous provided that the ground is flat and that front and rear wheel centres have parallel transverse lines (e.g. that wheels and tyres are the same size side-to-side). If the ground is not flat then some mean ground plane must be adopted. The roll angle and roll velocity are in practice fairly clear concepts. Asymmetries, such as the driver, mean that the roll angle is not necessarily zero under reference conditions.

Roll is geometrically equivalent to bump of one wheel and droop of the opposite one, relative to the body. Thus roll speed generally results in a scrub speed of the tyres relative to the ground, causing temporary changes to slip angles and hence to tyre forces. In a rolled position, suspension geometry is generally such that there are changes of wheel steer angles relative to the body. This is roll steer, dealt with in Section 5.14. It is equivalent to bump steer for independent suspension, but not for solid axles. The roll gradient is the rate of change of roll angle ϕ with lateral acceleration A:

$$k_\phi = \frac{\mathrm{d}\phi}{\mathrm{d}A}$$

The total roll gradient is the sum of suspension and axle roll gradients:

$$k_\phi = k_{S\phi} + k_{A\phi}$$

The rolled position also results in wheel camber relative to the body, and more importantly relative to the road, introducing camber forces. In general, in the rolled position the spring stiffnesses on the two sides are different, usually greater on the lower side, so there is some heave of the body in consequence of roll. This is spring jacking. If a droop stop is engaged first, the body may be lowered. There are other jacking effects through the links in cornering and through damper action, which are discussed later.

The vehicle suspension roll couple is the sum of the two suspension roll couples arising because of body roll. The vehicle ride roll stiffness is the sum of the two ride roll stiffnesses. The total ride roll stiffnesses k_R at each end arises from the suspension roll stiffness k_S acting in series with the axle roll stiffness k_A arising from tyre vertical stiffness. Hence

$$\frac{1}{k_R} = \frac{1}{k_S} + \frac{1}{k_A}$$

The suspension roll stiffness is the rate of change of the suspension roll couple with respect to suspension roll angle. For small roll angles the roll couple depends on the main springs and anti-roll bars, whilst for larger angles the bump and droop stops come into play. Longitudinal leaf springs contribute some roll stiffness of their own because they are in torsion, as does the crossbar of the trailing twist axle, these effects being similar to an anti-roll bar. The roll couple may be found by assuming some angular position and calculating the couple. In general the body heave position varies to keep the total vertical force constant. The total vehicle roll couple may be found experimentally by applying a couple and observing the angle. If all four wheel vertical forces are measured on a flat surface, then an applied body roll angle will show the roll stiffness for each suspension. Whilst doing this the wheels should be free to scrub.

The suspension roll couple is most easily understood for a solid axle with linear spring stiffness K_s, at a spring spacing of S. At a body roll angle of ϕ radians each spring force changes by $\frac{1}{2}K_s\phi S$, so there is a couple of moment $\frac{1}{2}K_s\phi S^2$, and the suspension roll stiffness is

$$k_S = \tfrac{1}{2}K_s S^2$$

This depends on the spring spacing. Where coil springs act on links, it is the spacing of the link connection on the axle and the effective stiffness that the spring creates at the link end that are relevant. For non-linear springs, or if the bump or droop stops are engaged, it is necessary to find the body heave position first. There may be additional torques arising from torsion in leaf springs, or from the bushes of locating links. For a solid axle with coil springs $\frac{1}{2}K_sS^2$ is the correct basic expression, and hence the roll stiffness is less than the effect that the double-bump wheel rate would give if acting at the track spacing. However for leaf springs the extra torsional stiffness of the leaves is such that the track T may be used as an approximation for the effective spacing, instead of the actual spring spacing.

In the case of independent suspension, it is possible to calculate the roll moment by detailing the spring and link forces, but it is much easier to work directly with the wheel forces. Independent suspension can be considered as equivalent to a solid axle with wheel rate springs acting at the full wheel track, giving

$$k = \tfrac{1}{2}K_W T^2$$

Thus the actual position of the springs does not alter the relationship between roll and vertical stiffnesses, unlike the case of the rigid axle. Independent suspension is also subject to extra torques from rubber bushes in the links, or friction, but this can still be accommodated in terms of the wheel force. Because the spacing of springs on a solid axle is less than the track, the relationship between roll stiffness and heave stiffness is basically inferior to that of independent suspension.

For both solid and independent axles, the axle roll stiffness arising from tyre vertical stiffness is given by

$$k_A = \tfrac{1}{2} K_t T^2$$

Either a solid axle or an independent suspension may be fitted with an anti-roll bar to increase roll stiffness. The effective roll stiffness depends primarily upon the torsional stiffness of the active part of the rod, on the length of the lever arms, and on the point of connection of the drop links onto the suspension arms or axle. Secondary factors include the stiffness of the usual bushes, including those where the rod is mounted to the body, and bending of the rod, primarily in the lever arms. In some cases the anti-roll bar has been mounted by strapping directly to the trailing arms giving no contact with the body, i.e. as a nascent trailing twist axle, achieving the anti-roll effect with less noise transmission to the body.

The roll couple or stiffness of each end of the vehicle can be summarised by a roll couple graph, Figure 5.3.1. This is usually roughly symmetrical in roll, but not necessarily exactly so because of suspension asymmetries such as a Panhard rod. The initial position is approximately zero roll, the gradient at A being the initial roll stiffness. There is then typically a constant or rising stiffness until B, where there may be a discontinuity of stiffness where a bump stop is engaged. There may be a further stiffness discontinuity at C where the opposite droop stop is engaged, especially for solid axles. At D the inner wheel is

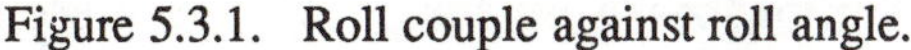
Figure 5.3.1. Roll couple against roll angle.

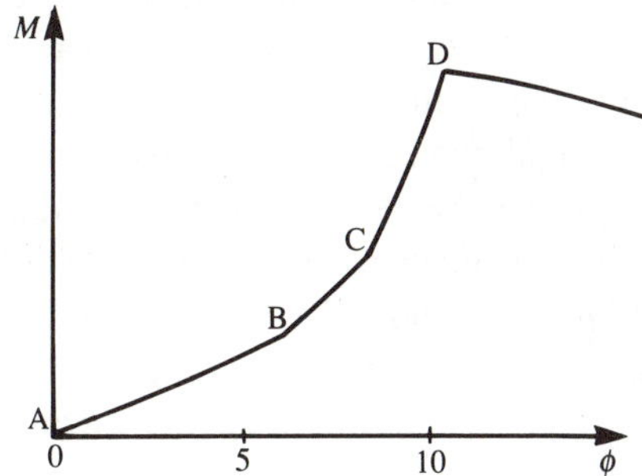

lifted from the ground.

There is one roll couple graph for each end of the vehicle, and one for the whole vehicle. At any roll angle, the roll couple distribution is the distribution of the total between front and rear, usually expressed as percentages. The roll stiffness distribution is similarly defined. Such distributions may be quoted for the suspension alone, or for the ride roll couple including tyre effects. There may be considerable differences, particularly if the suspension bump or droop stops are acting. In fact, the roll stiffness distribution is usually quoted for the zero-roll position, but this is not always satisfactory because the influence of the distribution of roll couple on handling mainly comes into play at medium to large lateral accelerations, when a substantial roll angle has developed, and the roll couple distribution may then be different from the initial value.

5.4 The roll-centre

This section presents the basic theory of the roll-centre. With detailed computer simulations that consider the forces in the individual suspension links it is not necessary to use the roll-centre concept. However, the roll-centre is a very useful idea, because the roll-centre height concisely summarises the effect of the links. With known roll-centre heights it is easy to calculate the roll angle and the load transfer at each of the front and rear axles. This is important in handling at large lateral accelerations because it affects the individual tyre vertical forces, and therefore their lateral forces. Hence the roll-centre height may be used as a summary of the load transfer characteristics of a suspension found by a detailed suspension analysis, or as the input specification for a simple handling simulation. An explanation of the use of roll-centres in lateral load transfer calculation is given in Sections 5.10 and 5.11.

The S.A.E. defines the roll-centre in terms of forces, despite its kinematic name. A definition based on forces will be presented and used here. However, many authors introduce the roll-axis as an axis about which the vehicle actually rolls during cornering, the roll axis being the line joining the front and rear roll-centres. When a vehicle is actually moving on a road, the concept of a kinematic roll axis is difficult to justify in a precise way, especially for large lateral accelerations. Therefore the idea of the vehicle rolling about such an axis, although useful as a qualitative idea, should be treated rather cautiously, except in the special case of a stationary vehicle subject to loads in the laboratory. From a practical point of view, its kinematic significance is that it permits calculation of the lateral movement of the sprung mass relative to the axle, which affects the load transfer (Figure 5.4.1), and is related to the scrub rate in bump.

The roll-centre is defined in S.A.E. J670e *Vehicle Dynamics Teminology* in the following way:

> "The [S.A.E.] roll-centre is the point in the transverse vertical plane through any pair of wheel centres at which lateral forces may be applied to the sprung mass without producing suspension roll."

This does not call for the roll-centre to be in the centre plane and is therefore ambiguous, although it is usually taken to be there. This is a convenient definition when the roll effect of a force applied externally to the body is required, for example an aerodynamic force. However, the usual application of the roll-centre is in evaluating the roll angle and the front-to-rear distribution of lateral load transfer during cornering; i.e. it is primarily concerned with the application of forces by the axle to the body, or by the body to the axle. These are action and reaction, which by Newton's third law are equal and opposite and have the same line of action.

The following basic definitions for roll-centre and roll-centre height will be adopted initially here:

> The roll-centre is a point in the centre plane and in the vertical transverse plane of the wheel centres, at roll-centre height.
>
> The roll-centre height is the height at which lateral forces may be applied to the sprung mass without producing suspension roll.

This is illustrated in Figure 5.4.1 (over page), showing the rear-view free-body diagram of an axle in left-hand cornering.

Figure 5.4.1 actually shows a solid axle; it is permissible for the current purpose to treat the parts of an independent suspension as equivalent to a solid axle. We are here assuming a small lateral acceleration. This is a notional single-axle two-dimensional vehicle; alternatively we could interpret it as a vehicle with identical suspensions front and rear. The real double-axle case is dealt with in Section 5.11. Figure 5.4.1 distinguishes the sprung and unsprung masses m_S and m_U at heights H_S (unrolled) and H_U. The total weight is

$$W = mg = m_S g + m_U g$$

There is a steady-state cornering lateral acceleration A, and the free body is shown in vehicle-fixed axes, so centrifugal compensation sideforces are included. The sprung mass with centre of mass at the

Figure 5.4.1. Solid axle free-body diagram in vehicle-fixed axes; rear view in left-hand cornering.

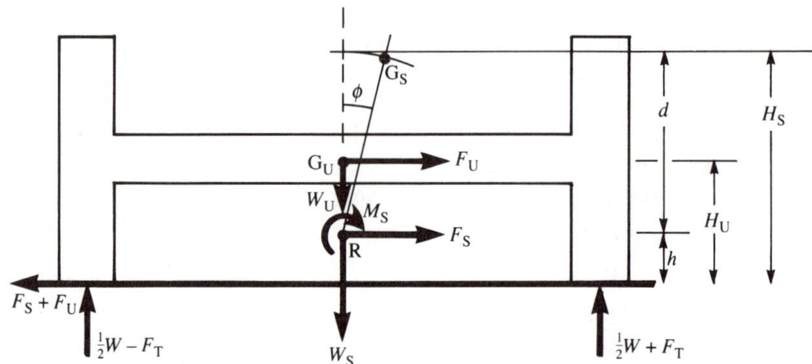

point G_S has rolled by ϕ about the roll axis at point R. There is a total lateral load transfer F_T, so the wheel vertical forces are $\frac{1}{2}W + F_T$ and $\frac{1}{2}W - F_T$. Acting at the unsprung centre of mass is the unsprung weight force, and also the unsprung compensation sideforce

$$F_U = m_U A$$

The sprung mass compensation sideforce

$$F_S = m_S A$$

really acts at G_S, but it is resisted at the roll-centre, so the sprung compensation sideforce F_S is shown acting at the roll-centre along with a moment M_S. The sprung mass weight force is also transferred to the roll-centre, and also contributes to M_S, giving a moment

$$M_S = m_S A d \cos\phi + m_S g d \sin\phi$$

where

$$d = H_S - h$$

is the distance of G_S from the roll-centre. It is the moment M_S that is resisted by the springs and anti-roll bar and creates the suspension roll angle. The sprung mass force F_S acting at the roll-centre is resisted directly by the links and creates no suspension roll, by definition. Hence the total load transfer can be considered in three factors: that from unsprung sideforce (F_{TU}), that from sprung sideforce at the roll-centre, which is transferred through the links of the idealised spring model (F_{TL}), and that from the sprung moment (F_{TM}). By taking moments about the mid-point between the tyre contacts, the load transfer is

$$F_T = F_{TU} + F_{TS}$$

$$= F_{TU} + F_{TL} + F_{TM}$$

$$= \frac{m_U A H_U}{T} + \frac{m_S A h}{T} + \frac{M_S}{T}$$

Thus a high roll-centre increases the load transfer through the links whilst reducing that through the springs and anti-roll bar, reducing the roll angle.

In the application of roll-centres in handling analysis, the load transfer contribution caused by the sprung-mass force at the roll-centre is calculated from the roll-centre height, so the relevant equation is

$$F_{TL} = \frac{m_S A h}{T} = \frac{F_S h}{T}$$

This provides an alternative mathematical definition of the roll-centre height:

$$h = \frac{F_{TL}}{F_S} T$$

To some extent this is a circular definition, because in practice F_{TL} is usually found from h. We could say that F_{TL} is the part of the total sprung-mass load transfer left after F_{TM} is subtracted. For independent suspension, in practice the springs are placed in a way that causes a complex combination of link and spring forces that to a large extent cancel out. However if we consider the idealised spring model of the suspension, then F_{TL} is simply the load transfer through the links, and F_{TM} is simply the load transfer through the springs and anti-roll bar; these are called the net link load transfer and the net roll load transfer. The idealised spring position model was introduced in Section 5.2, and is further explained in Section 5.10; it has the spring that directly resists F_1 in Figure 5.2.3. Using the idealised spring model we can define the roll-centre height using the above equation where F_{TL} is the net link load transfer, or as:

> The roll-centre height is the load transfer through the links of the idealised spring model, divided by the sprung mass sideforce, times the track.

We could alternatively think in terms of a non-dimensionalised roll-centre height, expressed as a fraction of the track. This is simply equal to the load transfer as a fraction of the sprung mass sideforce, which we can call the load transfer factor, f:

$$f = \frac{h}{T} = \frac{F_{TL}}{F_S}$$

Hence the load transfer factor is defined in words as follows:

The load transfer factor is the load transfer through the links of the idealised spring model divided by the sprung-mass sideforce, and also equals the roll-centre height divided by the track.

From the above we may deduce two obvious but important equations:

$$h = fT$$

$$F_{TL} = fF_S$$

Because, as discussed in Section 5.10, a high roll-centre reduces the roll angle, the load transfer factor could also be termed the 'anti-roll', analogous to anti-dive or anti-squat in longitudinal dynamics. The anti-roll coefficient is

$$J_{AR} = \frac{h}{H_S}$$

usually quoted as $J_{AR} \times 100\%$.

5.5 Independent suspension roll-centres – Part 1

The load transfer characteristics, and hence the roll-centre height, may be found by detailed force analysis of the suspension. Figure 5.2.3 showed the force exerted by the ground on the wheel for the case of a double-wishbone suspension. This acts at F, which we can approximate in the usual way as at the centre of the contact patch. Given appropriate tyre data (overturning moment), a better point could be chosen for F. Here we have already subtracted the part of the force associated with unsprung weight force, unsprung-mass sideforce and unsprung-mass load transfer, which can be dealt with separately. The force F is resolved into F_1 and F_2 perpendicular and parallel to EF. All further discussion is about the idealised spring model, hence eliminating the internal forces with no net effect. F_1 is the part of the force that has a moment about E, which therefore cannot be transmitted by the suspension links; it is therefore transmitted to the body by the idealised spring. The force F_2 acts through E and has no moment about E, and is therefore the total force exerted by the links of this side of the suspension on the body when the idealised spring is used. This side net total link force therefore has a line of action along EF. This is the basis of the construction of the line EF for a force roll-centre.

Figure 5.5.1 shows both wheels, with left and right link forces F_{L2} and F_{R2}. Their resultant must act through the intersection of F_{L2} and F_{R2} because they have no moment about this point. Hence this intersection point is the roll-centre. For a symmetrical vehicle this point is on the centre-line. A lateral force applied here can be reacted directly by the

Figure 5.5.1. Independent roll-centre and forces, rear view.

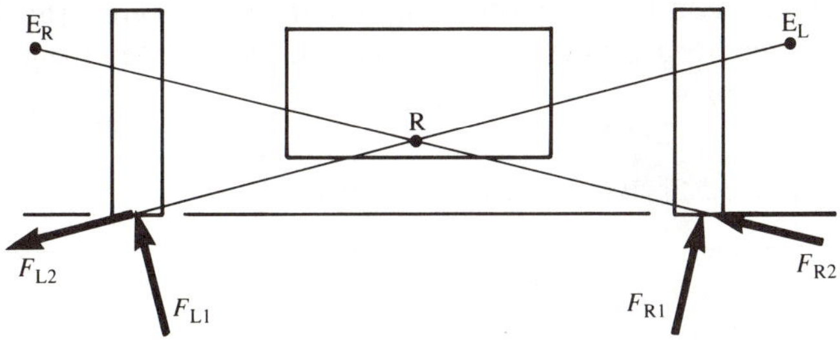

links, giving no force in the springs and therefore without resulting in any roll.

The above development shows that the roll-centre can be deduced by an argument based on forces only, without mention of a kinematic roll-centre. It also shows the real approximations that are usually involved, for example small friction and small bush stiffness, which are quite different from the approximations of the kinematic roll-centre. In a fully detailed analysis, usually by computer, the roll-centre height, with the definition adopted here, provides a convenient summary of the load transfer characteristics of the suspension.

From Figure 5.5.1 we can see that a lateral force at the geometric roll centre could be exactly equilibrated by suitable magnitudes of F_{L2} and F_{R2}, there being unique values to do this. However, there is no guarantee that this will actually occur, because the magnitudes of F_{L2} and F_{R2} depend on the distribution of tyre lateral force on left and right wheels. If they are unequal, with the correct horizontal total, then there will be a residual jacking component.

The presence of friction in the joints, tending to be greater with Macpherson or Chapman strut type suspensions than with wishbones, introduces an interesting point of principle. There is actually a range of heights over which a sideforce will cause no roll. The actual line of action of the link forces is uncertain within this band. Thus the force roll-centre height is also uncertain within this friction band range.

Figure 5.5.2 summarises the initial roll-centre constructions (i.e. for small lateral acceleration) for various independent suspensions. With the assumption of symmetry, it is sufficient to draw one side only. For the strut and link (Macpherson) suspension, Figure 5.5.2(b), the upper construction line is perpendicular to the slider, not to the steering axis AD; these are generally different. For the essentially obsolete slider (pillar) suspension, Figure 5.5.2(c), E is at infinity; there is a single construction line from the wheel contact patch perpendicular to the slider direction. For the true swing axle, Figure 5.5.2(d), the roll-centre is very high. For the transverse rigid arm, the swing arm, Figure 5.5.2(e), sometimes also called a swing axle or pseudo swing axle, it can be made lower.

Figure 5.5.2. Basic independent roll-centres.

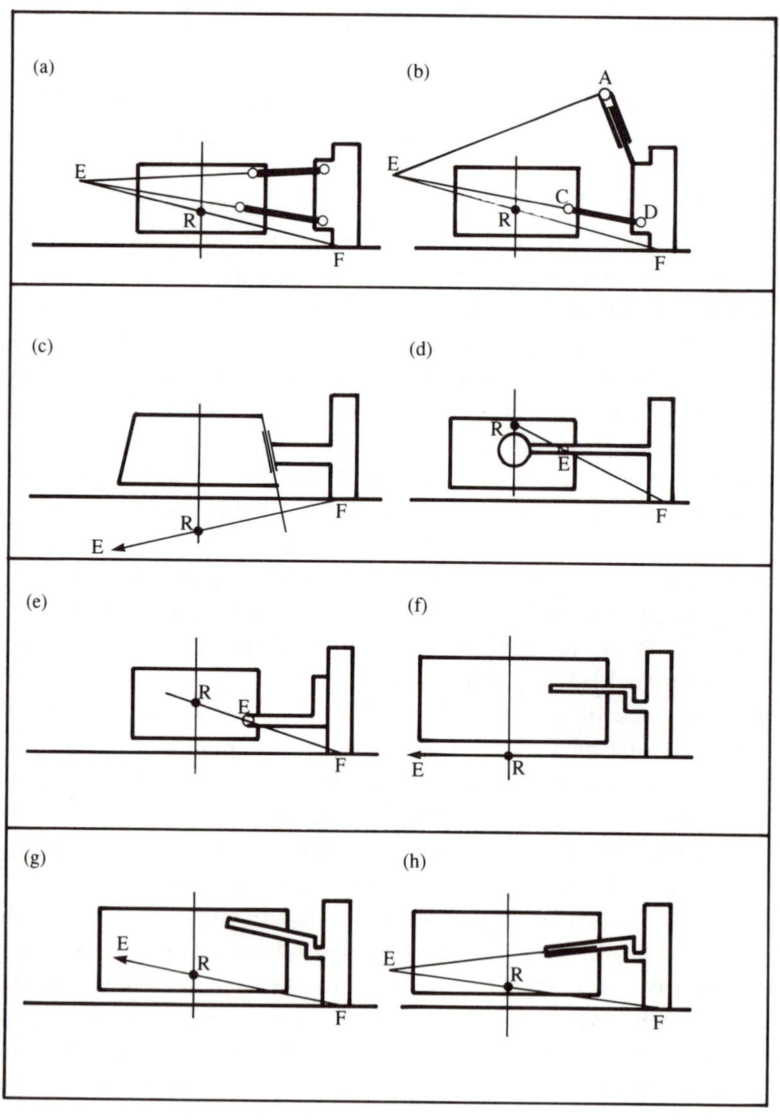

For a trailing arm or semi-trailing arm suspension, in general the pivot axis should be projected into the vertical transverse plane of the wheel centres. This provides the effective E point because the link cannot exert a moment about this axis, neglecting bush friction and stiffness. If the pivot axis is perpendicular to the vehicle centre plane in plan view, i.e. pure trailing arm, then the E point is at infinity, even if it is inclined in front view, Figures 5.5.2(f) and (g). For small lateral acceleration, on the simple trailing arm of Figure 5.5.2(f) the roll-centre is at ground level, even if the trailing arms are inclined in side view.

If the trailing arm pivot is inclined in front view only, Figure 5.5.2(g), then E is at infinity parallel to the pivot axis. For a semi-trailing arm the pivot axis is not perpendicular to the centre-line in plan view, so the axis must be projected into the vertical transverse plane of the wheel centres to find the actual E point which will not be at infinity, Figure 5.5.2(h).

An interesting and unusual case of double wishbones that has been used is with the pivot of the upper arm perpendicular to the vehicle centre-line in plan view. In front view this gives a straight path of the outer end, at any required angle to the vertical, so that the front view is geometrically rather like a strut suspension, but with the advantage of prospectively lower friction.

With the roll-centres determined by the above method, combined with the equations of the previous section, the net link load transfer can be calculated. A high roll-centre increases the net link load transfer, reducing the net roll load transfer and so helping to reduce the roll angle and roll steer effects. However, a high roll-centre means a high centro, which means more scrub in bump. Early independent suspension roll-centres were at ground level, but the compromise usually found nowadays is a height of 50 to 200 mm.

5.6 Independent suspension roll-centres – Part 2

The previous section presented the roll-centre concept in the degree of detail and accuracy normally applied. However the load transfer is most important at large lateral accelerations, in which case the body is rolled and the symmetry approximation may not be good. Therefore this section presents extended analysis of the roll-centre concept at large lateral accelerations. The explanation is still based on the idealised spring model.

We should make a distinction now between the height at which the link force acts, and the height at which the next increment of force acts, for an increment of lateral acceleration. Such a distinction is routinely made in related fields, such as the centre of pressure and the aerodynamic centre of a wing. The corresponding terms 'roll-centre' and 'incremental roll-centre' will be used here. The S.A.E. definition does not explicitly address this distinction. The formal definition of the incremental roll-centre height h_i that will be used here is:

> The incremental roll-centre height is the rate of change of lateral load transfer through the links of the idealised spring model with respect to the sprung mass sideforce, multiplied by the track.

Hence

$$h_i = \frac{dF_{TL}}{dF_S} T$$

Differentiating $F_{TL} = hF_S / T$ with respect to F_S and neglecting track change (which is not important in this case), it may be shown by the usual method for a product of variables that

$$h_i = h + F_S \frac{dh}{dF_S}$$

It is the incremental roll-centre heights that determine the front-to-rear distribution of the incremental lateral load transfer. Again, we can define a non-dimensional incremental roll-centre height, as a fraction of the track, which is the incremental load transfer factor f_i:

> The incremental load transfer factor is the incremental roll-centre height divided by the track, and equals the rate of change of lateral load transfer through the links of the idealised spring model with respect to the sprung-mass sideforce.

Hence

$$f_i = \frac{h_i}{T} = \frac{dF_{TL}}{dF_S}$$

for which

$$f_i = f + F_S \frac{df}{dF_S}$$

The position of the incremental force roll-centre is associated with movement of the force roll-centre. A high incremental roll-centre implies a rising roll-centre. For small lateral acceleration, the roll-centre height remains approximately constant, and the incremental roll-centre height equals the roll-centre height, but this is not so for large lateral acceleration.

In cornering at large lateral accelerations the body is rolled, in which case a symmetry approximation is not appropriate. For example, in the

Figure 5.6.1. Roll-centre at high lateral acceleration.

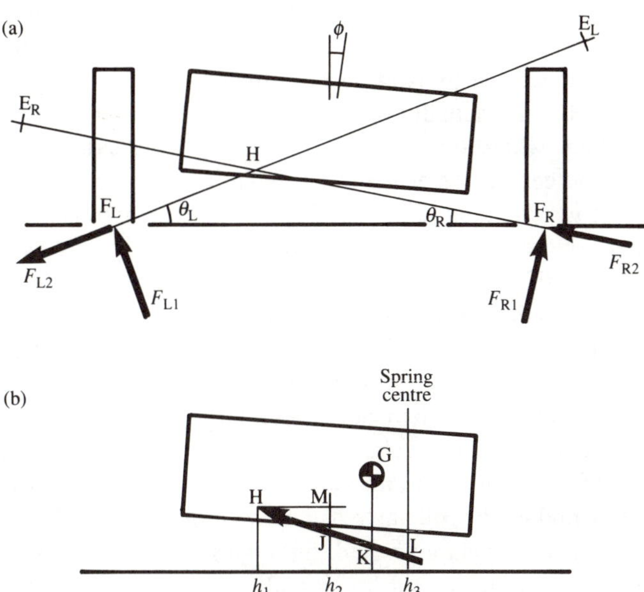

idealised spring model of Figure 5.6.1(a), representing any kind of independent suspension, the positions of E_L and E_R depending on the particular geometry, with a total vehicle radial acceleration to the left, the forces resolve as shown, with the link forces directed F_R to E_R and E_L to F_L. The intersection of the EF lines gives a point H on the line of action of the total link force on the body; in general this is not on the vehicle centre-line. This total force generally has a vertical upward component, because F_{R2} is different from F_{L2} and θ_R different from θ_L; this is the jacking force, which partially supports the body weight, relieving the springs and raising the body. From Figure 5.6.1(a), where the total link forces are F_{L2} and F_{R2}, the total jacking force is

$$
\begin{aligned}
F_J &= F_{R2} \sin \theta_R - F_{L2} \sin \theta_L \\
&\approx 2f(F_{YR} - F_{YL})
\end{aligned}
$$

Because the intersection point H is not on the centre-line, the jacking force itself is generally distributed unevenly on the two sides, and therefore has some influence on the body roll angle. Actually, for an asymmetric vehicle, such as some front-drive vehicles with unequal spring stiffnesses side-to-side, even a central jacking force will result in roll. The vertical line on which a vertical force will cause no roll is called the spring centre (S.A.E.). Here we should go further and distinguish between the incremental spring centre for a small additional vertical force applied in a given cornering trim, and the (ordinary) spring centre which is the place where the vertical component of the total link force can be considered to exert no roll effect.

The point H is not actually the kinematic roll-centre; that is, it is not the point about which the body rolls relative to the axle. Consider a point M on the centre-line at the same height as H. A rotation about H is equivalent to a rotation about M plus a height change. However, the height changes are actually governed by other factors such as jacking force, spring stiffness etc., not just by the suspension geometry. Therefore the rotation of the body relative to the axle is not constrained to be about H (geometrically there are two degrees of freedom). This is particularly obvious when H moves a long way off to one side, when a roll about H would then demand a very large height change, which simply does not occur. The actual centre of roll of the body relative to the axle is offset if there is any jacking, but its position can only be found by first finding the roll angle and heave motion.

Within the approximations stated, the total force exerted on the body by the links acts at H, but not horizontally, Figure 5.6.1(b). There are points J on the vehicle centre-line, K below G, and L on the spring centre, all on the line of action of the total link force. If we resolve the total link force into horizontal and vertical (jacking) components at L instead of H, then by definition the jacking component no longer has a rolling effect, and instead the horizontal component now has a different moment because of its new height h_3. In the simplified symmetrical case H, J, K and L are all on the centre-line at J.

Where then is the force roll-centre? Consider the roll-centre as the point of application of the suspension force to the body. According to this concept, anywhere on the line of action HJKL would do. However the roll-centre height is used to calculate the load transfer and the rolling moment, and usually in this process the rolling effect of the jacking force is neglected. In this case we really should use the height of the point L on the spring centre. The difference of height between H and L depends on the inclination of the suspension forces, which may be substantial for high lateral acceleration. If we use a point other than L as the roll-centre, then the moment of the jacking force should be dealt with as a separate item. We may consider the roll-centre to be on the centre-line at the height of L; it need not be considered to be at L itself.

The distinction between H, J, K and L is important in principle, and may also be in practice for large lateral accelerations. In particular, H may move sideways and up or down because of the suspension geometry, and L may move sideways because of non-linear springs and bump stops.

With the definition of the roll-centre proposed here, the roll-centre height in Figure 5.6.1(a) may be determined by force analysis. The definition of link lateral load transfer is:

> The load transfer through the links is one half of the difference in the vertical upward components of the net forces exerted through the links.

Hence, in Figure 5.6.1(a),

$$F_{TL} = \tfrac{1}{2}(F_{R2} \sin \theta_R + F_{L2} \sin \theta_L)$$

where a plus sign is used because F_{L2} acts downwards here with θ_L positive. The roll-centre height is then found from $h = (F_{TL}/F_S)T$. Of course, having found the load transfer directly, the roll-centre height need not be found, other than as a matter of record or interest, or as a convenient summary of the suspension characteristics.

Two specific examples will be considered here. Figure 5.6.2 shows a parallel equal-arm double-wishbone suspension. In this case the body roll lowers the inside end of the outer (right hand) side links, the link angle and hence θ depending on the link length and the lateral spacing between the link–body mounting points. In this case

$$F_{TL} = \tfrac{1}{2}(F_{R2} \sin \theta_R - F_{L2} \sin \theta_L)$$

By definition, the total link force from both sides exerts no moment about the roll-centre, for the idealised spring model. In general, at high lateral acceleration $F_{R2} > F_{L2}$, so the roll-centre will be nearer to the line of action of F_{R2} than that of F_{L2}, i.e. with increasing lateral acceleration and roll the roll-centre falls for this kind of suspension; this is generally true of double-wishbone suspensions.

Figure 5.6.2. Rolled equal parallel wishbones.

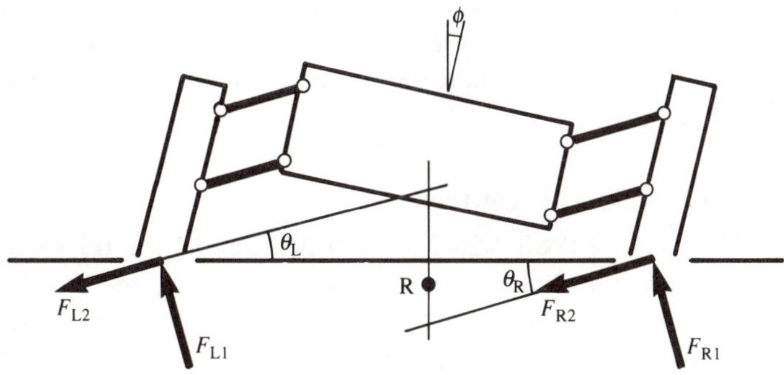

Figure 5.6.3. Rolled plain trailing arms.

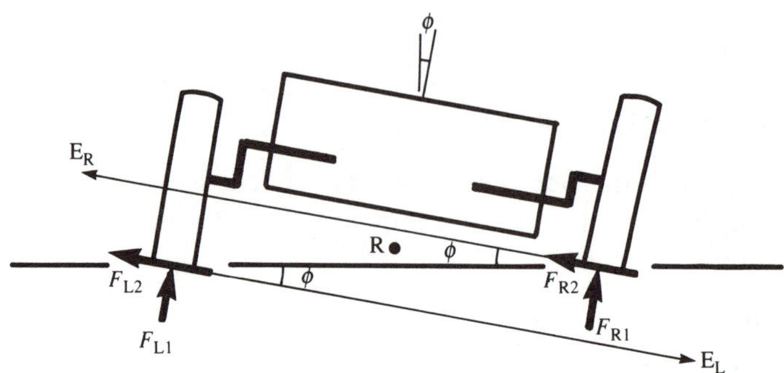

Figure 5.6.3 shows a plain trailing arm suspension. For this case, with roll angle ϕ,

$$F_{TL} = \tfrac{1}{2}(F_{R2} \sin \phi - F_{L2} \sin \phi)$$

Again, considering the point on the centre-line with no moment from F_{L2} and F_{R2} combined, and with $F_{R2} > F_{L2}$, we see that in this case the roll-centre rises above ground level. It approaches the line of action of F_{R2} as F_{L1}, and hence F_{L2}, tends to zero, and therefore may reach 100 mm or more, for large lateral accelerations. This is in marked contrast to the usual assumption that it remains at ground level for this kind of suspension. In other words, there is a link load transfer and jacking force in contrast to the usual assumption that these are zero. This is easily seen for the extreme case when the inner wheel vertical force goes to zero, leaving a jacking force $F_{R2} \sin \phi$ and link load transfer $\tfrac{1}{2}F_{R2} \sin \phi$. These are not apparent from the simple theory of Section 5.5 because they develop in a non-linear way.

Because the effect of a rising or falling roll-centre only comes into play as roll angle develops, it would be possible to treat the roll-centre height as fixed and to have an extra equivalent anti-roll bar. However this would have to be non-linear, probably with moment proportional to roll angle squared, and its properties could only be established by investigating the roll-centre movement. For a falling roll-centre it would have negative stiffness.

5.7 Solid axle roll-centres

This section deals with the determination of the roll-centre for solid axles with rigid link location. Leaf-spring axles are considered in the next section. Figure 5.7.1 shows a general four-link axle. The method is based on studying the support links to find two points A and

Figure 5.7.1. Roll-centre for four-link solid axle.

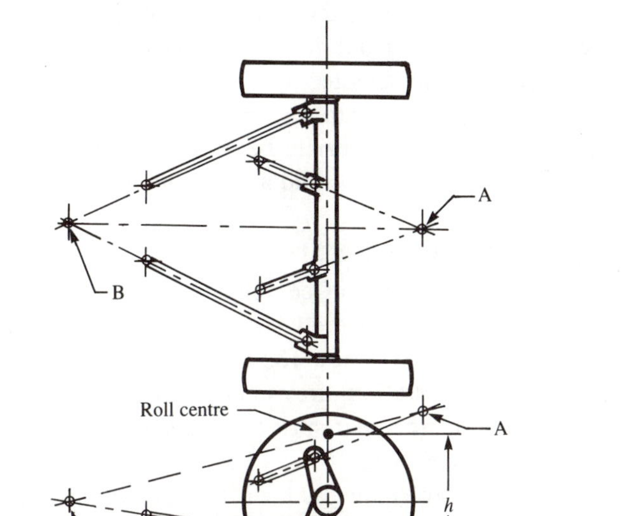

B where forces are exerted by the axle on the body, the roll-centre necessarily lying on the line joining A and B, at the point where this line penetrates the transverse vertical plane of the wheel centres. Consider idealised springs at the wheels, or at least springs that do not act on the links. One link pair has an intersection point at A, so the combined force exerted by these links on the body must act through A (neglecting bush torques and link weight). Similarly the other link pair exerts a force through B. The resultant of the two forces at A and B acts through a point somewhere on AB. However, the cornering force acts in the vertical transverse plane of the wheel centres, neglecting pneumatic trail, so the roll-centre is where AB intersects this plane. The torque due to trail can be dealt with separately.

Suitable points A and B can be found for other axle link layouts. For example, if the lower links are parallel then the point B is at infinity, so AB is parallel to the bottom links. If the bottom link pair is replaced by a torque tube or similar system, Figure 4.6.1(f), then point B is the front ball joint. If transverse location is by a Panhard rod, then point A is the point at which the rod intersects the vertical central plane. A characteristic of the Panhard rod is that the roll-centre rises for roll in one direction, and falls for the other, because of vertical motion of the point of connection to the body. For other axle lateral location systems it is similarly necessary to find point A where the line of action of the force intersects the central plane. In the systems described in Section 4.3, this generally corresponds to the pivot pin point C.

The forces exerted on a solid axle are as in Figure 5.4.1. The resolution into components directed along and perpendicular to the line from the contact to the roll-centre, as in Figure 5.5.1 for independent suspension, is no longer appropriate because the wheels exert forces directly on each other. Hence the solid axle has no link jacking force.

5.8 Compliant-link roll-centres

Finding the roll-centres of compliant-link systems is especially problematic. In the cases covered so far, the locating links were physically distinct from the springs. The functions of horizontal location and vertical force generation are sometimes combined in single elements, for example the leaf-spring or trailing twist axle. This complicates the issue considerably.

In the case of location by longitudinal leaf-springs, the load transfer properties depend on the bending and torsional stiffness of the springs and bushes, which only comes into play once the body rolls, so it is necessary to separate out the roll stiffness effects as an equivalent anti-roll bar, to leave the effective roll-centre height. As an example to find a first approximation to the roll-centre position, consider the simplified case of Figure 5.8.1.

Figure 5.8.1. Idealised leaf-spring axle.

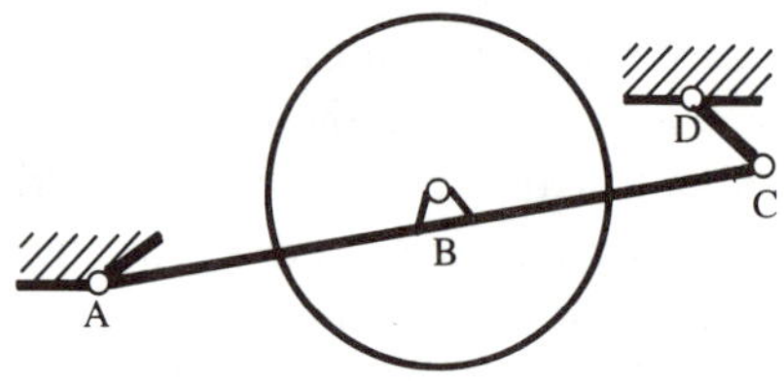

Firstly, consider the case of perfect torsional rigidity of the straight spring ABC, and complete rigidity of link DC and the bushes at A, C and D, other than their basic design motion. In such a case, a force could be applied to the body at any height without any roll rotation being possible. This is equivalent to a perfectly rigid anti-roll bar. To completely eliminate anti-roll bar effects, which must be done to find the roll-centre, compliance must be introduced into the system, in particular for the bushes about axes AC or AD. In practice, because of the torsional compliance of the leaf-spring, the effective stiffness of bush C will be less than that of D. If we consider bushes A and C to have complete compliance about AC, but D to be rigid, then a sideforce applied to the body on the line AC can be transmitted to the suspension because DC will act as a rigid cantilever. In this simplified case, the roll-centre will be at the height of the line AC where it passes the vertical transverse plane of the wheel centres. For other more complex cases, such as curved springs ABC, it is still usual to take this approximation, i.e. to use the line joining the front and rear spring eyes. This is clearly only an approximation, because even with a sideforce applied at roll-centre height there will be torsion of the spring, and additional torque at A and C. In any case, the roll stiffness because of spring torsion with body roll must be treated as an equivalent anti-roll bar.

Because of the lateral compliance of the leaf-spring location system, it is possible to supplement it with a Panhard rod or other lateral location device, as is sometimes done for racing. If the stiffness of the rod with its bushes is sufficient, i.e. much greater than the leaf system, then the rod will be decisive. Otherwise the result will be some intermediate position. If the rod is at a different height from B, this calls for increased lateral deflection of the leaf-springs, and therefore results in an additional roll stiffness, which must be treated as an equivalent anti-roll bar.

In the case of the trailing twist axle the cross-member acts in torsion giving an anti-roll bar effect, so to eliminate this we must consider a zero torsion stiffness cross-member, still with bending stiffness. Figure 5.8.2(a) shows a rear view of a simplified case with horizontal arms and small roll angle. The tyre lateral forces are L and L'. The cross-beam has the free-body diagram in rear view of Figure 5.8.2(b). The wheel

Figure 5.8.2. Trailing twist axle: (a) rear view, (b) beam free-body diagram, (c) bending-moment diagram.

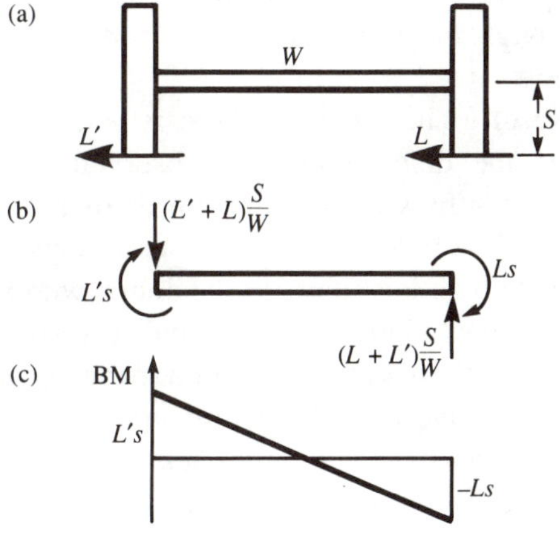

vertical forces are reacted separately by springs on the trailing arms. The tyre sideforces exert moments *Ls* and *L's* on the ends of the beam. For beam rotational equilibrium there must be vertical end forces as shown. This gives the beam the bending-moment diagram of Figure 5.8.2(c).

When $L \approx L'$ the central bending moment is small in this view. A similar argument applies in plan view. In conjunction with the zero torsion specification, this means that each half of the suspension exerts negligible moment about an axis from its own front bush through the mid-point of the cross-beam. Therefore this axis can be projected into the transverse plane of the wheel centres to find the force centre, as for a trailing arm. The approximations are increasingly in error as the lateral acceleration increases, so this is only the initial position of the roll-centre.

5.9 Experimental roll-centres

Various methods have been suggested at times for the experimental measurement of roll-centre position, none of which are really very satisfactory. One method is the observation of body displacement with application of a lateral force at the centre of mass, for example using a laterally slanted ramp, but this is definitely not suitable. The large lateral displacement at the tyres is a problem which may be eliminated by using solid wheels, but there will still be inappropriate lateral displacements within the suspension compliances. A correct method of finding the S.A.E. roll-centre would be to apply lateral forces at various heights. With the lateral force applied at the height of the roll-axis, there will be a lateral displacement of the body because of compliances, and a vertical displacement because of the jacking force, but no rotation. In this sense, the force roll-centre is directly analogous to a shear centre, which would probably be a better name for it, although the former name is no doubt too entrenched by use to change now.

As in the case of a shear centre, there is a more direct method of measurement. If the displacement is observed when a moment is applied, the shear centre will be the centre of rotation, i.e. the point of zero displacement. The most convenient way to apply a moment is by a couple, for example by a joist through the doors, with a load first on one end then on the other. This does involve a vertical load which is acceptable if the vehicle is to be tested in a loaded condition. Otherwise it is necessary to arrange for a vertical upward force to be applied as half of the couple, say by a hydraulic jack or a weight over a pulley. The actual deflection has been analysed by double-exposure photographs, but in the interests of accuracy it is much better to take accurate measurements of the displacement of specific points by dial gauges. If these points are not in the vertical transverse plane of the wheel centres, then the data must be used to find the roll-axis, and this then used to find the roll-axis position at each suspension.

This method is still subject to serious errors and problems. Under real cornering conditions there are actually large forces in the suspension links, causing geometric changes through bush distortions. These are largely absent from the test. Any small inappropriate lateral displacement of the body will give a large error in the measured roll-centre height. The applied couple will change the load on the tyres, giving an additional rotation of the complete chassis, including the wheels, about a point at ground level roughly midway between the tyres. The result is a false lateral motion of the body at roll-centre height. This must be guarded against by using solid disc wheels, although this exacerbates the problem of allowing scrub, because when finding the roll-centre for large angular displacements, there will be a consequent track change. This must be permitted, perhaps with air bearing pads, but without allowing any inappropriate lateral motion of the body, which is critical. It is difficult to say how the body should be located.

In summary, it is fairly easy to observe the displacement of a vehicle when it is subject to loads or couples, but it is very much more difficult to obtain results that have a worthwhile roll-centre interpretation, especially for large lateral accelerations. Probably the only satisfactory experimental method is to use an instrumented vehicle in real cornering to find the actual load transfer for each axle, and deduce the net link load transfer and force roll-centre height from that, using the equations of Section 5.4.

5.10 Suspension load transfer

During cornering, the vertical forces on the outer tyres increase at the expense of those on the inner ones. This is vertical force transferred laterally, commonly called lateral load transfer. This section describes lateral load transfer on a single solid or independent axle. A complete vehicle with two axles is examined in the next section.

As perceived in vehicle-fixed axes, the load transfer F_T is accomplished by several simultaneous compensation forces on the axle, referring back to Figure 5.4.1. These are:

(1) The centrifugal force $F_U = m_U A$ on the unsprung end-mass at its G height, giving load transfer F_{TU}.

(2) The centrifugal force $F_S = m_S A$ on the sprung end-mass acting at roll-centre height through the links, giving the net link load transfer F_{TL}.

(3) The moment M_S resisted by the idealised springs and anti-roll bar, because of roll angle, giving the net moment load transfer F_{TM}.

This gives

$$F_T = F_{TU} + F_{TL} + F_{TM}$$

Actually, the springs normally act on the links in a way that creates additional internal forces in both that have no net effect on the body or axle. Of course, in the investigation of internal effects, e.g. bush distortions, it is necessary to consider the real spring position. However the important effect of the spring and anti-roll bar is to produce the wheel vertical force, and for handling their exact position is not important provided that the end result is the same. For example, for a given wheel force a spring acting on a wishbone will need to exert a smaller force if it acts closer to the wheel end. Different spring positions for a given wheel force give different spring and link forces, but do not alter the net effect, so it is very convenient to reduce a suspension to the idealised spring model. This has the same links, but the spring and anti-roll bar are deemed to act in a way that creates no link forces; e.g. in Figure 5.2.3 this will mean that the spring force acts directly along the line of action of F_{R1}. In the idealised spring model, the conflicting internal forces are eliminated, leaving only the important net effects.

In Figure 5.5.1, the net link load transfer is the difference of vertical components of F_{L2} and F_{R2}. The net roll load transfer from the springs and anti-roll bar is the difference of vertical components of F_{L1} and F_{R1}. The jacking force is the total vertical force exerted by the links of the model, and so is the sum of the vertical components of F_{L2} and F_{R2}.

Figure 5.10.1(a) shows the rear-view free-body diagram of a highly simplified notional two-dimensional vehicle in steady-state left-hand cornering, in Earth-fixed (inertial) coordinate axes. There are tyre lateral forces giving the vehicle a radial acceleration A to the left. Figure 5.10.1(b) shows the free-body diagram in the vehicle-fixed axes. The appropriate compensation force mA has been added at the centre of mass, so that the vehicle has no acceleration in this coordinate system. Bearing in mind that the total tyre lateral force $F_{YL} + F_{YR}$ equals mA, that the total tyre vertical force $F_{VL} + F_{VR}$ equals W, and that there is no roll acceleration, summing moments about the centre of mass G for either of these figures gives

$$\Sigma M = \tfrac{1}{2}F_{VL}T - \tfrac{1}{2}F_{VR}T - (F_{YL} + F_{YR})H = 0$$

Figure 5.10.1. Free-body diagrams: (a) Earth-fixed axes, (b) vehicle-fixed axes.

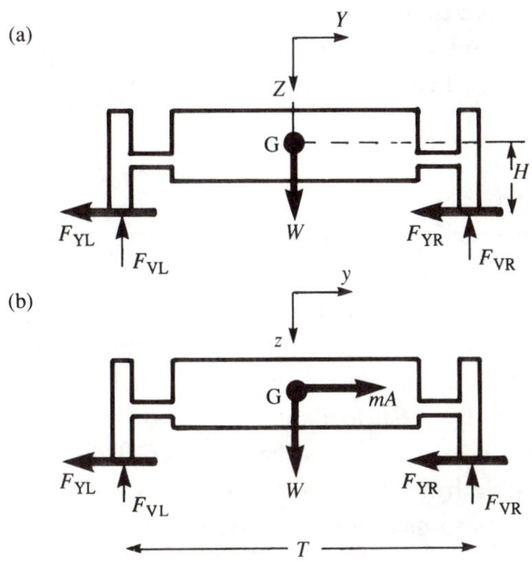

Hence

$$\tfrac{1}{2}(F_{VL} - F_{VR})T = (F_{YL} + F_{YR})H = mAH$$

The total load transfer is half of the difference of vertical tyre forces:

$$F_T = \frac{F_{VL} - F_{VR}}{2} = \frac{mAH}{T}$$

The wheel vertical forces are therefore

$$F_{VR} = \frac{W}{2} + \frac{mAH}{T}$$

$$F_{VL} = \frac{W}{2} - \frac{mAH}{T}$$

The total load transfer depends only on the lateral force mA, the centre of mass height H and the track T. This demonstrates that for this model the total load transfer cannot be influenced by adjustment to internal characteristics such as the suspension.

Actually, this is not exactly true for a body that rolls, because the roll angle moves the centre of mass out slightly relative to the wheels, depending on the height of G_S above the roll axis, as shown in Figure 5.4.1. At maximum lateral acceleration this lateral motion of G gives a load transfer of about 5% of the total weight, compared with a load transfer of about 35% of the weight from the other factors. It is sometimes stated that a vehicle that does not roll has no load transfer;

this is false. Zero roll will eliminate the small factor because of lateral G movement, but will have no effect on the major factors shown in the zero-roll model above.

It is often convenient to express the lateral load transfer in terms of a non-dimensional vertical force lateral transfer factor e_V. This is defined as

$$e_V = \frac{F_T}{F_{Vi0}}$$

where F_{Vi0} is the static vertical force on the inner wheel. Hence $e_V = 1$ corresponds to inner wheel lift-off.

It is desirable for safety reasons that overturning should not occur before the tyre friction sliding limit, placing limits on the relationship between H and T – limits that are met by normal cars but often not met by trucks, especially when loaded. Because of the tyre characteristics, the greatest total lateral force can be achieved if the vertical force is as uniform as possible from side to side, giving a strong incentive for racing and sports vehicles to have a low centre of mass and a wide track.

Roll-over would occur at a lateral acceleration at which the load transfer reduces the inner reaction F_{VL} to zero:

$$A_R = \frac{WT}{2mH} = \frac{gT}{2H}$$

If the maximum lateral acceleration achievable, limited by grip, is A_M, then the safety factor against roll is defined to be

$$S_R = \frac{A_R}{A_M}$$

Using the above expression for the value of A_R, then

$$S_R = \frac{\frac{1}{2}gT/H}{A_M}$$

which is around 1.5 for a conventional passenger car. The simple expression for S_R immediately above can be refined by a more accurate calculation of A_R, including load transfer from body roll (lateral G position), gyroscopic effects, and tyre lateral distortion. This gives:

$$A_R = \frac{\frac{1}{2}gT/H}{1 + \dfrac{m_S g(H_S - h)\,k_\phi}{mH} + \dfrac{m_{4w}\;k^2}{m\,H\,R_e} + \dfrac{mg}{2H\,k_y}}$$

where m_{4w} is the mass of the four wheels, k is their radius of gyration, and k_y is the tyre lateral stiffness relating the offsetting of F_V to the sideforce. This more accurate value of A_R is about 20% lower than the simply calculated value, giving a correspondingly much lower S_R.

Figure 5.4.1 showed the forces exerted on an axle seen in vehicle-fixed axes, including consideration of separate unsprung and rolling sprung masses. The total tyre lateral force is

$$F_S + F_U = m_S A + m_U A$$

The unsprung mass centrifugal compensation force $m_U A$ is applied at the unsprung centre of mass height H_U. The sprung-mass centrifugal compensation force, really applied at the sprung centre of mass, is shown instead transferred to the roll-centre, along with a moment. The sprung weight force, now offset because of roll, is also transferred to the roll-centre, and also contributes to the moment. Using

$$d = H_S - h$$

then the moment is

$$M_S = m_S A d \cos\phi + m_S g d \sin\phi$$

In the vehicle-fixed axes, all moments are in equilibrium, so setting moments about the mid-point between the tyre contacts to zero and dividing by the track T gives the total load transfer:

$$\begin{aligned} F_T &= F_{TU} + F_{TL} + F_{TM} \\ &= \frac{F_U H_U}{T} + \frac{F_S h}{T} + \frac{M_S}{T} \end{aligned}$$

Thus the load transfer arises from three factors: unsprung sideforce at the unsprung centre of mass height, sprung sideforce through the links at roll-centre height, and the moment associated with the roll angle. It is usual to treat the two unsprung parts of an independent suspension as having a combined centre of mass. For independent suspension, the unsprung load transfer actually results in link forces that affect the sprung mass roll angle slightly, depending on the details of the linkage geometry. This is usually neglected for simplicity.

Part of the total load transfer from the sprung mass is taken by the roll angle acting against the roll stiffness giving F_{TM}, and part is exerted directly by the links giving F_{TL}. The allocation of load transfer to the two methods depends on the roll-centre height. Usually $0 < h < H_S$, in which case the vehicle rolls outwards, and the load transfer by both methods is positive. If $h = H_S$ there will be zero roll angle, and all the load transfer is through the links. For a roll-centre higher than the sprung centre of mass ($h > H_S$), not normally met in practice, the links would transfer too much load, which would be balanced by a negative load transfer in the springs, and the body would lean into the turn. For a roll-centre below ground level ($h < 0$), sometimes met in practice, the links have a negative load transfer, so the springs and anti-roll bar must have a load transfer exceeding the total. However, whatever the roll-centre height, the total load transfer is not altered, other than that a high roll-centre helps to reduce the roll angle, which reduces the relatively small effect of the lateral movement of the centre of mass.

The roll moment produced by the stiffness elements in roll depends on the body roll relative to the axle, i.e. relative to a line joining the wheel centres. This is the suspension roll angle. Because of load transfer on the tyre vertical stiffness, typically 250 N/mm, the axle itself, solid or independent, has a small roll angle called the axle roll angle. The body roll is the suspension roll plus the axle roll. Axle roll is typically one-eighth of suspension roll, reaching 1–2°.

5.11 Vehicle load transfer

Figure 5.11.1 shows a two-axle vehicle model. This distinguishes the sprung mass m_S from the front and rear unsprung masses m_{Uf} and m_{Ur}, each with its own centre of mass. There are different front and rear roll-centre heights, tracks, etc. The vehicle is in steady-state left-hand cornering. It is convenient to perform this analysis in the accelerating coordinate system xyz attached to the vehicle (vehicle-fixed axes), so the centrifugal compensation forces are included at each mass centre. In this coordinate system there are no linear or angular accelerations. The sprung-mass weight force is shown; other weights and tyre forces are omitted for clarity of the figure. The lateral acceleration considered here is that perpendicular to the vehicle centre-line. This is not quite the same as the cornering radial acceleration, because the vehicle body has an attitude angle to its direction of travel arising from steering, the slip angle of the rear tyres and roll and compliance steer effects, so the vehicle radial acceleration resolves into a lateral acceleration and a longitudinal one. The longitudinal acceleration component causes a front-to-rear load transfer that can be calculated separately. The linear analysis will be presented here in order to illustrate the principles. In practice, the load transfer is important at higher lateral accelerations, in which case non-linearities may well come into play (for example bump stops). A good computer simulation can, of course, deal with the non-linearities.

In these vehicle-fixed coordinate axes, the vehicle is held in equilibrium, with zero acceleration. The forces $m_{Uf}A$ and $m_{Ur}A$ act directly on the front and rear axles respectively, each transferring load only between its own pair of tyres. The effect of the sprung-mass forces $m_S A$ and $m_S g$ may be determined as before. The distance of the

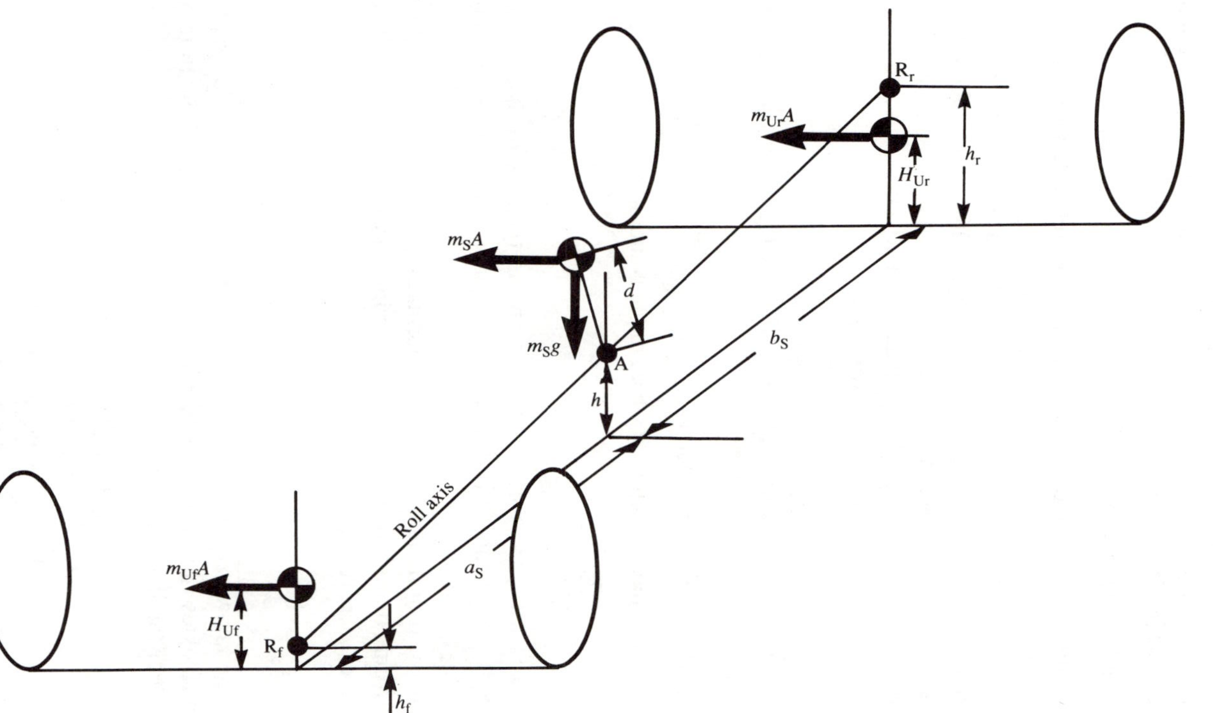

Figure 5.11.1. Vehicle load transfer model, vehicle-fixed axes.

sprung centre of mass from the roll-axis is $d = H_S - h$. The roll angle is ϕ. The forces $m_S A$ and $m_S g$ may be replaced by forces at A plus a moment, which is

$$M_S = m_S A d \cos\phi + m_S g d \sin\phi$$

This moment is reacted by the springs and anti-roll bars, distributed front-to-rear appropriately (according to the distribution of roll stiffness for the linear case for a torsionally rigid body). This determines the suspension roll angle. To find this roll angle, this total applied moment must be matched against the total vehicle roll couple characteristic from Figure 5.3.1. The roll angle is usually small enough for a small ϕ approximation for the applied moment, giving

$$M_S = m_S A d + m_S g d\phi$$

If the resistance moment is also linearised, which is a more restrictive approximation, then

$$M_\phi = k_S\,\phi = (k_{Sf} + k_{Sr})\phi$$

in which case the roll angle has the simple solution

$$\phi = \frac{m_S A d}{k_S - m_S g d}$$

Here k_S is used, the roll compliance on the tyres being neglected for simplicity. In practice it is desirable to include this, especially for higher roll stiffness suspensions.

Consistent units must be used, it normally being necessary to convert $m_S g d$ from Nm/rad to Nm/deg to agree with k, this then giving roll angle in degrees. The roll-angle gradient is

$$k_\phi = \frac{\mathrm{d}\phi}{\mathrm{d}A} = \frac{m_S d}{k_S - m_S g d}$$

From the known roll angle and the front and rear roll couple graphs or roll stiffnesses, the front and rear load transfer from roll angle may be found. For the linearised case

$$F_{\mathrm{TfM}} = \frac{k_f \phi}{T_f} = \frac{k_f m_S A d / T_f}{k - m_S g d}$$

$$F_{\mathrm{TrM}} = \frac{k_r \phi}{T_r} = \frac{k_r m_S A d / T_r}{k - m_S g d}$$

This completes the calculation of roll stiffness load transfer.

The effect of the sprung mass force $m_S A$, which has been transferred to point A on the roll axis, is found by redistributing the sideforce between the front and rear roll-centres according to the position of the unsprung centre of mass along the wheelbase. The wheelbase is

$$l = a + b = a_S + b_S$$

where the sprung centre of mass is a_S behind the front axle. The front and rear sprung end-masses are:

$$m_{Sf} = \frac{m_S b_S}{l} = m_f - m_{Uf}$$

$$m_{Sr} = \frac{m_S a_S}{l} = m_r - m_{Ur}$$

where m_f and m_r are the total front and rear end-masses:

$$m_f = \frac{mb}{l}$$

$$m_r = \frac{ma}{l}$$

where m is the total mass. At the front and rear axles the sprung end-mass forces are

$$F_{Sf} = m_{Sf} A$$

$$F_{Sr} = m_{Sr} A$$

The sprung end-mass forces F_{Sf} and F_{Sr} each act on the appropriate suspension and cause load transfer on their own end only, according to the roll-centre height for that end, giving the net link load transfer:

$$F_{TfL} = \frac{m_{Sf} A h_f}{T_f} = m_{Sf} A f_f$$

$$F_{TrL} = \frac{m_{Sr} A h_r}{T_r} = m_{Sr} A f_r$$

where $f = h/T$ is the suspension load transfer factor. This completes calculation of the load transfer from the sprung mass.

The unsprung mass load transfer is simply caused by the height of the centrifugal compensation force on the unsprung centre of mass (an approximation for independent suspension), so

$$F_{TfU} = \frac{m_{Uf} A H_{Uf}}{T_f}$$

$$F_{TrU} = \frac{m_{Ur} A H_{Ur}}{T_r}$$

where H_U is the unsprung centre-of-mass height.

The total load transfer from sprung and unsprung effects at each end is then

$$F_{\mathrm{Tf}} = F_{\mathrm{TfU}} + F_{\mathrm{TfL}} + F_{\mathrm{TfM}}$$

$$F_{\mathrm{Tr}} = F_{\mathrm{TrU}} + F_{\mathrm{TrL}} + F_{\mathrm{TrM}}$$

For the linear case, all this can be summarised by

$$\phi = \frac{m_{\mathrm{S}}Ad}{k_{\mathrm{S}} - m_{\mathrm{S}}gd}$$

$$F_{\mathrm{Tf}} = \frac{m_{\mathrm{Uf}}AH_{\mathrm{Uf}}}{T_{\mathrm{f}}} + \frac{m_{\mathrm{Sf}}Ah_{\mathrm{f}}}{T_{\mathrm{f}}} + \frac{k_{\mathrm{f}}\phi}{T_{\mathrm{f}}}$$

$$F_{\mathrm{Tr}} = \frac{m_{\mathrm{Ur}}AH_{\mathrm{Ur}}}{T_{\mathrm{r}}} + \frac{m_{\mathrm{Sr}}Ah_{\mathrm{r}}}{T_{\mathrm{r}}} + \frac{k_{\mathrm{r}}\phi}{T_{\mathrm{r}}}$$

In this way the front and rear lateral load transfer may be found. The distribution of lateral load transfer should be carefully distinguished from the distribution of roll stiffness, even in the linear case. Where required, the calculation may be refined by including tyre overturning moment, the gyroscopic effect of the wheels, and possibly engine, and aerodynamic effects. These may have a substantial cumulative influence.

The procedure for finding the cornering load transfer and its distribution may be summarised as follows:

(1) Find the total sprung mass applied rolling moment in terms of ϕ.

(2) Find ϕ using the total roll couple curve or roll stiffness.

(3) Find each end roll load transfer using the end roll couple curves or roll stiffnesses.

(4) Find the front-to-rear distribution of sideforce from the sprung mass.

(5) Find each end sprung-mass force load transfer using the roll-centre heights (net link load transfer).

(6) Find each end unsprung-mass load transfer using unsprung-mass heights.

(7) For each end, sum the load transfers by unsprung-mass sideforce, net link sideforce, and the roll stiffness moment.

It is sometimes suggested that account should be taken of the inclination ρ_{ra} of the roll axis, which is usually 6° or less, and 12° in the extreme. Then instead of transferring the sprung sideforce directly down to the roll axis, it is moved perpendicularly to the roll axis; in practice this means slightly backwards and with a slightly smaller moment arm. However, if this is done then one should also allow for the fact that the moment about the roll axis has a component about a vertical axis, which compensates for the rearward movement. Also, for realistic values of inclination the cosine is so close to 1.0 that the change of moment arm is negligible. Allowing for axis inclination therefore simply complicates the equations without usefully improving the accuracy.

The above analysis assumes that the body is torsionally stiff, to the extent that the torsion angle is much less than the roll angle, i.e. the front and rear suspension roll angles are taken to be equal, which is generally true enough for cars, but not so for trucks. If this is not the case, then the moment about the roll-axis is distributed front-to-rear in a more complex way. For a very torsionally compliant body, adjusting the distribution of roll stiffness ceases to be an effective tool in handling adjustment. Also, additional effects may sometimes need to be considered. For example, the centre of mass may be off-centre, or the ground may be sloping, or aerodynamic forces may be significant.

Aerodynamic effects may be added, calculating the forces and moments according to the methods of Chapter 3. The effect of lift and pitch is considered in detail in Section 5.13. Roll, yaw and side lift have some effect on lateral load transfer; this is usually small, but may be significant for some vehicles. For left-hand cornering, as in the rear view of Figure 5.10.1, positive forces act to the right, so the compensation forces are positive and the tyre forces are negative, and clockwise moments will be taken as positive. Taking the attitude gradient as positive, the attitude is negative for low speed with front steering, becoming positive for large lateral acceleration. C'_S, C'_R and C'_Y are usually simply given positive values, in which case the side lift, roll moment and yaw moment are

$$F_{SAe} = -C'_S \beta_{Ae} q A_F$$

$$M_{RAe} = -C'_R \beta_{Ae} q A_F l$$

$$M_{YAe} = -C'_Y \beta_{Ae} q A_F l$$

where β_{Ae} is the aerodynamic attitude angle, which in still air is

$$\beta_{Ae} = \beta = -\frac{b}{R} + k_\beta A$$

The aerodynamic effect on roll angle depends on the aerodynamic roll moment about the roll axis. Considering that the sideforce acts at the mid-point of the wheelbase, and that the yaw moment has a component about the roll axis, which has inclination

$$\rho_{ra} = \frac{h_r - h_f}{l}$$

the aerodynamic moment on the sprung mass about the roll axis is

$$M_{\mathrm{SAe}} = R - \frac{S(h_{\mathrm{f}} + h_{\mathrm{r}})}{2} + \frac{Y(h_{\mathrm{r}} - h_{\mathrm{f}})}{l}$$

Hence

$$C'_{\mathrm{Rra}} = C'_{\mathrm{R}} - \frac{C'_{\mathrm{S}}(h_{\mathrm{f}} + h_{\mathrm{r}})}{2l} + \frac{C'_{\mathrm{Y}}(h_{\mathrm{r}} - h_{\mathrm{f}})}{l}$$

with a representative value of 0.003 for a saloon car. Including consideration of the sprung-mass weight force, this gives an aerodynamic roll angle

$$\phi_{\mathrm{Ae}} = \frac{M_{\mathrm{SAe}}}{k_{\mathrm{S}} - m_{\mathrm{S}} g d}$$

$$= - \frac{C'_{\mathrm{Rra}}\, \beta_{\mathrm{Ae}} q A l}{k_{\mathrm{S}} - m_{\mathrm{S}} g d}$$

which when significant is generally negative, although not invariably so because the attitude angle may be negative with front steering. This roll moment is distributed front-to-rear according to the roll stiffnesses acting against the roll angle.

The side lift force F_{SAe}, acting at the mid-point of the wheelbase, is distributed equally to the front and rear roll-centres. Hence the front and rear aerodynamic load transfers are:

$$F_{\mathrm{TfA}} = \frac{F_{\mathrm{SAe}}\, h_{\mathrm{f}}}{2T_{\mathrm{f}}} + \frac{k_{\mathrm{f}} \phi_{\mathrm{Ae}}}{T_{\mathrm{f}}}$$

$$F_{\mathrm{TrA}} = \frac{F_{\mathrm{SAe}}\, h_{\mathrm{r}}}{2T_{\mathrm{r}}} + \frac{k_{\mathrm{r}} \phi_{\mathrm{Ae}}}{T_{\mathrm{r}}}$$

where F_{SAe} and ϕ_{Ae} are generally negative when significant. Hence, for positive attitude angle the aerodynamic roll is generally negative, reducing the normal roll slightly, and the aerodynamic roll load transfers are negative. The aerodynamic roll angle is generally less than 1°, and the aerodynamic load transfer reaches 2–3% at high speed for an average saloon car.

5.12 Pitch

Vehicle longitudinal dynamics are not of relevance in themselves here, but sometimes need to be considered in conjunction with lateral dynamics, for example during combined braking and cornering, or during strong acceleration and cornering, and in racing. Because of aerodynamic pitch moment, or hill climbing, or braking or accelerating, or steady-state cornering with an attitude angle, there will be longitudinal load transfer, i.e. longitudinal transfer of tyre vertical force.

It is considered positive when the rear reactions are increased. It generally also results in ride height changes at the front and rear suspensions, which may alternatively be considered as a change of height at the centre of mass plus a change of pitch angle. This has geometric effects on the wheels, such as camber and steering angle changes, and also affects the caster angles significantly. As in the case of lateral load transfer, the longitudinal load transfer may be achieved partly through the springs and partly through the links. When the linkage arrangements are such as to transfer some of the load through the links rather than through the springs, this is called anti-dive at the front and anti-rise at the rear in the case of braking, and anti-lift at the front and anti-squat at the rear in the case of traction. Having a roll-centre above ground could correspondingly be described as 'anti-roll'.

The principle of anti-dive for a twin-wishbone front suspension is shown in Figure 5.12.1. The lines where the planes of the wishbones (i.e. the planes of the pivot points) intersect the plane of the wheel are arranged to converge to a point E, possibly at infinity. With the usual approximation of zero moment about the pivot axis, the suspension links can exert no moment about E. When, resulting from brake action, a horizontal force is applied by the road to the wheel at the bottom, this can be resolved into components along and perpendicular to AE. The component F_2 perpendicular to AE must be reacted by the springs: at the front it acts downwards on the wheel, extending the spring and so opposing the usual compression of the front spring in braking. This is anti-dive. Figure 5.12.1 assumes that the brakes are outboard. If they are inboard then the torque transferred by the driveshafts means that the brake force is effectively applied at the wheel centre height, so it is necessary to arrange for appropriate inclination of line CE instead. Similar principles can be applied to the rear suspension, but in this case the force centre must be in front of the wheel, to discourage the usual

Figure 5.12.1. Anti-dive and anti-squat geometry (only braking forces shown).

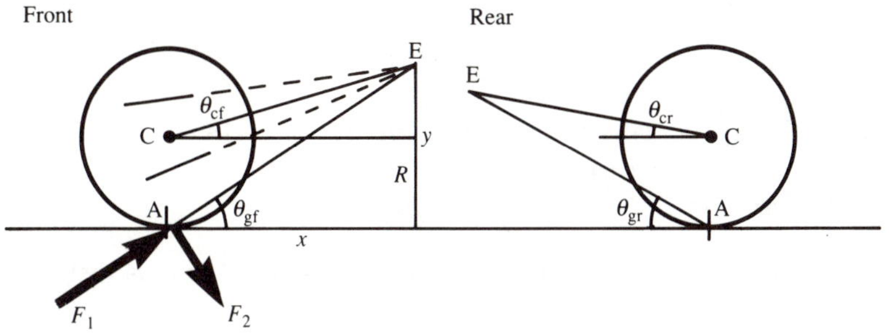

extension of the rear springs in braking or compression in acceleration.

The notation for angles in Figure 5.12.1 is that subscripts c and g mean from the centre and ground respectively, and subscripts f and r mean front and rear as usual. The anti coefficients will be denoted by J (in practice usually expressed as $J \times 100\%$). For example J_{al} is the anti-lift coefficient. Table 5.12.1 summarises the parameters, showing which angle is relevant to which coefficient.

Table 5.12.1. Anti-dive/rise/lift/squat parameters

Action	Symbol	Direction	End	Relevant angle Outboard brakes	 Inboard brakes
anti-dive	J_{ad}	braking	front	θ_{cf}	θ_{gf}
anti-rise	J_{ar}	braking	rear	θ_{cr}	θ_{gr}
anti-lift	J_{al}	driving	front	θ_{cf}	θ_{cf}
anti-squat	J_{as}	driving	rear	θ_{cr}	θ_{cr}

The total deceleration depends on the total force at both ends, but the horizontal suspension force on each end depends on the braking proportion at that end. Accurate calculation of the anti-dive requires consideration of the proportion of brake force at the front, the sprung and unsprung masses, and the sprung centre-of-mass height. The analysis is usually simplified by neglecting the inertial effect of the unsprung mass, as in the following. As an example, consider the braking force

$$F = F_f + F_r = pmA_B + (1-p)mA_B$$

where p is the proportion of braking at the front, which may be acceleration-dependent, for example if there is a rear pressure limiter. The resulting vehicle deceleration is

$$A_B = \frac{F_f + F_r}{m}$$

The total longitudinal load transfer to the rear is

$$F_{TX} = -\frac{mA_B H}{l}$$

The vertical force exerted by the ground on the front axle is therefore

$$F_{Vf} = W_f - F_{TX}$$

$$= W_f + \frac{mA_BH}{l}$$

The forces on the suspension from the body have no moment about E, so taking moments about E for the front wheel shows the idealised front spring force (two wheels) to be

$$F_S = W_f - W_{Uf} + \frac{mA_BH}{l} - \frac{pmA_By}{x}$$

Hence the front anti-dive against forces exerted on the wheel at ground level is

$$J_{gf} = \frac{p(y/x)}{(H/l)}$$

$$= \frac{\tan\theta_{gf}}{\tan\theta_{gfi}}$$

where θ_{gfi} is the 'ideal' angle for full anti-dive, given by

$$\tan\theta_{gfi} = \frac{H}{pl}$$

In braking, with outboard brakes, at the rear the anti-rise is correspondingly given by

$$J_{gr} = \frac{\tan\theta_{gr}}{\tan\theta_{gri}}$$

where

$$\tan\theta_{gri} = \frac{H}{(1-p)l}$$

In the case of traction, with tractive force

$$F = F_f + F_r = tmA + (1-t)mA$$

where t is the tractive force fraction on the front wheels, the tractive force is produced because of torque in the driveshafts, totalling for both wheels $tmAr$ at the front and $(1-t)mAr$ at the rear (neglecting rotational inertia of the wheels). This means that the effective line of action of the tractive force is transferred to the centre of the wheel. The front anti-lift is

$$J_{cf} = \frac{t\,[(y-r)/x]}{(H/l)}$$

$$= \frac{\tan\theta_{cf}}{\tan\theta_{cfi}}$$

where θ_{cf} is the angle of CE at the front:

$$\tan \theta_{cf} = \frac{y-r}{x}$$

and θ_{cfi} is the angle for full anti-lift, given by

$$\tan \theta_{cfi} = \frac{H}{tl}$$

The traction anti-squat at the rear is

$$J_{cr} = \frac{(1-t)\,[(y-r)/x]}{(H/l)}$$

$$= \frac{\tan \theta_{cr}}{\tan \theta_{cri}}$$

where

$$\tan \theta_{cri} = \frac{H}{(1-t)l}$$

In the case of inboard brakes, the braking force is associated with a driveshaft torque, so the inclination of CE gives the relevant angle. All the above equations can be extended to include the effect of the translational and rotational inertia of the unsprung masses. For example, during linear vehicle deceleration, the wheels also have angular deceleration, with angular momentum change that must be provided by the moment of a longitudinal load transfer.

In practice there are objections to anti-dive geometry. It tends to lead to harshness of the front suspension on rough roads because the wheel moves forward as it rises, attacking the bump; it may also cause steering kick-back and wander under braking. It also becomes more difficult to achieve good-quality steering geometry, so instead of full anti-dive a proportion is often used, usually expressed as a percentage. Up to 50% anti-dive has been used on passenger cars; more can be used in racing because of the smooth tracks and low centres of mass. However large anti-dive may be problematic on small radius turns, where the large steer angle plus cornering force results in significant jacking forces. The most successful applications of anti-dive seem to be those in which the geometry is arranged to minimise changes of caster angle.

In the case of the pitch-up caused by traction forces, with rear-wheel-drive it is quite common to have some anti-squat at the rear. It is not possible to provide anti-rise at the front of a rear-drive vehicle because there is no associated horizontal force applied to the wheel. Anti-squat may have detrimental effects on traction on rough surfaces.

Anti-squat effects may also be achieved by other arrangements of the rear suspension, such as a lift bar, which is rigidly attached to the rear axle, protruding forward and acting upward on a rubber block on the body; the axle reaction torque therefore provides an upward force on the rear body, depending on the length of the lift bar, relieving the rear springs.

It is sometimes convenient to express the longitudinal load transfer in acceleration in terms of a vertical force longitudinal transfer factor

$$e_X = \frac{F_{TX}}{N_f}$$

where N_f is the constant speed front axle reaction, so $e_X = 1$ is front axle lift off. Correspondingly, the vertical force braking transfer factor is

$$e_B = -\frac{F_{TX}}{N_r}$$

5.13 Wheel vertical forces

The four wheels of a conventional vehicle constitute a statically indeterminate system for the vertical forces at those wheels. There are only three equilibrium equations available: heave, roll and pitch. As a consequence, the distribution of the vertical forces at the wheels depends upon the system's internal characteristics. For example, a symmetrical vehicle with centre of mass at 46% of the wheelbase from the front, on level ground, would have 27% of the weight supported at each front wheel and 23% at each of the rear wheels, giving 54% total on the front, 46% total on the rear and 50% on each side, Figure 5.13.1(a). The diagonals D_1 and D_2 are equally loaded, being 50% each. However, because of the indeterminacy, the load on one diagonal can be increased at the expense of the other by adjusting the suspension; this is static diagonal load transfer, usually called diagonal bias. Figure 5.13.1(b) shows a diagonal bias of 6% of the total weight, i.e. the diagonals D_1 and D_2 are 56% and 44%, whilst the front total, rear total, and side totals remain as before. The convention will be adopted here that a positive diagonal bias means more reaction on the right-front diagonal (D_1).

Diagonal bias can be caused by adverse accumulation of production tolerances, operating distortion, incorrectly repaired crash damage, suspension friction or engine torque reaction. Competition vehicles that predominantly run on one-handed tracks are sometimes deliberately adjusted in this way (called weight-jacking in the U.S.A.), achieved by adjusting the height of a spring seat or the length of an anti-roll bar drop-link.

Figure 5.13.1. Wheel vertical forces (static): (a) symmetrical, (b) 6% diagonal bias.

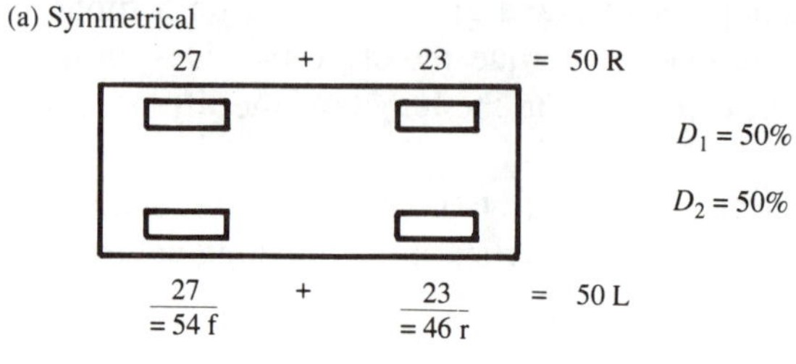

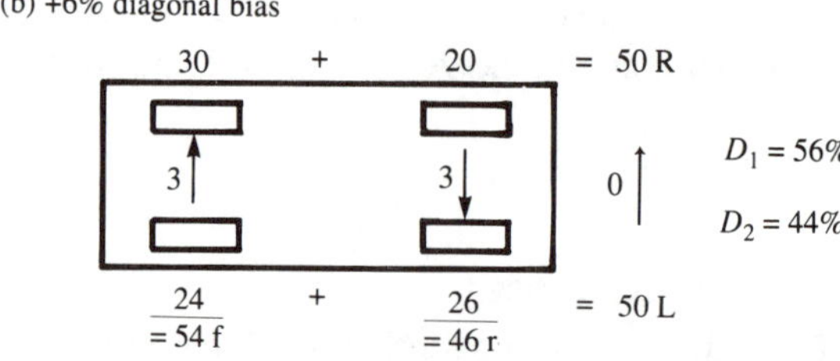

The distribution of vertical wheel forces is important in handling at higher lateral accelerations. The desired result is achieved by controlling the front-to-rear distribution of lateral load transfer. Figure 5.13.2(a) shows the same car with symmetrical trim, statically as in Figure 5.13.1(a), in left-hand cornering with lateral load transfer equal to 20% of the total weight, where the distribution of lateral load transfer is shared equally between front and rear. This still leaves the diagonal sums D_1 and D_2 at 50% each. Increasing the lateral load transfer at the front to three-quarters of the total load transfer gives Figure 5.13.2(b), where the diagonal sums are now 60% and 40%.

These effects may be summarised by the following equations, using subscripts L for left, R for right, f for front and r for rear. The diagonal bias is the force transferred diagonally:

$$F_{TD} = \tfrac{1}{2}[(F_{Rf} + F_{Lr}) - (F_{Lf} + F_{Rr})]$$

For zero (static) diagonal bias, the cornering effects give front and rear load transfers

$$F_{Tf} = \tfrac{1}{2}(F_{Rf} - F_{Lf})$$

$$F_{Tr} = \tfrac{1}{2}(F_{Rr} - F_{Lr})$$

Hence the diagonal bias is

$$F_{TD} = F_{Tf} - F_{Tr}$$

The total lateral load transfer is

Figure 5.13.2. Wheel vertical forces (cornering): (a) load transfer distribution 50/50 front/rear, (b) distribution 75/25 front/rear.

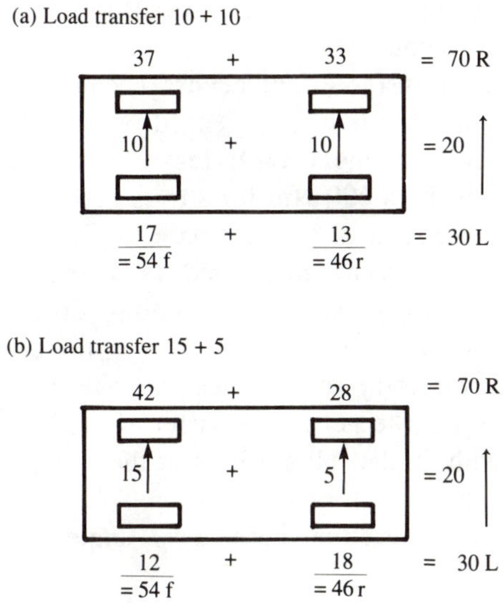

(a) Load transfer 10 + 10

37 + 33 = 70 R

10 + 10 = 20

17 + 13 = 30 L

= 54 f = 46 r

(b) Load transfer 15 + 5

42 + 28 = 70 R

15 + 5 = 20

12 + 18 = 30 L

= 54 f = 46 r

$$F_T = \tfrac{1}{2}[(F_{Rf} + F_{Rr}) - (F_{Lf} + F_{Lr})]$$
$$= F_{Tf} + F_{Tr}$$

The aerodynamic effects on wheel vertical forces may be found as follows, with forces and moments calculated by the methods of Chapter 3 using the coordinate axes of Figures 3.5.1 and 3.5.2. The lift force is

$$F_{LAe} = C_L q A_F$$

and this acts vertically upwards at the centre of the wheelbase, this being the aerodynamic coordinate centre. Therefore each wheel vertical force is changed by

$$\Delta F_{V\,lift} = \frac{-F_{LAe}}{4} = \frac{-C_L q A_F}{4}$$

For passenger vehicles, the lift is usually positive, thus reducing the wheel vertical forces, typically by 5 to 10% at full speed.

A positive pitch moment is defined to give a longitudinal load transfer increasing the rear wheel vertical forces. The pitch-up moment is

$$M_{PAe} = C_P q A_F l$$

so the total longitudinal aerodynamic load transfer is

$$F_{TXAe} = \frac{M_{PAe}}{l} = C_P q A_F$$

and half of the above for each wheel. The pitch coefficient usually causes a longitudinal load transfer of 1% to 2% of total weight at high speed. There may also be a small lateral load transfer from the aerodynamic roll moment.

A diagonal load transfer effect occurs because of the driveshaft torque on a conventional live axle. The driveshaft torque, basically engine torque times the gear ratio less friction (except for rear gearboxes), is as much as 600 Nm for strong acceleration in low gear when the effect is worst. It acts on the axle to give a load transfer. Its reaction acts on the sprung mass and is distributed front-to-rear according to the roll stiffnesses, compensating partially for the direct propshaft torque on the axle. In the case of a three-wheeled vehicle with rear axle, all the body roll stiffness is at the rear, so there is a resulting roll angle but the net load transfer effect is zero. For a four-wheeled vehicle with all the roll stiffness at the front, the full diagonal loading would occur.

The propshaft torque reaction causes a roll angle

$$\phi_{\mathrm{Pr}} = \frac{M_{\mathrm{Pr}}}{k - m_{\mathrm{S}} g d}$$

which is normally positive in left-hand cornering, i.e. it then increases the roll. The propshaft front load transfer is then

$$F_{\mathrm{Tf\,Pr}} = \frac{k_{\mathrm{f}} \phi_{\mathrm{Pr}}}{T_{\mathrm{f}}}$$

The rear load transfer is the negative torque on the axle, plus the body load transfer through the roll stiffness, giving

$$F_{\mathrm{Tr\,Pr}} = -\frac{M_{\mathrm{Pr}}}{T_{\mathrm{r}}} + \frac{k_{\mathrm{r}} \phi_{\mathrm{Pr}}}{T_{\mathrm{r}}}$$

For a vehicle with a live solid rear axle the result is a significant net lateral load transfer at the rear during strong acceleration. This acts to limit traction forces and hence to limit forward acceleration. It is a relatively small asymmetric effect in steady-state cornering, i.e. it has differing effects for different directions of turning. This effect is absent for most front-drive vehicles with the driveshafts from the side of the differential on the sprung mass, or for de Dion rear axles.

For manual computation these effects can best be summarised as in Table 5.13.1, which shows the result corresponding to the diagonal bias of Figure 5.13.1(b) plus the cornering load transfer of Figure 5.13.2(b), and 4% longitudinal load transfer, plus example aerodynamic lift and pitch effects.

Table 5.13.1. *Example load transfer effects (percentages of mg)*

Wheel	Weight	D. Bias	Lat. LT	Long. LT	Lift	Pitch	Total
Lf	27	–3	–15	–2	–2	+1	6
Rf	27	+3	+15	–2	–2	+1	42
Lr	23	+3	–5	+2	–2	–1	20
Rr	23	–3	+5	+2	–2	–1	24
Totals	100	0	0	0	–8	0	92

It is often convenient to express the vehicle or axle lateral load transfer in terms of the vertical force lateral transfer factors e_V, e_{Vf} and e_{Vr}. These are defined as the lateral load transfer divided by the vertical force on the inner wheel(s) before the load transfer. Thus for the case of Table 5.13.1,

$$e_{Vf} = 15/21 = 0.714$$

$$e_{Vr} = 5/25 = 0.200$$

$$e_V = 20/46 = 0.435$$

The vertical force longitudinal transfer factor is defined in a similar way.

5.14 Steering

Directional control is normally achieved by steering the front wheels, i.e. by rotating them about a roughly vertical axis. Rear-wheel steering is inherently unstable at high speeds, but because of its convenience in manoeuvring it is sometimes used on specialist low-speed vehicles, such as dumper trucks. Recently some interest has been shown in variable rear steering for road cars, coupled with conventional front steering, and this is now commercially available.

In the early days of motoring, various hand controls were tried for the driver. It was Benz who introduced the steering wheel, and this was almost universal by 1900. Tests in other control applications show that the hand wheel is the best way to combine rapid large movements with fine precision. For cars, the road wheels steer through a total of about 70°, and the steering wheel through three and a half turns, requiring a gear ratio of about 18. The average is actually about 17 for power-assisted steering and 21 for unpowered steering. For trucks the steering-wheel movement and gear ratio are about twice those for cars. More steering-wheel movement means a smaller steering force requirement, but it is less favourable for rapid response in emergencies.

The steering system must connect the steering wheel to the road wheels with the appropriate ratio, and also meet other geometric constraints, such as limits on bump steer. It is desirable for the forward efficiency of the system to be high, in order to keep the steering forces low. On the other hand, a low reverse efficiency helps to reduce the transmission of road roughness disturbances back to the driver, at the cost of some loss of the important feel that helps a driver to sense the frictional state of the road.

For independent suspension there are two principal steering systems in use, one based on a steering box, the other on a rack and pinion. In the typical steering box system, Figure 5.14.1, known as the parallelogram linkage, the steering wheel operates the Pitman arm A via the steering box. The box itself is nowadays usually a cam and roller or a recirculating ball worm-and-nut system. The gear ratio of the box alone is usually somewhat less than that of the overall ratio, because of the effect of the links. Symmetrical with the Pitman arm is an idler arm B, connected by the relay rod C, so that the whole linkage is geometrically symmetrical, although the forces are introduced on one side. From appropriate points on the relay rod, the track-rods (tie-rods) D connect to the steering arms E. The length and alignment of the track-rods are critical in controlling bump steer effects. The steering box system has the advantage of a suitable reverse efficiency, but this has become less important than in the earlier days of motoring because of the improved quality of roads.

Figure 5.14.1. Steering box system.

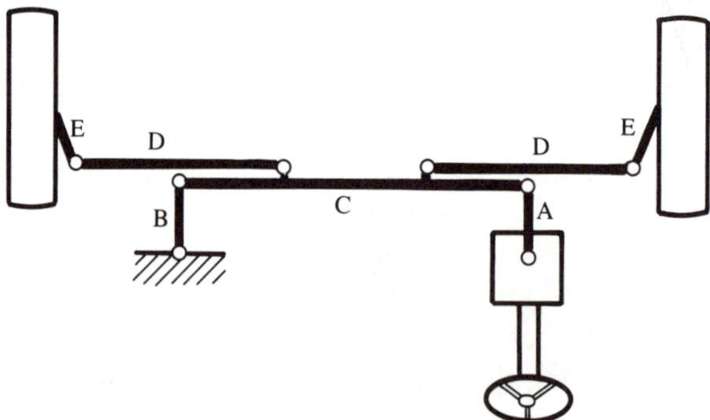

In the steering rack system, the steering column is connected directly to a pinion acting on a laterally moving rack. The track-rods may be connected to the ends of the rack, or they are sometimes attached close to the centre where the geometry favours this, for example when the rack is high up with strut suspension. Road shock feedback can be controlled to some extent by choosing a suitable gearing helix angle, minimising wheel offset, or increasing handwheel inertia. Flexible mounting of the rack has also sometimes been used, but this causes loss of steering precision.

Precision is of great importance in the steering system, and the rack system has a superior reputation, although it is quite difficult to observe any substantial difference between a rack and a good box system in comparative driving tests. To prevent play in the various inter-link ball joints, they are spring-loaded.

On trucks it is still common to use a rigid axle at the front, mounted on two longitudinal leaf-springs. Usually the wheel steering arms are connected together by a single tie-rod, Figure 5.14.2. Steering is effected by operating a second steering arm A on one of the wheels, by the horizontal drag link B from a vertical Pitman arm C. This acts from the side of the steering box, which is mounted on the sprung mass.

For all systems, the wheels and hubs are pivoted about the kingpin axis, Figure 5.14.3. Nowadays, on cars at least, kingpins are no longer

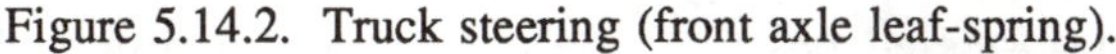
Figure 5.14.2. Truck steering (front axle leaf-spring).

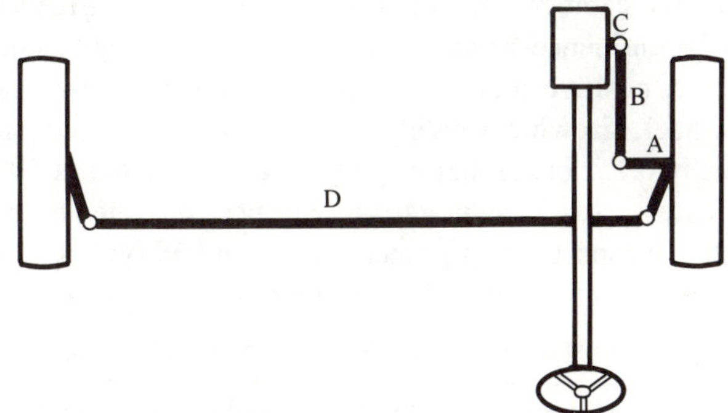

Figure 5.14.3. Wheel steering axis geometry.

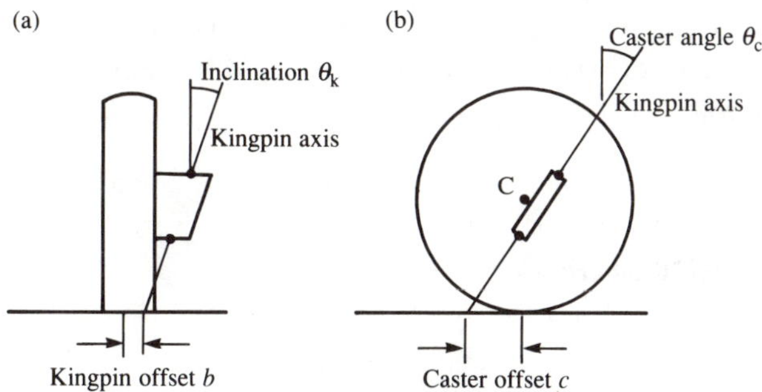

used, the steering axis now being defined by a pair of ball joints. In front view the axis is at the kingpin inclination θ_k, usually from zero to 20°, giving a reduced kingpin offset b at the ground. The inclination angle helps to give space for the brakes. Where the steering arms are forward (rack in front of the wheel centres) it also gives room to angle the steering arms for Ackermann geometry. Sometimes a negative offset is used, this giving straighter braking when surface friction varies between tracks. Zero offset is called centre-point steering. If the inclination angle is also zero, it is called centre-line steering.

In side view, Figure 5.14.3(b), the kingpin axis is slanted at the caster angle θ_c, usually zero to 5°. This introduces a mechanical trail, the caster trail (or caster offset), that acts in concert with the tyre pneumatic trail. On cars and trucks it is usual for the kingpin axis to pass through the wheel spin axis C in side view, but this is not essential, and some offsetting of the axis from the centre enables the caster angle and trail to be independently varied. The road wheel steer angle δ is the angle between the vehicle longitudinal axis and the line of intersection of the wheel plane and the ground. This is approximately the same as the angle of rotation of the wheel about the inclined kingpin axis. Once the wheel is in a steered position, the caster and kingpin inclination angles affect the camber angle, and this can therefore influence turn-in, and especially influence handling in small-radius corners. For realistic steer angles, a positive inclination angle causes a positive camber on the outer wheel, growing roughly with the steer angle squared, and being typically 0.15° of camber per degree of inclination at 30° of steer. Positive caster causes a negative camber on the outer wheel, approximately proportional to steer angle, and is typically –0.50° of camber per degree of caster at 30° of steer. The actual camber angle is

$$\gamma = \gamma_0 + \arccos(\sin\theta_k \cos\delta) + \theta_k + \arccos(\sin\theta_c \sin\delta) - 180°$$

The steering wheel angle δ_s is the angular displacement of the handwheel from the straight-ahead position. The overall steering ratio G is the rate of change of steering-wheel angle with respect to the average steer angle of the steered wheels, with negligible forces in the steering system, or assuming a perfectly rigid system, and with zero suspension roll:

$$G = \frac{d\delta_s}{d\delta}$$

The mean overall steering ratio is

$$G_m = \frac{\delta_s}{\delta}$$

For a linear system G and G_m are equal and constant. In this case it is sometimes convenient to introduce the reference steer angle

$$\delta_{ref} = \frac{\delta_s}{G}$$

This is not the same as δ because it incorporates the effect of steering compliance. However, it has the advantage over δ_s that it is based on the road wheel angles and therefore it is not very sensitive to G, unlike δ_s.

The steering-wheel angle gradient is the rate of change of steering-wheel angle with steady-state lateral acceleration ($d\delta_s/dA$) in rad/m s^{-2}, deg/m s^{-2} or deg/g, and is a measure of the position control sensitivity. The steering wheel torque gradient is the rate of change of torque T_s with steady-state lateral acceleration (dT_s/dA) typically in Nm/m s^{-2} (Ns^2) or Nm/g, and is a measure of the force control sensitivity.

Steering systems are dynamically complex combinations of components, and subject to various modes of vibration. Fortunately, handling effects are essentially low-frequency phenomena, so simple dynamical models usually suffice for handling analysis. The essential inertias are those of the wheels about the kingpin axis, about 1 kg m^2 each, and that of the steering wheel, which for cars is usually in the range 0.025 to 0.065 kg m^2, averaging 0.040 kg m^2. However they are connected by the overall gear ratio of typically 20, and inertias are factored by the gear ratio squared, so the steering wheel angular inertia referred to the road wheel motion is about 16 kg m^2, which is much greater than that of the road wheels. The dominant compliance is the torsional compliance of the steering column, being typically 25 Nm/rad. Depending on the design, sometimes other compliances should be included, such as the long tie-rod D on a rigid axle, Figure 5.14.2.

The torque required to steer the road wheels is greatest for static vehicle conditions. Provided that the kingpin axis is not too far from the centre of tyre contact, for example if it is in the footprint, as it usually is, the following empirical equation gives a fair estimate of the torque:

$$T_R = \frac{\mu F_V^{1.5}}{3\sqrt{p_i}}$$

where p_i is the inflation pressure.

Angular compliance in the steering column proves to be of importance in handling analysis, so it is important to be able to calculate the steering torque about the kingpin axis in dynamic conditions. This comprises a torque to provide wheel angular acceleration, usually small, plus torque to balance the forces exerted by the ground on the tyre, and to balance any driveshaft torque, plus the spring torsional stiffness in the case of Macpherson struts.

Driveshaft torque occurs for driven wheels or for inboard brakes. The consequent steering torque is simply the component of the driveshaft torque along the kingpin axis. These ideally balance from side-to-side, but will not do so when the driveshafts are at different angles for any reason, for example because they are of different lengths or because of body roll, or engine torque rock, or when the shaft torques are different as may occur when a limited slip differential is fitted.

The consequences of the tyre forces may be found by considering the three forces and moments at the centre of tyre contact. The rolling resistance moment and the overturning moment are negligible in this context. The aligning torque, acting about the vertical axis, may easily be resolved into its component about the kingpin axis, Figure 5.14.3. The sum for the two wheels is

$$M_A = (M_{ZL} + M_{ZR}) \cos(\phi^2 + \theta^2)^{0.5}$$

This moment attempts to rotate the steering in such a way as to restore straight running. The tyre lateral force acts at a distance $c \cos \theta$ from the axis, so the wheel pair moment for small θ is

$$M_L = (F_{YL} + F_{YR})\, c$$

For positive caster offset c this acts in the same sense as the aligning torque moment, and therefore also has a stabilising effect.

The tractive force acts at a moment arm of $d \cos \phi$ giving a wheel pair moment for small ϕ of

$$M_T = (F_{XL} - F_{XR}) d$$

The two tractive forces balance each other in the symmetrical condition but, as for the driveshaft torques, may be unbalanced with a limited slip differential, especially on variable surfaces, this leading to the characteristic steering fight of limited-slip front-drive vehicles. Imbalances may also occur for braking on asymmetrical surfaces or for tyre deflation, there being some advantage here in avoiding large kingpin offsets.

The influence of the tyre vertical force, less the wheel weight, is more difficult to see, and is best considered in two separate parts, one the consequence of kingpin inclination angle, the other of caster angle. As a result of the kingpin inclination angle, there is a component $F_V \sin \phi$ that acts on a moment arm of $b_a \sin \delta$, for small ϕ, when the wheel is steered at angle δ, where b_a is the kingpin offset at wheel axis height, perpendicular to the axis. For the two wheels the moment is

$$M_{Vi} = -(F_{VL} + F_{VR})\, b_a \sin\phi \, \sin\delta$$

This is a steering restoring moment independent of load transfer, the wheels acting in the same sense. This moment is often also correctly explained as resulting from the tendency of this geometry to lift the vehicle when steering is performed. It is significant for large steer angles.

As a result of the caster angle, and now neglecting the offset c, there is a force component $F_V \sin\theta$ that, for small θ, acts on a moment arm of $d\cos\delta$. The moment for the pair of wheels is

$$M_{Vc} = (F_{VL} - F_{VR})\, d \sin\theta \, \cos\delta$$

In this case the moments from the two wheels oppose each other, and the net moment depends directly on the load transfer.

To compare the actual size of these torques, consider a medium saloon of total mass 1400 kg with a front weight of 7 kN. The maximum drive thrust is about 7 kN on the two wheels at a radius of 0.35 m, so the shaft torque is 1225 Nm. With 70° between shaft and kingpin axis, the torque component is 420 Nm. Angle asymmetries by roll will cause imbalances of about 40 Nm. With a limited slip differential, in an extreme case there could be the full 420 Nm difference. The aligning moment will peak at a value of approximately 5000 N on 30 mm giving 150 Nm for the pair of wheels. At 5° caster with the standard geometry, the caster trail c is 30 mm, giving a further 150 Nm from the lateral force. The tractive force, with a limited slip differential, would give up to 2500 N on a kingpin offset of say 30 mm, giving 75 Nm. For the vertical force effect because of kingpin inclination, an inclination of 15°, b_a is 60 mm, and a maximum steer angle of 35° gives a moment of 60 Nm. Finally, the caster contribution from vertical force, with 5° caster and complete load transfer, at zero steer, is 18 Nm.

These figures illustrate, firstly, the strong disruptive effect of a limited slip differential, mainly arising from the direct torque component. With a plain differential, the dominant effect is that of the self aligning torque and the lateral force. In practice the steering feel is adjusted by the caster offset through the caster angle, to obtain a desired relationship between lateral force and total aligning torque, so that the experienced driver can tell from the steering torque when the tyres approach their lateral force limit. Adding caster trail moves the maximum steering torque closer to the maximum lateral force, i.e. the steering goes light later.

During cornering, the steering must also support the centrifugal compensation forces on the steering mechanism, for example the rack and the track-rods. This is called centrifugal caster, and reaches a typical moment about the kingpin of 10 Nm.

On a bumpy road, as the wheel rolls the loaded radius constantly varies. This causes changes of wheel angular speed, with associated longitudinal forces on the tyre. These forces, of magnitude about 1 kN, act on the hub at wheel axis height, and hence because of the offset b_a disturb the steering. This can be eliminated only by centre-line steering.

Another steering disturbance, for front drive, is the side-to-side difference of the component of the driveshaft torque along the kingpin axis; this is a problem where the driveshafts have different inclinations, because they have different lengths, or where they are momentarily differently inclined because of rough roads. This is worst for large torques, and hence during acceleration.

5.15 Turning geometry

When a vehicle moves in a curved path at a very low speed the lateral acceleration is very small, so roll and axle lateral force are negligible. Thus the wheel angles are those arising geometrically, not from the need to produce lateral force. There may however be opposing slip angles on the two ends of an axle, giving zero net force.

For a normal vehicle with front steering, to turn with zero slip angles means that the turning centre C must be in line with the rear axle, Figure 5.15.1. The front wheels must be steered by different amounts, the inner wheel more, in order for both of them to have zero slip angle. The difference between the steer angles for both wheels to have zero or equal slip angles equals the Langensperger angle λ subtended at the

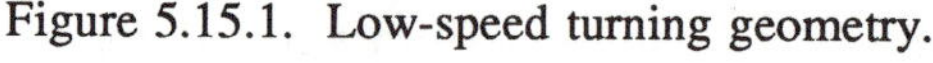

Figure 5.15.1. Low-speed turning geometry.

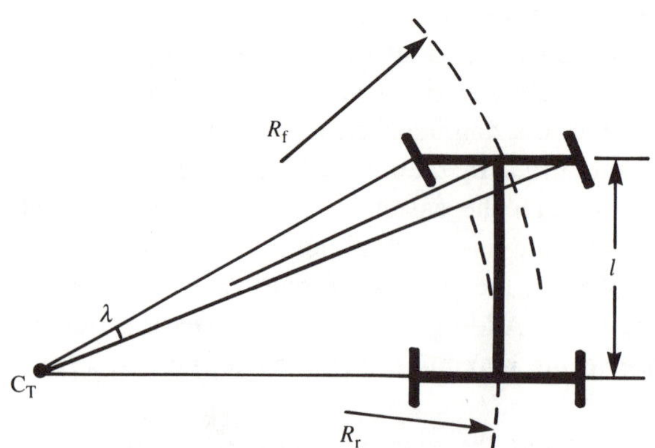

turning centre by the axle. This is known as the Ackermann steering concept, although actually invented by Langensperger, which we would expect to be desirable for low-speed manoeuvring to avoid tyre scrub, squeal and wear. Various geometries are used in practice, more or less related to Ackermann geometry. For a given vehicle with a given mean steer angle δ_m there would be an Ackermann difference of steer angles equal to the Langensperger angle at zero attitude angle, λ_0. For the real steering, at this mean steer angle the difference of steer angles is actually $\delta_L - \delta_R$. The Ackermann factor is

$$K_A = \frac{\delta_L - \delta_R}{\lambda_0}$$

which is zero for parallel steering, and 100% for true Ackermann. For a real steering arrangement it is only approximately a constant.

For vehicles that do a great deal of turning or need a very small turn radius, such as purpose-built taxis and urban delivery vehicles, then full Ackermann is used. The 'Lunar Rover' vehicle used on the Moon by Apollo astronauts had four-wheel steering, with close to Ackermann at both front and rear, giving a minimum turn radius equal to the wheelbase of 2.286 m. The traditional London Taxi has almost perfect Ackermann over its full 60° of inner road-wheel steer angle. There is less case for Ackermann steering under dynamic cornering conditions. This is because as attitude angle develops, the Langensperger angle λ subtended by the front axle at the turning centre reduces. Also, the outer tyre has greater vertical force, hence needing a greater slip angle than the inner tyre for maximum cornering force. Even anti-Ackermann has been used on occasion, and various steering geometries are used in practice. If the steering is not perfect Ackermann, then at low speed each wheel pair must adopt equal and opposite slip angles to give zero net force.

The most convenient way to obtain the different steer angles of the Ackermann layout is to angle the steering arms inwards (for a rack behind the kingpins), as in Figure 5.14.1, so that as steer is applied there is a progressive difference in the effective moment arms. It is widely believed that aligning the steering arms so that their lines intersect at the rear axle will give true Ackermann steering (the Jeantaud diagram). However, this is far from true, the actual Ackermann factor varying in a complex way with the arm angle, rack length, rack offset forward or rearward of the arm ends, whether the rack is forward or rearward of the kingpins, and with the actual mean steer angle.

The two ends of the vehicle corner at different radii, Figure 5.15.1. This is one reason why four-wheel-drive vehicles need a centre differential or front over-run clutch. At low speed, with zero slip angles

$$R_r^2 = R_f^2 - l^2$$

The difference of radii is known as the offtracking:

$$R_f - R_r \approx \frac{l^2}{2R_f} \approx \frac{l^2}{2R}$$

where R is the turn radius of the centre of mass. The offtracking can be about 0.4 m for cars, leading to occasional kerbing of the rear wheels, or worse. It is more problematic for long trucks, and especially for trailers. It can be obviated by four-wheel steering as is now available for some cars.

The low-speed mean steer angle will here be called the kinematic steer angle:

$$\delta_K = \arctan\left(\frac{l}{R_r}\right) \approx \arctan\left(\frac{l}{R}\right) \approx \frac{l}{R}$$

At significant speed, when a tyre slip angle is required, then attitude angle develops, and the offtracking changes. The turning centre C_T has now moved forward from the axle line. The condition of zero offtracking will occur when C_T has moved forward half of the wheelbase. This is possible for some particular speed that depends on the cornering radius, according to tyre characteristics etc., and is typically 10 m/s at 30 m radius. For a greater speed, the offtracking becomes negative, the rear following a path of greater radius than the front.

5.16 Bump steer, roll steer, toe

The term bump steer means changes of wheel steer angle when the wheel is moved relative to the body in bump and droop. The term roll steer refers to changes of steer of the pair of wheels, i.e. of the axle, when the body rolls. These are obviously related, but bump steer tends to be the basic form of data for independent suspension, and roll steer for solid axles. Some forms of bump steer can cause poor straight-line stability, and very unpredictable and unpleasant vehicle behaviour, the handling being sensitive to small changes of wheel steer angles.

The bump steer coefficient β_S is the rate of change of wheel steer angle with vertical wheel position, usually in deg/m. This will be taken as positive for toe-in with a rising wheel. The bump camber coefficient β_C is positive for positive-going camber with a rising wheel, again usually in deg/m.

The roll steer coefficient ε_S is the rate of change of axle mean wheel steer angle with respect to suspension roll angle, and is usually expressed as deg/deg, i.e. it is dimensionless. Hence for small roll, the steer angle is $\varepsilon_S \phi$. For a single wheel it is closely related to the bump steer coefficient, but for independent suspension it is also influenced by the body width between the suspension mountings. For a solid axle it relates to the steer angle of the complete axle, whereas for independent suspension it relates to the mean value for the pair of wheels.

Derived from the roll steer coefficient, the roll steer gradient $k_{\varepsilon S}$ in deg/m s^{-2} or deg/g is sometimes used, defined by

$$k_{\varepsilon S} = \varepsilon_S \frac{d\phi}{dA} = \varepsilon_S \, k_\phi$$

where k_ϕ is the roll angle gradient, ϕ is the roll angle and A is the lateral acceleration.

The roll camber coefficient ε_C for a single wheel is the camber change with suspension roll, in deg/deg, related to the bump camber coefficient but again depending on the body width between the suspension mountings. This is the mean camber angle for the two wheels, relative to the road. Hence for small roll the camber is $\varepsilon_C \phi$. The roll camber gradient is

$$k_{\varepsilon C} = \varepsilon_C \, k_\phi$$

A solid axle is not subject to suspension roll camber, but it does camber in roll because suspension roll leads to axle roll as a result of load transfer on the tyre vertical stiffness. Arguably then, a solid axle might be said to have an axle roll camber coefficient, typically of about 0.12 (deg/deg), this being additionally applicable to independent suspensions.

For a twin-wishbone front suspension, Figure 5.16.1, considering the wheel to move in bump ideally with no steer angle change, then the track-rod end B should move in an ideal arc with centre at A, where the position of A depends on the wishbone geometry. If the track-rod to rack joint, or track-rod to relay-rod joint, is actually at A then there will be no bump steer. In practice, there are often deliberate or accidental discrepancies of height of the actual joint C, or of length of the track-rod BC. The possibility of achieving an accurate steering motion is an important advantage of independent suspension over a steered solid axle. Predictable and precise handling is particularly important for competition and high-performance vehicles, and in such cases it is considered to be of paramount importance that the rack is mounted at exactly the right height, within 1 or 2 mm, and that the track rods should be of the appropriate length. For ordinary road vehicles such complete accuracy is often not attempted. The errors consequently introduced

Figure 5.16.1. Wishbone suspension and track-rod in rear view.

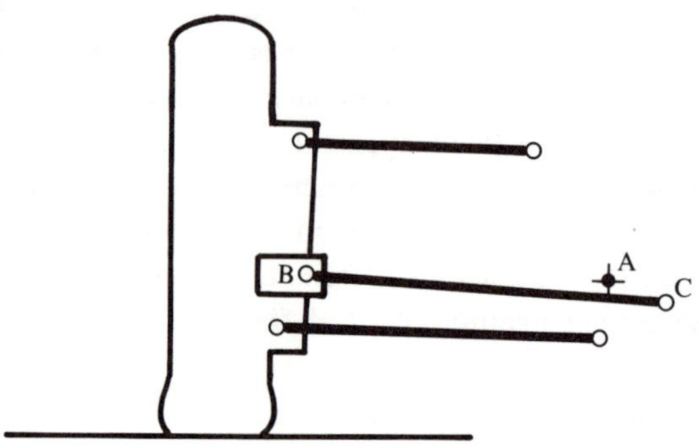

are sometimes claimed to give less wheel response to rough roads or to have handling advantages, although this is a controversial issue.

If the actual wheel steer angle is plotted against bump, a typical result such as Figure 5.16.2 is obtained.

Figure 5.16.2. Example bump steer graph.

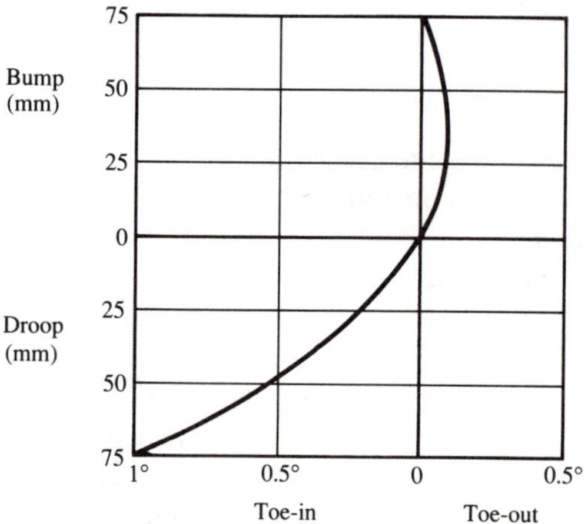

Actually there is normally also a toe angle at zero bump; this is the static toe, and does not usually appear in bump steer plots because it is readily adjustable and is measured separately; bump steer is usually measured as the change of angle. It can be characterised in two ways. At zero bump there is a gradient of steer change with bump, in this case about 0.5° toe-in in 75 mm bump, i.e. a bump steer coefficient of 6.7 deg/m. This results from the rack height, say about 7 mm too low in this case with the rack behind the wheels, or high for the rack in front. Secondly there is a curvature of the graph. In this case the curvature is towards toe-in, which would result from a track-rod longer than ideal for a rack behind the wheels, and shorter than ideal for a rack in front of the wheels, as can be seen by imagining the ideal and actual arcs of the track-rod end. The initial toe-out tendency in bump means that the wheel itself tends to recede from the bump, and so possibly reduces the steering fight on rough roads. The curvature toes-in the inner wheel in cornering, reducing its slip angle. This can help prevent undue wear because of excessive slip on the lightly loaded tyre, an example of the abandonment of Ackermann steering under dynamic conditions.

Because of the nominal symmetry of the whole steering and suspension system, even if these deliberate errors are adopted, body heave should not cause net steering effects, which is an advantage over a steered solid axle. Nevertheless, the toe-in varies with load which can affect steering feel and tyre wear, for which reason if these effects are used it is best to design them around a light load position. Where present, bump steer is often a palliative for some other fault such as bad weight distribution, and is likely to give poor straight-line stability and tyre wear. At best such effects compromise the basic handling in order to gain some rough road or other small advantages, and so should be used with caution.

Many cars are designed with zero theoretical bump steer, but positioning of the steering rack is critical and sufficiently close tolerances are rarely held in production, particularly if this aspect was not considered in the design. Even for a given design, bump steer varies from car to car and even from side to side of one car, sometimes to the extent that one wheel toes-in with bump and the other toes-out, which is especially bad for straight-line stability.

Calculation of the ideal pivot centre A (Figure 5.16.1) is a purely geometrical problem. In the case of strut and link suspension, the ideal link length is highly sensitive to the vertical position of the ball joint B, tending to infinity when it is at the strut top. A common solution is to use a rack with centre-mounted track-rods, and then to choose a rack height for which these are the correct length, which roughly matches up with the main spring seat. For twin wishbones, the usual rack end connection is suitable. There are various other independent front suspensions, not often seen nowadays, for which there exist suitable ideal steering layouts free of bump steer. The bibliography gives details.

The trailing arm is a common form of rear suspension. For pure trailing arms there is no bump steer. For semi-trailing arms there is often considerable bump steer, depending on the angle of the pivot axis to the vehicle centre-line. This is of the second-order (curvature) kind relative to the mid-position, but relative to a loaded position there is usually both static toe-in and a first-order dependence of toe on bump.

Although this is a disadvantage, it is often considered an acceptable sacrifice in order to gain the advantage of favourable camber in roll.

Rear solid axles can be considered in two groups: those with link location and those with longitudinal leaf springs. In the case of link location, the roll-centre was found by identifying the lateral location points A and B according to the particular linkages (Section 5.7). Because these points define the lateral location of the axle relative to the body, they also define an axis about which the axle will roll relative to the body if the 'road' is rolled (subject to the approximations discussed in Section 5.7). If we consider the body to be rolled about a horizontal axis, then if the front point A is lower than the rear point B the different sideways movements of A and B will result in a steer rotation of the axle, such that the axle tries to increase its slip angle. This is at the rear, so it is an understeer effect. If the axle axis is inclined at ρ_A radians, a roll angle of ϕ results in an axle steer angle of $\phi\rho_A$. The roll steer coefficient (the rate of change of roll steer with suspension roll angle) is therefore equal to ρ_A in radians. This is often expressed as a percentage roll steer, i.e. $100\rho_A\%$.

The variation of roll steer coefficient with load is important, and can be examined easily by considering the change of ρ_A from the motion of A and B with increasing load. If they move equally in the same direction then there is a change of roll-centre height, but no change of roll steer. In some cases, for example the convergent four-link suspension of Figure 5.7.1, when the body moves down then A moves up and B moves down, giving a small change of roll-centre height but a large change of roll steer coefficient. Some positive sensitivity, i.e. increasing ρ_A, may be desirable to help to compensate for the otherwise general trend towards oversteer with increasing load that occurs because of the tyre characteristics. This can help with primary understeer but does not help with final understeer or oversteer.

It might appear in the above discussion that the axle axis angle should be measured relative to the vehicle roll-axis rather than to the horizontal. However, as discussed in Section 5.5, the vehicle does not in a real sense roll about its roll-axis. Rolling the body about the inclined roll-axis implies a roll about a horizontal axis plus a yaw movement which will affect front and rear suspensions equally, and so will have no net result on the steering angle required.

In the case of longitudinal leaf springs, the roll steer depends on the inclination of the equivalent link AB that describes the motion arc of the wheel centre, Figure 5.16.3 (considering the figure without the steering system). This equivalent link is directed towards the unshackled end, is about 3/4 of the length of that end of the spring, and is roughly parallel to it. When the body rolls, A rises on the inner side and falls on the outer, thus tending to steer the axle. Horizontal equivalent links give no steer, because both sides move forward equally. Having point A higher than B gives roll oversteer for a rear axle, i.e. reduced slip angle tendency, B higher gives roll understeer. For a front axle the effects are opposite. The roll steer coefficient is equal to the AB axis inclination ρ_A, independent of the spring length or separation. On the other hand, the spring length affects the influence of load variation on roll steer coefficient. Too high a coefficient, apart from being bad for handling, also leads to harshness on rough roads because of the wheel path in bump, and other problems. Early Hotchkiss axles, pre-1930, were given a negative coefficient because of their favourable response to roughness. In the early 1930s a positive coefficient was first used, and a dramatic improvement in directional stability was found.

When the solid axle with longitudinal leaf-springs is used at the front, as on many trucks, it is subject to all the effects described above, plus additional effects because of the steering linkage. The critical steering link is always the one that connects the sprung and unsprung parts of the steering. Figure 5.16.3 shows a typical arrangement with front steering and a rear-spring shackle. Here D is the drag link connection to the wheel hub, and C is its connection to the Pitman arm on the sprung mass. When the axle moves, D has an ideal no-steer arc centred on E. If E and C do not coincide, there will be steering errors. As early as the 1920s, it was attempted to match the E and C positions for roll motions, but with disappointing results. The reason is that the arc of D is different for roll, single-wheel bump, and heave, and different again with braking because of axle wind-up. Also there are differences because of production variability of springs, and variation between spring options. One improvement that is sometimes adopted is to use an unsprung steering box, the sprung to unsprung connecting linkage being through a splined steering column.

The trailing twist rear axle (Figure 4.7.1) is a special case, as usual.

Figure 5.16.3. Steered leaf-spring axle.

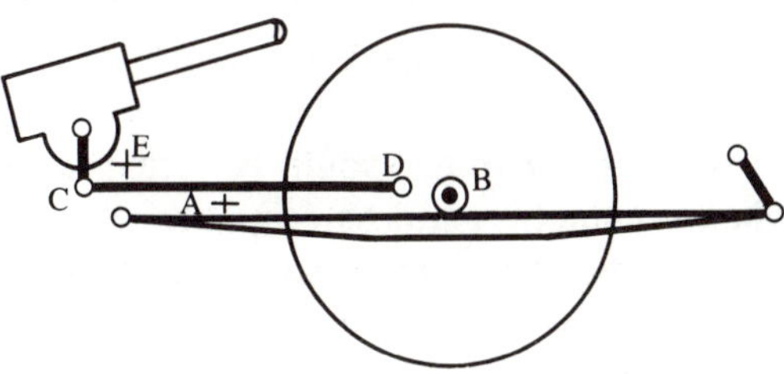

In heave the whole axle rotates about the front pivots and there is no steer effect. In single-wheel bump, the other trailing arm remains approximately level, so the pivot axis is from the pivot bush of the bumped side to the intersection of the cross-beam with the unbumped arm, equivalent to a semi-trailing arm. For roll, the mid-point of the cross-beam remains approximately stationary, so the pivot axes are from the bush to the centre of the beam – a different equivalent semi-trailing arm. The bump and roll steer values follow from these semi-trailing arm equivalents.

Even if the smooth road ideal of zero roll steer is abandoned, there are definite limitations to the degree of roll steer that is acceptable. Roll oversteer of independent front suspension, i.e. wheel toe-in in bump, gives severe wheel-fight on rough roads. Roll oversteer of the rear gives increased body attitude angles, a very unpleasant uncertain feeling for the driver, and bad directional response to rough roads and to side winds. It does not seem to be effective to balance an oversteering rear with an understeering front, and there is little reason to try to do this. In short, there is good reason to avoid roll oversteer at either end. Some would argue for rear roll understeer for its reduced attitude angle and hence possibly faster response, especially for large cars which tend to have a larger dynamic index (I_z/mab), but this can lead to problems with rough roads or wind, and may result in engine torque steering because of body roll in strong acceleration, unless the power is transmitted through independent or de Dion axles or an offset lift bar is used.

A small static toe-in at the rear often has a surprisingly large effect on handling, increasing understeer; front-wheel-drive vehicles, being lightly loaded at the rear, and therefore having a large rear tyre cornering stiffness coefficient, are especially sensitive to this. Front toe-out might be expected to have the same effect. However, practical experience shows the reverse: front toe-out gives a vague steering feel whereas front toe-in gives a favourable feel and leads to increased understeer that can be measured on the skid pad. This seems to be caused by the combination of load transfer, aligning torque, lateral force on caster trail, and steering compliance. Too much front toe-in affects corner turn-in, giving an unprogressive and imprecise steering feel.

Static toe settings are governed within quite narrow bands by tyre wear. For least wear, toe settings should be arranged to give minimal steer angles when running, regardless of camber. This means a small static toe-in for undriven wheels, and a small toe-out for driven ones, so that forces and compliances act to bring the toe angles to zero when running. Within the allowable band for low wear (a range of about one degree), there is limited scope to use toe angles to tune handling characteristics.

5.17 Compliance steer

Changes of wheel angles, as a result of the tyre forces and moments, are known as compliance steer and compliance camber. The compliance may be in the suspension or in the steering linkage. Change of steer and camber angles may in principle result from any of the tyre forces or moments. In quoting and using these coefficients it is therefore important to specify the particular force or moment being considered. The effects are sideforce compliance steer η_S (deg/N), aligning moment compliance steer η_A (deg/Nm), sideforce compliance camber (η_C) and overturning moment compliance camber (η_O). The last of these is usually neglected, and often the last two or three. The actual compliance angles are

$$\alpha_{\eta S} = \eta_S F_y$$

$$\alpha_{\eta A} = \eta_A M_z$$

$$\gamma_{\eta C} = \eta_C F_y$$

$$\gamma_{\eta O} = \eta_O M_x$$

The total effect of compliance in the linear regime may be summarised by the compliance steer gradient k_η in deg/m s^{-2} or deg/g. Because the sideforce effective in creating distortions is the sprung mass force rather than the total tyre force, we have

$$k_{\eta S} = m_{Sf}\eta_{Sf} + m_{Sr}\eta_{Sr}$$

$$k_{\eta A} = m_{Sf}t_f\eta_{Af} + m_{Sr}t_r\eta_{Ar}$$

where t is the tyre pneumatic trail. Typical values of compliance understeer gradient are 1.0 deg/g at the front and 0.2 deg/g at the rear. Sometimes rear compliance oversteer is deliberately introduced. Under tractive or braking forces, there may also be compliant changes of caster angle. These will augment the caster change relative to the ground caused by vehicle pitch due to longitudinal load transfer.

For the design of ordinary passenger vehicles, handling must be seen in the context of the ride/handling compromise. To provide comfort and to avoid noise, vibration and harshness it is necessary to have considerable compliance of wheel motion, not just vertically but also in longitudinal and lateral directions, achieved by the springs and by the extensive use of rubber bushes. Unfortunately this has generally led to considerable angular compliance of the wheels, resulting in unfavourable or unpredictable handling because of deflection camber and especially because of deflection steer. This conflict is now largely resolvable, by allowing the wheel relatively generous movement in translation, but little angular movement in steer or camber for the forces that it actually experiences. The method is to bring the shear centre of an independent suspension, in plan view, close to the centre of tyre contact, or for a solid axle near to the mid-point between the tyres. Tyre forces then have little moment about the shear centre, and so although the wheel still has angular compliance, there is little angular response, so that steer angle changes are controlled. Careful location of the shear centre can even be used to introduce favourable small deflections. A significant characteristic of compliance steer is that unlike roll steer it occurs almost immediately.

Consider a typical driven semi-trailing arm, supported by two bushes, Figure 5.17.1. If the bushes A and B are of equal stiffness then the shear centre lies mid-way between them. If one is harder then the shear centre is closer to that one, but always between A and B. A tractive force on the tyre will have a clockwise moment about the shear centre, so the wheel will respond by toeing in. Braking forces will cause toe-out, resulting in some instability under braking. Sideforce will also cause significant toe changes. If a driver enters a corner under power, the tractive force will contribute a toe-in; if he lifts off the accelerator to

Figure 5.17.1. Conventional trailing arm.

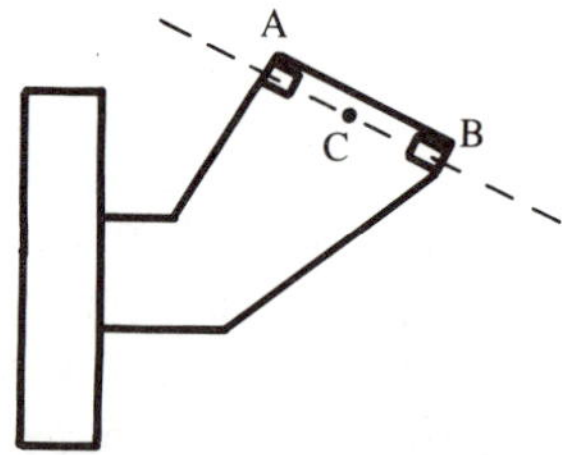

Figure 5.17.2. Equivalent plan view linkage of Weissach axle.

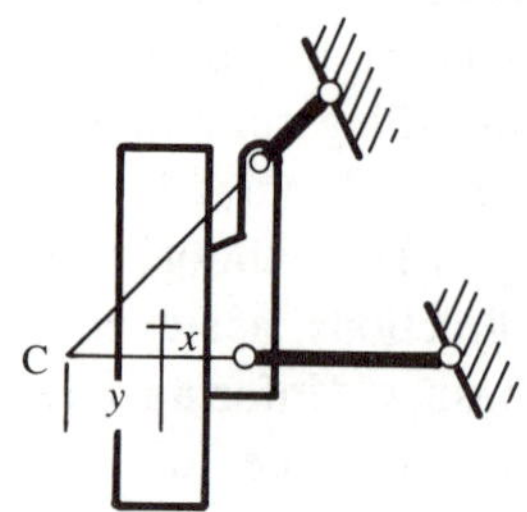

slow down, the wheel will make a toe-out change, tightening the curve in a disconcerting manner. This is known as lift-off tuck-in, and is similar to power understeer.

One solution to this problem is the Weissach axle, which has the equivalent plan-view linkage of Figure 5.17.2. With all the usual approximations, the centro of wheel motion relative to the body is at C, and this is also the shear centre. Actually wheel motion about this centre is resisted by suitable stiffnesses, e.g. the rubber bushes at the link ends. Tractive and braking forces now have the opposite effect to the ordinary semi-trailing arm of Figure 5.17.1, giving toe-out and toe-in respectively. The stiffnesses are such that the total range of angular motion under cornering and braking forces is a little less than one degree. For outboard brakes, C is the point where the pivot axis intersects the ground plane. For traction, because of driveshaft torque, the total drive force on the suspension acts at wheel centre height, so C is the point where the pivot axis intersects the horizontal plane at wheel centre height. This distinction permits separate tuning of the response for braking and traction.

Response to fore–aft forces depends on the lateral position of the shear centre relative to the centre of tyre contact (y), whereas response to lateral forces depends on the fore–aft position of the shear centre (x). It is not possible to achieve completely zero sideforce steer because the tyre pneumatic trail varies with sideforce. Actually some sideforce understeer may be desirable on vehicles of large dynamic index (I_z/mab), usually large vehicles, to improve response time and stability in lane change manoeuvres, but sideforce oversteer may give better limit controllability because of more progressive tyre breakaway and more controlled response to road roughness or friction variations.

Similar principles are applicable to solid rear axles. For a Panhard rod supported axle, the rod being DE in Figure 5.17.3, the position at which the rod crosses the centre-line is the shear centre. A sideforce is

Figure 5.17.3. Compliance steer by Panhard rod, plan view.

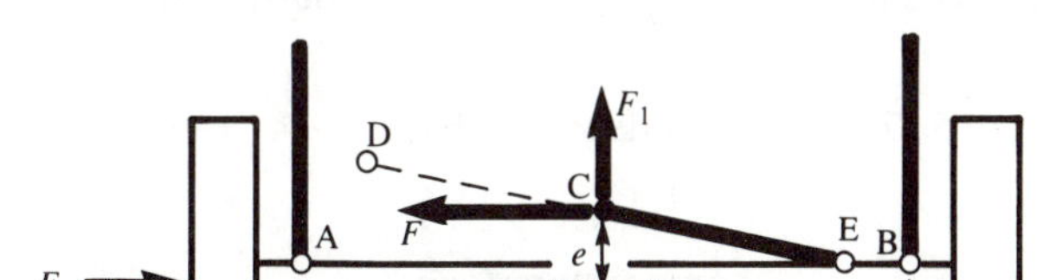

opposed by lateral and forward components of force in the rod, for a tilted rod. The forward component F_1 is then reacted by equal forces at the longitudinal locations at A and B, which cause no steer effect. The couple from the offset force F and the wheel sideforce F gives a moment Fe, requiring equal and opposite forces Fe/S at A and B, leading to a steering rotation of the axle. Thus the sideforce steer depends on the centre-line offset e of the Panhard rod forward from the tyre sideforce, and on the effective total stiffness k of the longitudinal arms, which is basically that of the bushes at A and B and the bushes at the front of the arms. The sideforce understeer coefficient for such an arrangement is

$$\eta_s = -\frac{2e}{kS^2}$$

For a rod forward of the axle the result is sideforce oversteer. This can be used to give progressive limit state handling. The offset e actually varies with the lateral force because of change of pneumatic trail. If the axle is located by a torque tube or equivalent instead of fore-and-aft links at the side, then there is almost inevitably sideforce oversteer regardless of the lateral location system position.

One disadvantage of the solid axle compared with an independent system is that different tractive or braking forces side-to-side, e.g. from a limited slip differential, will cause steer. This will be minimised for a given sideforce steer if the bushes are stiff and the rod offset large. The solid axle has been very successful for the rear of front-drive vehicles, but these effects suggest that independent suspension may be superior for rear drive in this respect.

Conventional trailing twist axles have a marked sideforce oversteer because of compliance of the bushes at the front mountings, the shear centre being between them. This has been adjusted by using wedged bushes that react to a side displacement with a fore–aft displacement too, essentially as in Figure 5.17.4, where the sideforce F causes a y displacement of the wedged pivot bolt.

By mounting the trailing twist axle in such bushes, it can be arranged

Figure 5.17.4. Radially deflecting bush.

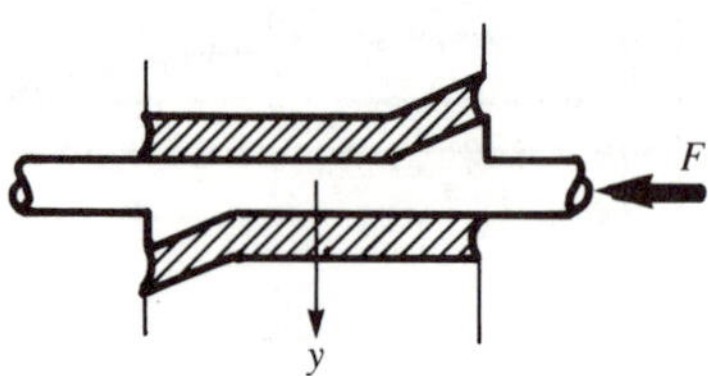

that a sideforce turns the whole axle in a direction that attempts to oppose the side displacement at the wheels. In effect the shear centre is therefore moved rearward from the bushes, and can be placed between the wheels, or as required. A similar effect could be achieved by using conventional bushes with their axes approximately perpendicular to the line from the desired shear centre.

With longitudinal leaf springs parallel to the vehicle centre-line there is no sideforce steer effect. However the spring deflection tends to be perpendicular to the spring, so if the springs are convergent towards the front there will be sideforce oversteer for a rear axle, and sideforce understeer for a front axle. Longitudinally split springs have been tried, increasing the lateral compliance, but this leads to a vague steering feel even for parallel springs.

The conventional swing axle is essentially free of sideforce steer, but is likely to have significant tractive force toe-in, i.e. power understeer and brake oversteer.

The situation at the front of the vehicle is complicated by the steering. The outside wheel in cornering is subject to an inward lateral force, putting the lower wishbone in compression and the upper one in a smaller tension. Because of the bushes this results in a wheel camber. There will be no resultant motion of the wheel at some height between the wishbones, depending on the bush stiffnesses. If the steering arm is at this height then there will be no steering effect from this cause, whether the steering tie-rod is in front or behind the kingpin axis. In plan view the lateral force acts behind the kingpin because of pneumatic and caster trail. For front or rear tie-rods, the steering compliance then results in a sideforce understeer. This is the most significant of all sideforce steer effects because of the considerable compliance of the steering column. We can express this relatively high compliance by saying that for sideforce the shear centre (the point where a force will not cause a steer deflection) is a rather small distance behind the kingpin axis. For tractive forces the shear centre is again rather closely aligned with the kingpin axis, so in practice steering not far removed from centrepoint steering is used. For front drive, careful tuning of the system is required to prevent power steer, although the situation is complicated by the dominant influence of the driveshaft torque component (see Section 5.14).

On low-friction surfaces such as ice or snow, the tyre characteristics are different, and there is a low limit to lateral acceleration, so load transfer distribution becomes less significant, and roll steer and sideforce steer become more critical in determining limit handling, especially at the rear.

5.18 Problems

Q 5.2.1 For a suspension stiffness of 20 N/mm and a tyre vertical stiffness of 240 N/mm, what is the ride stiffness?

Q 5.2.2 In Figure 5.2.1, with EH = 300 mm and HF = 2000 mm, what is the rate of track change in heave? What is the value of the S.A.E. equivalent swing arm radius?

Q 5.2.3 Sketch the constructions to find the centre E for double-wishbone, Macpherson, trailing arm and semi-trailing arm suspensions.

Q 5.2.4 Draw a typical graph of vertical force against bump position. Describe its features.

Q 5.2.5 Sketch, and justify, the construction to find the force centre E for double-wishbone, Macpherson, trailing arm, semi-trailing arm and swing axles.

Q 5.2.6 A wheel has a scrub rate in bump of 0.167, and rises at 3 m/s over a bump on a vehicle travelling at 15 m/s. Calculate the effect of scrub on slip angle.

Q.5.2.7 Find an expression for the wheel rate in terms of the idealised spring stiffness. Discuss the result in general terms for a slider and spring for deflections at an angle to the slider motion. (Note that the wheel rate is the rate of change of vertical force with vertical deflection.)

Q 5.3.1 Discuss the problems with the kinematic concept of the roll-centre.

Q 5.3.2 Explain the difference between body roll, suspension roll and axle roll.

Q 5.3.3 Explain how body jacking can occur because of springs or stops. (After studying the rest of the chapter, compare with link jacking.)

Q 5.3.4 Describe a typical graph of suspension roll moment versus roll angle.

Q 5.3.5 Describe a typical graph of vehicle roll moment versus roll angle (i.e. for two suspensions acting in parallel).

Q 5.4.1 An axle of track 1.48 m has a roll-centre height of 160 mm. What are the load transfer factor and the net link load transfer at a sprung end-mass sideforce of 2000 N?

Q 5.4.2 An instrumented vehicle axle is found to have a net link load transfer of 320 N at a sprung sideforce of 2200 N. The track is 1.44 m. What are the load transfer factor and the roll-centre height?

Q 5.4.3 Compare and contrast the roll-centre definitions used here with the S.A.E. definition.

Q 5.4.4 At a lateral acceleration of 7 m/s^2 a suspension of track 1.82 m has a net link lateral load transfer of 680 N for a sprung mass sideforce of 7 kN, and at 7.5 m/s^2 has 740 N on 7.5 kN. For each case find the roll-centre height and the load transfer factor.

Q 5.4.5 According to Steeds (1958), if the tyre lateral forces are unequal, then a lateral force at roll centre height causes a roll angle. Do you agree? Quantify this (if not zero) in terms of the sideforce transfer factor, i.e. where the inner and outer sideforces are $\frac{1}{2}(1 - e_S)F$ and $\frac{1}{2}(1 + e_S)F$, for a very small roll angle. Do this by considering an independent suspension, with compensation centrifugal force F at the roll centre. Draw free body diagrams of the two sides of the unsprung mass, and taking moments about the instantaneous centres find the effect on the idealised spring forces. Hence find how the effective moment and jacking force on the sprung mass depend on e_S, if either is non-zero.

Q 5.4.6 A vehicle has a sprung centre-of-mass height of 0.642 m, 1.422 m back on a 3.141 m wheelbase. The roll centre heights are 0.082 m and 0.284 m front and rear. Evaluate the anti-roll coefficient.

Q 5.5.1 Justify the construction of the EF line for a double-wishbone suspension.

Q 5.5.2 Justify each of the constructions of Figure 5.5.2.

Q 5.5.3 Explain with suitable figures how to find the initial roll-centre of a trailing-arm suspension with pivot axes not perpendicular to the centre plane in either front or plan view.

Q 5.5.4 Compare Figure 5.5.2 with other such published figures. Note any discrepancies. Satisfy yourself that Figure 5.5.2 is correct.

Q 5.5.5 Consider a vehicle with three axles, with roll centres that are not co-linear. Where is the roll axis? Is this a problem for the kinematic concept of the roll axis? Is it a problem for the force concept of the roll axis?

Q 5.5.6 Draw front views of double-wishbone independent suspensions with a roll-centre below ground level, one with positive and one with negative S.A.E. swing arm radius.

Q 5.6.1 Explain the concept of the incremental roll-centre, and compare it with the roll-centre.

Q 5.6.2 Show that $h_i = h + F_S(\mathrm{d}h/\mathrm{d}F_S)$ and $f_i = f + F_S(\mathrm{d}f/\mathrm{d}F_S)$. Also explain physically why h_i is approximately equal to h for small lateral acceleration.

Q 5.6.3 Explain how the suspension jacking force arises.

Q 5.6.4 For each of the suspensions of Figure 5.5.2, does the roll-centre rise or fall with increasing lateral acceleration? Can this be stated with confidence without knowing the geometric details?

Q 5.6.5 For the vehicle of Q 5.5.5, find the mean load transfer factor and the mean incremental roll-centre height for this increment of acceleration.

Q 5.7.1 Explain with diagrams how to find roll-centres for rigid link located axles.

Q 5.7.2 In an axle roll-centre analysis, with track 1.44 m, point A is found 1.20 m behind the suspension plane at a height of 420 mm, and point B is 2.15 m in front at height 220 mm. What are the roll-centre height, net link load transfer factor and net link load transfer at 3200 N sprung mass sideforce?

Q 5.7.3 Analyse the effect of vertical load on roll-centre height for the axles of Figure 4.6.1.

Q 5.7.4 'A solid axle is not subject to the link force jacking effect of independent suspensions.' Discuss, with diagrams.

Q 5.8.1 Explain the roll-centre position of a leaf-spring mounted solid axle.

Q 5.8.2 Explain the roll-centre position of a trailing twist axle.

Q 5.9.1 Discuss various methods of experimental roll-centre measurement.

Q 5.9.2 For the best method of experimental laboratory roll-centre measurement, consider the possible inaccuracies in relation to the force roll-centre for real cornering conditions.

Q 5.10.1 Explain why the total lateral load transfer cannot be altered by suspension adjustments, for a given height of G and track width T, neglecting roll angle.

Q 5.10.2 In load transfer analysis, the axle unsprung masses are treated separately from the sprung mass. Justify the treating of the unsprung mass as a single mass in this context, for an independent suspension.

Q 5.10.3 In steady-state cornering there may be a longitudinal load transfer. List all the possible causes of this, and explain them.

Q 5.10.4 Explain, with equations, how the total lateral load transfer is distributed front and rear.

Q 5.10.5 A vehicle has total mass 1600 kg, lateral acceleration 6.5 m/s^2, $H = 550$ mm, $a = 0.46l$, track 1.46 m, $h_f = 70$ mm, $h_r = 320$ mm, roll stiffness 350 Nm/deg front, 150 Nm/deg rear. Find the roll angle neglecting the sideways movement of G, the percentage load transfer distribution, and the tyre vertical reactions. Neglect the unsprung mass and treat the body as torsionally rigid.

Q 5.10.6 Taking other data from the last question, with front unsprung mass 140 kg at height 280 mm and rear unsprung mass 200 kg at 300 mm, sprung mass 1260 kg at 550 mm height, with $a_s = 0.44l$, find the roll angle allowing for sideways movement of G, and the percentage load transfer distribution.

Q 5.10.7 With the same data as Q 5.10.5, but with a driveshaft torque of 400 Nm exerted by the body on the rear axle, find the roll angle and load transfer because of this.

Q 5.10.8 Repeat Q 5.10.5, allowing for a tyre vertical stiffness of 200 N/mm, also finding the two axle roll angles.

Q 5.10.9 Explain the idealised spring model of the suspension, and discuss its advantages.

Q 5.10.10 For a vertical slider pillar suspension, obtain an expression for the real load transfer in the links with the spring acting at the pillar. Compare with the idealised spring model. Where is the roll-centre? Discuss.

Q 5.10.11 'The roll axis inclination can be neglected in load transfer calculations.' Discuss, giving equations and considering the sprung mass weight and sideforce.

Q 5.10.12 'The rotation of the wheels gives a gyroscopic effect that affects lateral load transfer in steady state handling.' Explain with equations.

Q 5.10.13 Evaluate the gyroscopic effect of the wheels on lateral load transfer in steady state handling, as a percentage change, for an ordinary passenger car and for a racing car. See Appendix B for example data.

Q 5.10.14 A racing car designer proposes to improve the cornering performance by eliminating steady state lateral load transfer. This is to be done by installing a flywheel with axis parallel to the axles. Obtain an algebraic expression for the required angular speed of the flywheel, in terms of car mass, centre of mass height, flywheel second moment of mass, etc. State the required direction of rotation. What gear ratio is needed if the flywheel is driven from a rear wheel? Comment on the practicality of such a system (giving numerical values). Is the yaw balance of the car directly affected by the flywheel (i.e. is a yaw torque needed)?

Q 5.10.15 Analyse the lateral load transfer effect caused by rotation of a transverse engine. Give values, including the effect of gear ratio.

Q 5.10.16 A vehicle has track 1.440 m, centre of mass height 640 mm, and maximum lateral acceleration 8.10 m/s^2. Find the roll-over lateral acceleration, and the (simply calculated) safety factor against roll-over. Give a rough estimate of the probable real value that would be found by more accurate calculation.

Q 5.10.17 Define and explain the vertical force lateral transfer factor e_V.

Q 5.11.1 Derive an expression for the roll angle gradient of a vehicle, neglecting aerodynamics.

Q 5.11.2 Derive an expression for the roll angle gradient, including aerodynamics, for varying radius at constant speed.
[Use $\beta = -(b/R) + (k_\beta / k_\phi)\phi$.]

Q 5.11.3 Explain the effect of tyre overturning moment on lateral load transfer for: (1) a solid axle , (2) independent suspension.

Q 5.11.4 In cornering with an independent suspension, there may be a moment (M_X) on the unsprung mass from the couple of compensation force and tyre force, from tyre overturning moment or from wheel gyroscopic precession. Describe the corresponding lateral load transfer process in detail. Will this cause any roll angle of the sprung mass?

Q 5.11.5 In calculating cornering lateral load transfer from compensation force on the unsprung masses of an independent suspension, can the unsprung masses be treated as a solid axle? Justify your answer in detail.

Q 5.12.1 Describe how complete anti-dive may be achieved for inboard and outboard braked vehicles. What are the problems with such a system?

Q 5.12.2 Describe how anti-squat may be applied to a four-wheel-drive vehicle. Is this compatible with anti-dive, for inboard or outboard brakes?

Q 5.12.3 Analyse the jacking effect resulting from large anti-dive in small radius corners.

Q 5.12.4 Define, explain and give example values for the tyre vertical force longitudinal transfer factor in acceleration and braking, e_X and e_B.

Q 5.13.1 Explain why the vertical wheel reactions on a conventional vehicle are statically indeterminate.

Q 5.13.2 Find the percentage wheel reactions for a vehicle: (1) with centre of mass at 0.44 of the wheelbase, and plus 4% diagonal bias, (2) with centre of mass at 0.54 of the wheelbase, and minus 4% diagonal bias (linear springs).

Q 5.13.3 Find the percentage wheel reactions for a vehicle in right hand cornering with centre of mass at 0.48 of the wheelbase, with zero static diagonal bias, and 24% lateral load transfer distributed 75% at the front.

Q 5.13.4 Find the percentage wheel reactions for a vehicle in left-hand cornering with centre of mass at 0.44 of the wheelbase, plus 4% diagonal bias, and 20% lateral load transfer distributed 60% at the front, and 2% longitudinal load transfer.

Q 5.13.5 Apply the vertical force lateral transfer factor e_Y to a complete vehicle. Define e_Y for the vehicle, and for front and rear suspensions.

Q 5.13.6 For the vehicle of Q 5.13.4, evaluate e_Y, e_{Yf} and e_{Yr}.

Q 5.14.1 Discuss rear-wheel steering.

Q 5.14.2 Describe the most common types of steering system for independent suspension, with diagrams.

Q 5.14.3 Define, explain and give typical values of caster angle, caster offset (trail), kingpin inclination and kingpin offset.

Q 5.14.4 What is the significance of independently-varying caster angle and trail?

Q 5.14.5 Define and explain all terms relating to steer angle and steering gear ratio. Give example values.

Q 5.14.6 Estimate the steering wheel torque to statically steer a vehicle with mass 1500 kg, $a = 0.40l$, at 170 kPa inflation pressure with $\mu = 1.1$. With steering ratio of 16 and forward efficiency 85%, what is the steering-wheel torque requirement (unpowered)?

Q 5.14.7 Explain the origin of the most important steering system inertias and compliances, giving example values.

Q 5.14.8 Define and explain the terms steer angle, steering wheel angle, overall steering ratio, reference steer angle and any related terms.

Q 5.14.9 Sketch example curves of δ_s against δ with and without forces in the steering system, for non-linear steering ratio. Discuss the application of the concept of reference steer angle in such a case.

Q 5.14.10 Explain in detail how braking or driving driveshaft torques affect the steering. Give equations and example values.

Q 5.14.11 Explain in detail how the tyre lateral force and aligning moment affect the steering.

Q 5.14.12 Explain in detail how tractive or braking forces affect the steering, with and without a limited slip differential, on front drive.

Q 5.14.13 Explain in detail how tyre vertical forces affect the steering. Give suitable diagrams.

Q 5.14.14 Explain how caster trail is used to give a desired form to the Gough plot.

Q 5.14.15 Explain and give example values for centrifugal caster.

Q 5.15.1 Evaluate the maximum low-speed offtracking for a car, using representative dimensions.

Q 5.15.2 Describe in detail the low-speed offtracking of a two-wheeled trailer on a four-wheel vehicle.

Q 5.15.3 Define the Langensperger angle for an axle, and discuss its significance.

Q 5.16.1 Define and explain bump steer.

Q 5.16.2 Define and explain roll steer.

Q 5.16.3 Define, explain and give example values of bump steer coefficient, bump camber coefficient, roll steer coefficient and roll camber coefficient.

Q 5.16.4 Describe a typical bump-steer graph; explain how it might be measured and why certain features may be considered desirable.

Q 5.16.5 In Figure 5.16.2, what is the bump steer coefficient at 50 mm droop?

Q 5.16.6 Explain how the link geometry of an axle leads to roll steer, and how this may depend on the vehicle load in a favourable way.

Q 5.16.7 Discuss roll steer of leaf spring rear axles.

Q 5.16.8 Discuss the extra complications of roll steer on leaf-spring axles at the front rather than the rear.

Q 5.16.9 Explain roll steer of the trailing twist axle.

Q 5.16.10 Discuss the desirability of, and limitations on, deliberate static toe-in and toe-out.

Q 5.16.11 Analyse qualitatively the effect of load on the roll steer coefficient of the axles of Figure 4.6.1.

Q 5.16.12 Is it possible to arrange for no bump steer of a slider (pillar) suspension?

Q 5.17.1 Define and explain compliance steer coefficients and compliance camber coefficients.

Q 5.17.2 Explain the principle by which only desirable wheel compliant movements are allowed and undesirable ones minimised.

Q 5.17.3 Considering the two main types of independent suspension, with steering in front and behind the kingpin axis, describe how compliance steer arises. Which force or moment components of tyre force are likely to be the main cause of steer effects?

Q 5.17.4 Which components of tyre force and moment are likely to be the main cause of compliance steer?

Q 5.17.5 Explain the compliance steer effects that occur on a standard semi-trailing arm because of drive, braking and lateral forces.

Q 5.17.6 Explain the function of the Weissach axle.

Q 5.17.7 Explain how a Panhard-rod located dead rear axle can be tuned to give desired compliance steer characteristics. Give relevant equations.

Q 5.17.8 Explain in detail how wedged bushes may be used to eliminate excessive sideforce-steer from a trailing twist axle.

Q 5.17.9 A vehicle has a leaf-spring front axle with the spring divergent at 6° each in plan view, towards the front, spaced at 1.2 m at the axle. Each spring lateral stiffness is 62 kN/m. The front sprung end-mass is 1400 kg. Obtain an expression for the compliance understeer gradient ($d\eta/dA$) because of this, and evaluate it.

Q 5.17.10 A Panhard rod supported rear axle has a forward offset e of 120 mm, and longitudinal supports at spacing 1.100 m with bushes of stiffness 120 kN/m. The sprung end-mass is 900 kg. Estimate the compliance understeer gradient ($d\eta/dA$) neglecting pneumatic trail.

Q 5.18.1 For one of the vehicles specified in Appendix B, apply the theory of each section of this chapter to analyse the suspension.

Q 5.18.2 For vehicle G of Appendix B, calculate the load transfer distribution with and without allowance for tyre vertical stiffness.

Q 5.18.3 – 5.18.7

One set of published data for a modern racing car is for a Tyrell F1 on Signes corner of Paul Ricard Circuit in practice for the 1990 French Grand Prix (Curtis, 1990). (This author cannot vouch for the accuracy of the data, which when analysed gives rise to some doubts, but it is presented here for interest.) Signes is a very fast right-hander, with a recorded speed of 83.4 m/s (186.5 m.p.h.) and lateral acceleration of 39.2 m/s^2 (4.0g) at the path radius of 177 m (about 600 ft). The data are:

Mass = 600 kg

Aerodynamic downforce	– front wing	=	4500 N
	– body	=	4308 N
	– rear wing	=	3426 N
	Total	=	12 234 N
Tyre vertical force	– front inner	=	1667 N
	– front outer	=	5093 N
	– rear inner	=	4456 N
	– rear outer	=	6904 N
	Total	=	18 120 N
Tyre side force	– front inner	=	3182 N
	– front outer	=	6998 N
	– rear inner	=	6166 N
	– rear outer	=	7144 N
	Total	=	23 490 N
Typical F1 wheel track	– front	=	1.804 m
	– rear	=	1.626 m
Typical F1 wheelbase		=	2.800 m

Q 5.18.3 Calculate the front, rear and overall mean tyre side force coefficients.

Q 5.18.4 Calculate the longitudinal G position a/l using the side forces (neglect steer angle).

Q 5.18.5 Calculate the front, rear and total vertical force transfer, and the load transfer distribution, the front and rear load transfer factors, and the G height (the result is suspicious).

Q 5.18.6 Calculate the front and rear downforces and the aerodynamic downforce position (using the G position) and the total downforce area.

Q 5.18.7 Analyse the tyre maximum cornering force coefficient sensitivity to vertical force.

5.19 Bibliography

For a qualitative discussion of many example suspensions, see Norbye (1980). A more quantitative analysis with many practical examples is given in Bastow (1980 or 1987) and Campbell (1981). For more examples specifically on front-wheel drive see Norbye (1979). Ellis (1969 and 1989) provides a mathematical perspective on suspension analysis, while Segel *et al.* (1980) is a useful reference on the suspension of commercial vehicles.

A full definition of terms is given in S.A.E. J670e *Vehicle Dynamics Terminology* (June 1978) and M.I.R.A. 1965/1 *Definition of Handling Terms*.

A good deal of useful practical information is provided in the popular work by Puhn (1981). Many details of suspension design are discussed in unpublished General Motors reports by Olley (1961a, 1961b, 1962a, 1962b). Reimpell (1982) gives an extensive analysis of suspensions, in German. For a more detailed discussion of roll-centres for independent suspensions, see Dixon (1987a). Riede *et al.* (1984) have surveyed typical values of vehicle parameters needed for handling analysis.

Although there has been a good deal of discussion elsewhere about active suspension systems they have not so far found great commercial success or shown any consistent advantage in racing. They have the advantage of being able to eliminate roll, even with a ground-level roll-centre, and to eliminate dive and squat, freeing constraints on the link geometry. They are likely to be rather expensive. Bastow (2nd edition, 1987) gives an introduction. For details, see for example the proceedings of the International Conference on Advanced Suspensions (I.Mech.E. 1988).

6

Steady-state handling

6.1 Introduction

In a broad sense, the purpose of handling theory is to assist in the design of better vehicles, i.e. to determine if there are optimum or preferred vehicle characteristics. If such characteristics exist, then handling theory should show the designer how to achieve them. In a narrower sense the purpose is to predict the behaviour of a vehicle in response to control inputs, or to environmental disturbances such as road roughness or wind. This chapter deals with the theory of response to steady-state control inputs and disturbances.

Two examples of steady-state disturbances are a constant sidewind and a constant road camber, i.e. side slope. Later sections deal with the vehicle's response in such cases, and with the associated problem of the control inputs required to overcome the disturbance. More attention will be devoted to investigating the steady-state response to control inputs. In this case steady-state is taken to mean the absence of ground or wind disturbances, and fixed controls. This implies constant path radius of curvature, constant translation speed, constant angular speed and constant magnitude of lateral acceleration, although the velocity and acceleration are not constant in direction relative to the ground. The S.A.E. definition of steady-state is:

> "Steady-state exists when periodic (or constant) vehicle responses to periodic (or constant) control and/or disturbance inputs do not change over an arbitrarily long time. The motion responses in steady-state are referred to as steady-state responses."

By this definition, regular oscillatory control inputs, for example for a slalom test, would be included in steady-state. Ambiguously, the S.A.E.

definition of transient state is:

> "Transient state exists when the motion responses, the external forces relative to the vehicle, or the control positions are changing with time."

The definition adopted here will be that transient states are states that are not steady.

Following aircraft practice, steady-state theory is sometimes said to be a theory of trim states, i.e. of the vehicle response to given control trim conditions. Trim is formally defined by the S.A.E. as:

> "the steady-state condition of the vehicle with constant input which is used as the reference point for analysis of dynamic vehicle stability and control characteristics."

It is sometimes unfortunately said to be the 'equilibrium condition'; in general the vehicle is not in equilibrium in Earth-fixed axes, having a non-zero lateral acceleration. The question of the static and dynamic stability of the vehicle at any particular trim state will be considered in the next chapter. This is closely related to the widely used, and often abused, concepts of understeer and oversteer, which are examined in some detail.

Steady-state handling theory can be considered to fall into two main areas. The first area is the representation of the total vehicle handling characteristic, which may be by the sideslip, yaw and steer angles versus lateral acceleration, or by the moment-method graph. Such results may be found by testing. The second area is the prediction of the total vehicle characteristic from its design details such as the tyre and suspension characteristics.

6.2 Parameters

The properties of the chassis, the suspension, the tyres and the environment – i.e. the road surface and the atmosphere – combine to give the vehicle its handling characteristics. The parameters representing the motion state of the vehicle in steady conditions are its forward speed and the path curvature (reciprocal of the path radius of the centre of mass). 'Steady state' implies constant values for these two variables.

The speed and path curvature are the basic dependent variables that result from the driver control inputs to the system, Figure 1.1.1. These input variables are the steering-wheel position, the accelerator position, and the gear-lever position (for automatics the gear is a dependent variable). In a full computer simulation we would expect to find the steering wheel and accelerator positions as real variables, the gear as an integer variable, and the resulting speed and path curvature being deduced as real variables. These are the fundamental control inputs and responses of the system as perceived by the driver.

Even a cursory examination of the research literature quickly shows that handling theory is not actually presented in terms of the above three input variables. Certainly there are innumerable instances of the steering wheel position being considered, but this author cannot recall ever seeing a quantitative, or even qualitative, representation of the accelerator position. As an alternative to the accelerator and gear positions, we could use the effective torque at the rear wheels as an input variable. This is the result of the accelerator position and gear and engine characteristics, environmental conditions, gearbox ratio and friction, and differential and final drive ratio and friction. It also has the advantage of a reduction of the total number of input control variables from three to two, and the elimination of side-effects such as the influence of atmospheric conditions on the engine. Another possible advantage is that 'minor' features of the vehicle, such as the throttle linkage, are no longer directly relevant, although experience shows that the apparent response of a vehicle in ordinary driving is governed as much by the progressive action of the throttle linkage as by the maximum engine power; also, for high-powered or competition vehicles the controllability may critically depend upon the provision of a good design of throttle linkage, so this simplification is not achieved without some loss of reality. A definite disadvantage is that the driver does not perceive the torque at the rear wheels directly, i.e. it is not strictly a control input, although he does experience it indirectly through the vehicle's speed and acceleration. Again, however, examination of the literature shows that torque at the driven wheels is not normally considered the input variable.

Under steady-state conditions, driven-wheel torque primarily influences the forward speed. It is the steering-wheel position and the forward speed that are generally taken as the basic independent variables of steady-state handling theory. Even this is an oversimplification because the steering-wheel angle is often considered after gearing down to the reference steer angle

$$\delta_{\mathrm{ref}} = \frac{\delta_{\mathrm{s}}}{G}$$

In the results of testing, the vehicle speed, angular speed or lateral acceleration may be measured, although it is generally the last of these that is used in graphical results, e.g. Figure 1.13.1. For a speed V and path curvature

$$\rho = \frac{1}{R}$$

the yaw angular speed is

$$r = \rho V = \frac{V}{R}$$

and the lateral acceleration is

$$A = \rho V^2 = rV = \frac{V^2}{R}$$

Hence any two of these four variables suffice to find all four. In practice because the actual vehicle controls are the accelerator and the steering wheel, it is usual to take the speed, plus one of ρ, r or A. Arguably, the path curvature is the most fundamental of these, although each has interesting characteristics in its own right, and will be considered in more detail.

At any particular trim state, the sensitivity of any of the response measures to the steer angle is called a gain. The path curvature gain is

$$G_\rho \equiv \frac{\mathrm{d}\rho}{\mathrm{d}\delta}$$

the yaw velocity gain is

$$G_\mathrm{r} \equiv \frac{\mathrm{d}r}{\mathrm{d}\delta} = V\frac{\mathrm{d}\rho}{\mathrm{d}\delta} = VG_\rho$$

and the lateral acceleration gain is

$$G_\mathrm{A} \equiv \frac{\mathrm{d}A}{\mathrm{d}\delta} = V^2\frac{\mathrm{d}\rho}{\mathrm{d}\delta} = V^2G_\rho$$

By definition, these derivatives are taken at constant speed. The steer angle used may be the steering-wheel angle δ_s, the reference steer angle

$$\delta_\mathrm{ref} = \frac{\delta_\mathrm{s}}{\mathrm{G}}$$

or the mean of the front-wheel steer angles, and so should be specified. In principle there are also partial derivatives with respect to speed, keeping steer angle constant, which in practice do not find wide application.

Thinking in terms of force control rather than position control, the input parameter is the steering-wheel torque T_s instead of the steer angle. This leads to three new gains, less widely used than the previous ones; the force control lateral acceleration gain $G = \mathrm{d}A/\mathrm{d}T_\mathrm{s}$ is perhaps the most common. Sometimes the torque is instead expressed as a force at the steering-wheel rim, giving three more gains, e.g. $G = dA/dF$.

In practice various units are used for responses and gains. Because the basic S.I. unit of angle is the radian, with degrees as an allowable alternative, speed in S.I. is m s^{-1} or m/s, although km/hour and miles/hour (m.p.h.) sometimes appear, and acceleration in S.I. is m s^{-2} or m/s^2, but the technical unit of g (equal to 9.81 m/s^2) is convenient and often preferred. An interesting detail is that the units of path curvature are most clearly expressed as rad/m (radians per metre) but are sometimes simplified to m^{-1}, which has the same meaning. In summary the usual expressions are: speed in m s^{-1}, angle in radians or degrees, path curvature in rad/m, yaw velocity in rad/s, lateral acceleration in m/s^2 (m s^{-2}) or g, curvature gain in m^{-1}, yaw velocity gain in s^{-1}, and lateral acceleration gain in m s^{-2}/rad, m s^{-2}/deg or g/deg.

6.3 Basic handling curve

Figure 6.3.1 shows a minimal prototype vehicle without suspension. The properties of the pair of real wheels on each axle are compressed into single wheels on the centre-line, thereby simplifying the geometry. This is therefore sometimes known as the bicycle model, because it contains only two wheels, but this does not imply that the handling is in any way related to a real bicycle, which banks in corners. In its simplest form there is no suspension, although the front wheel can be steered. Aerodynamic forces are neglected.

Figure 6.3.1 actually shows two non-equilibrium conditions of the vehicle. In the first, there is a steer angle but no body yaw, the result being a lateral force and moment. In the second there is a yaw angle

Figure 6.3.1. Bicycle model vehicle: (a) with steer angle only, (b) with yaw angle only.

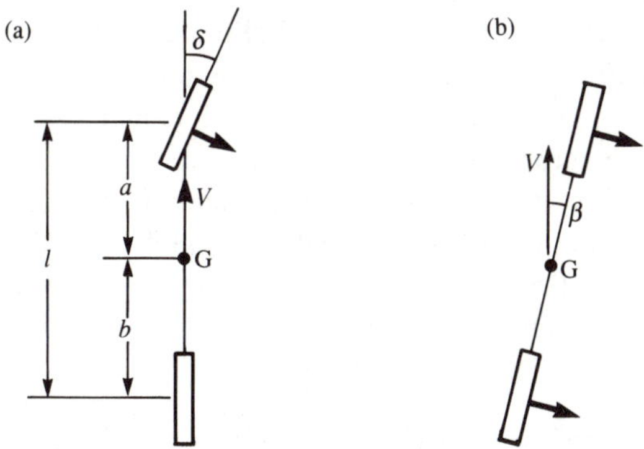

with no steer angle. Again there is a lateral force and a moment. In the latter case the total force acts close to the centre of mass, usually a little behind it. Thus the moment from yaw can be balanced by a steering moment from a relatively small steer angle. An alternative way to view this is that the yaw angle gives a rear-wheel slip angle, and the steering can be used to give a suitable front slip angle to give zero yaw moment and hence constant yaw speed, i.e. steady state. Thus handling depends on a balance between front and rear.

There are essentially two different approaches to representing the steady-state handling of a vehicle. The first, both the older and more common, we may call the kinematic method because, although forces are considered in the analysis, the result is to relate the response in kinematic terms to the control inputs. Thus path curvature, yaw speed and lateral acceleration are produced as functions of steer angle and forward speed. The second approach, at the same time more difficult and more comprehensive, is called the moment method, and basically considers the total vehicle lateral force and yaw moment as functions of yaw and steer. This gives a handling 'portrait' of the vehicle on a plot of moment versus lateral force, with separate plots required for different speeds when speed effects become significant, for example through aerodynamics.

The kinematic method is essentially a steady-state method, whereas the moment method can be used in the prediction of transient behaviour. Most of this chapter will be devoted to the kinematic method, although the moment method is introduced later in Section 6.22.

For the model of Figure 6.3.1, with no suspension, no aerodynamic forces and no tractive forces, and small steer angles, the handling characteristics may be deduced in a particularly simple way, from the front and rear tyre sideforce coefficient versus slip characteristics.

The axle characteristics are plotted together in Figure 6.3.2. A given specified lateral acceleration A requires the same F_y/m factor front and rear; the graph then shows the required slip angles α_f and α_r front and rear. For this simple model the vehicle yaw angle equals α_r at the rear axle. If the vehicle is describing a curved path then the yaw angle varies along the vehicle length, being α_r at the rear axle, but $\alpha_r - b/R$ at the centre of mass, the point at which it is normally defined. Because of the difference in slip angles required front and rear, the driver must provide an extra steer angle equal to the difference in slip angles. This is the understeer angle:

$$\delta_U = \alpha_f - \alpha_r$$

The total steer angle required in a corner follows from Figure 6.3.3.

Figure 6.3.2. Vehicle and tyre characteristics: (a) tyre characteristics, (b) resulting understeer angle.

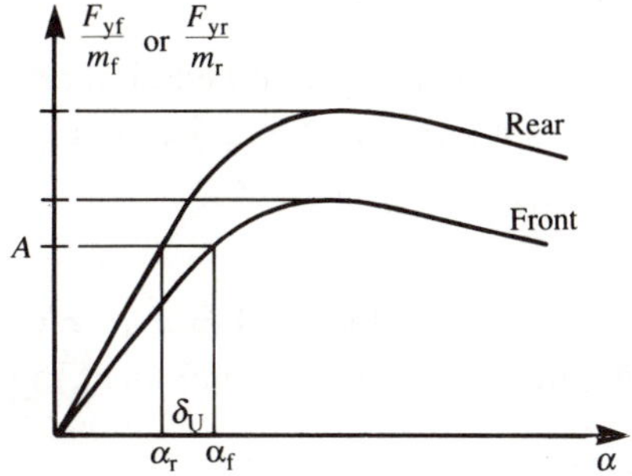

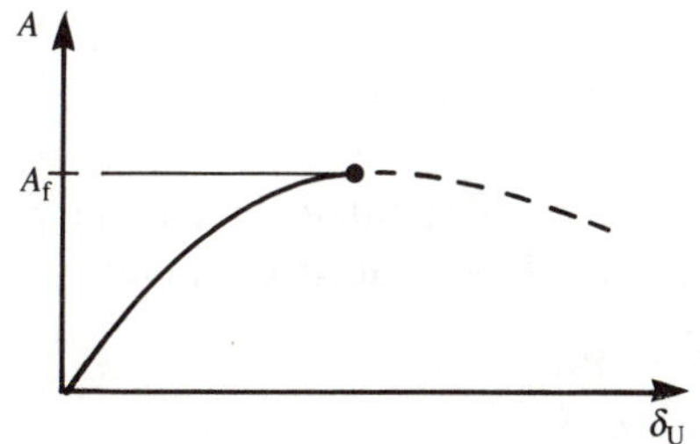

Figure 6.3.3. Steer angles and slip angles.

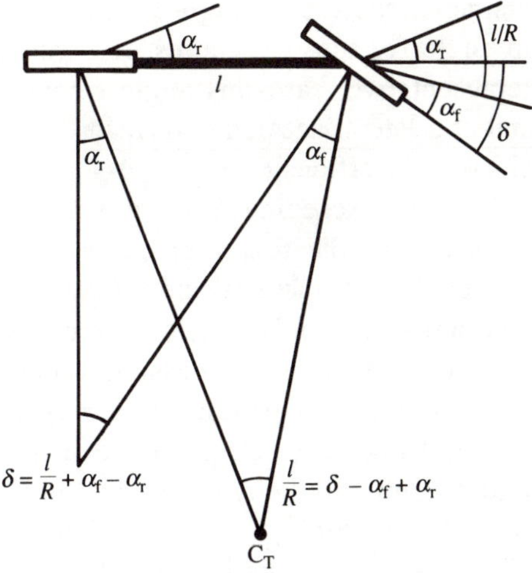

Here we see that the total steer angle δ required is a combination of a kinematic angle δ_K because of the path curvature, plus a dynamic angle δ_U because of the slip angles:

$$\delta = \delta_K + \delta_U$$

At very low speed, there is negligible lateral acceleration, and hence no slip angles, so there is a steer angle requirement of

$$\delta_K = \frac{l}{R}$$

Sometimes $\arctan(l/R)$ is used instead of l/R, but this is not really an improvement. The total steer angle requirement in general is therefore

$$\delta = \frac{l}{R} + (\alpha_f - \alpha_r)$$

Although it is convenient, and normal practice, to call $\alpha_f - \alpha_r$ the understeer angle, we shall see later that the S.A.E. definition of the understeer condition is not the same as having a positive understeer angle.

In some cases it is convenient to break the kinematic steer angle down into front and rear kinematic steer angles:

$$\delta_K = \delta_{Kf} + \delta_{Kr}$$
$$= \frac{a}{R} + \frac{b}{R}$$

Independently of the particular radius, from Figure 6.3.2(a) we can deduce a curve of lateral acceleration against understeer angle; this is the understeer characteristic, as in Figure 6.3.2(b). It is simply the difference of slip angle for the two curves of Figure 6.3.2(a). In this particular, and representative, case this angle smoothly increases and becomes horizontal at a lateral acceleration A_f for which the front tyres have their peak sideforce coefficient.

For lateral accelerations exceeding A_f there is no solution for a front slip angle, so the shape of the rear tyre curve at values above the corresponding force coefficient can have no influence on the understeer characteristic. In principle, if the front tyre characteristic is peaked, as shown in Figure 6.3.2(a), then a second possible value of α_f occurs for a given lateral acceleration, corresponding to a very large steer angle, giving the dashed line of Figure 6.3.2(b). If the rear tyre characteristic is highly peaked, falling again into the relevant region below A_f then there are further possible solutions for δ_U giving additional branches to the A versus δ_U curve. These are interesting in principle, although not of great practical importance. It is also possible for the rear tyre characteristic to lie beneath the front tyre one, but this is generally undesirable because the understeer angle then reduces with lateral acceleration, the condition known as oversteer.

Figure 6.3.2(b) showed lateral acceleration as the ordinate, and steer angle as the abscissa. This is convenient when deducing the curve from the tyre characteristics. However, when the steer characteristic is found experimentally it is usually the acceleration that is the abscissa, Figure 6.3.4, with the total steer angle as ordinate. Which way round to plot these curves is a matter of personal taste; more usually acceleration is taken as the abscissa.

Figure 6.3.4. Handling (understeer angle) characteristics.

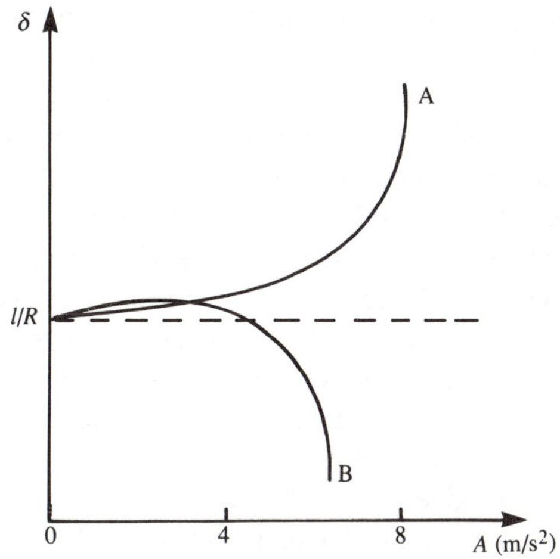

In the most common form of handling test, the vehicle is driven at various speeds around a fixed radius, with typical results as in Figure 6.3.4. Curve A is representative of a modern car. Curve B was common on older designs of cars; it would nowadays usually be considered unacceptable for a car, but it is still fairly typical of trucks. Curve A is fairly linear up to about 3 m/s^2 (0.3g) lateral acceleration, perhaps as high as 0.45g for a modern high-performance car, and curve B up to about 1 m/s^2. Thus the linear region covers most normal driving conditions, which we may call the primary handling regime. For a typical car, from 3 to 6 m/s^2 constitutes a non-linear region, the secondary handling regime, where effects such as lateral load transfer become significant. Beyond 6 m/s^2 is the final handling regime where tyre frictional effects are paramount. This is the province of desperate accident avoidance for the road vehicle, although entered as a matter of routine in deliberate testing and by competition vehicles. For commercial vehicles all these regions are at lower bands of lateral acceleration. These handling regimes will be considered separately in various sections of this chapter, beginning with the linear region, which is the simplest theoretically. From a design point of view, division of the handling curve into three regimes is useful, because the shape in each regime depends upon different variables.

6.4 Cornering forces

Figure 6.4.1 shows the simple 'bicycle' model in cornering without tractive forces, rotating about the turn centre C_T. Because of the constant angular speed, the slip angles must give no moment about the centre of mass G. Rolling resistance and aerodynamic forces are neglected. The tyre forces are perpendicular to the wheels, and intersect at the force centre C_F. Hence the line of action of the resultant force on the vehicle is through G towards the point C_F. This force has the desired centripetal component towards C_T, but also a tyre drag component so the tangential speed is not constant. Sometimes it is acceptable to neglect the tractive forces in this way.

Figure 6.4.2 shows a rear-drive vehicle with the required tractive force component F_{xr} to give steady tangential speed. To fit this requirement, the intersection of the total front and rear tyre forces F_f and F_r must intersect on the line GC_T. Figure 6.4.3 shows the equivalent force condition for a front-drive vehicle. In this case the intersection of the tyre forces occurs to the outside of the vehicle, tractive force F_{xf} having been added to give the required angle to F_f.

The vector force diagrams for rear- and front-drive vehicles appear in Figures 6.4.4 and 6.4.5. These are simply the forces taken from Figures 6.4.2 and 6.4.3. The general form of the diagram as shown corresponds to slip angles larger than a/R and b/R. In the case of small lateral acceleration at a small radius this will not be true, and the form of the vector diagram is significantly changed. The diagram shown assumes a positive understeer angle, i.e. $\alpha_f > \alpha_r$. For the rear-drive vehicle, the steer angle is the angle between F_{yr} and F_f, and for front-drive it is the angle between F_{yf} and F_r. In each case

$$\delta = \left(\alpha_f + \frac{a}{R}\right) - \left(\alpha_r - \frac{b}{R}\right) = \frac{l}{R} + (\alpha_f - \alpha_r)$$

$$= \frac{l}{R} + \delta_U$$

The diagrams of Figures 6.4.4 and 6.4.5 are easily extended to include additional effects such as aerodynamic forces and rolling resistance.

Figure 6.4.1. Plan view free-body diagram for no drive.

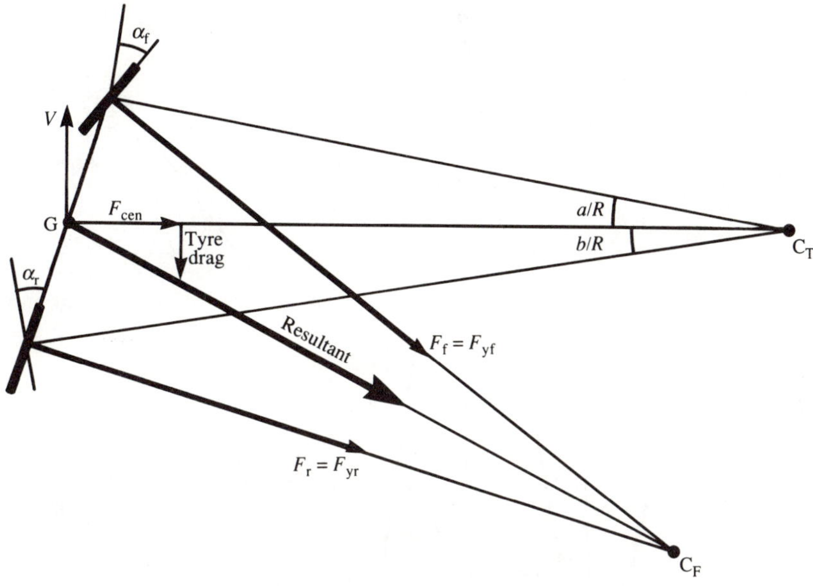

Figure 6.4.2. Plan view free-body diagram for rear drive.

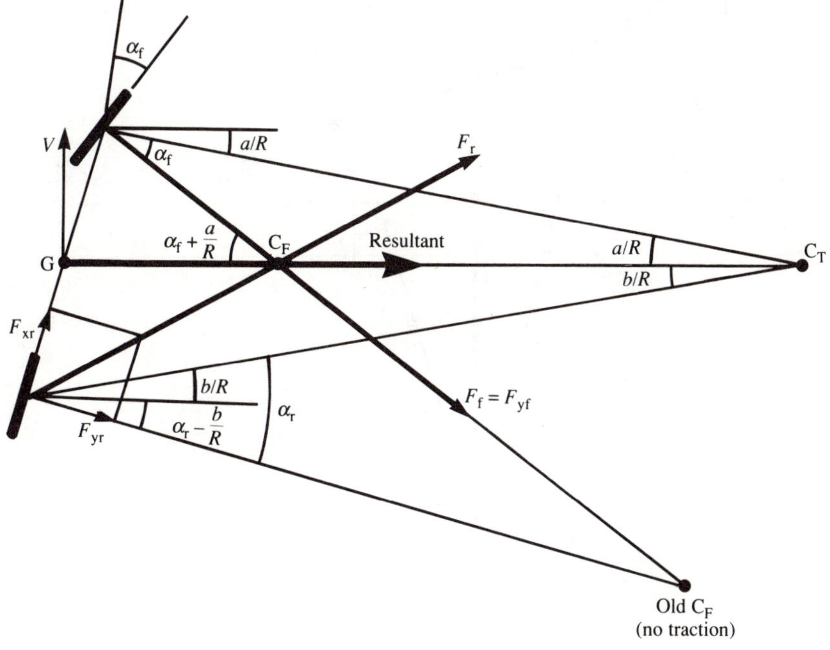

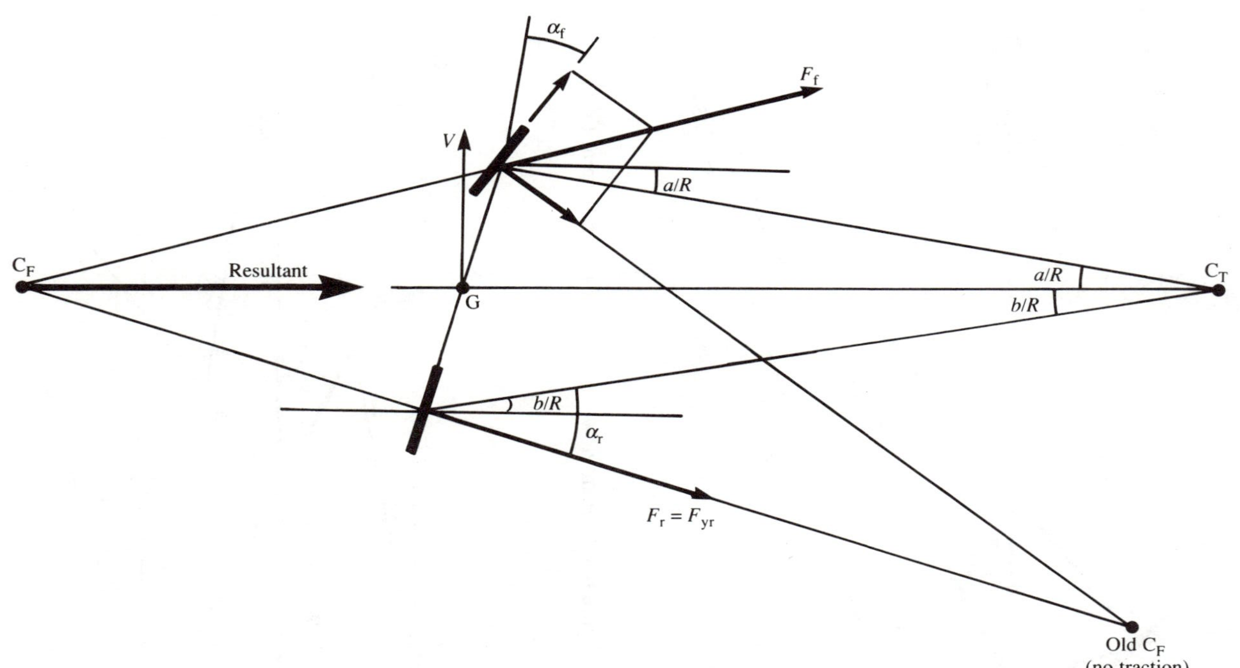

Figure 6.4.3. Plan view free-body diagram for front drive.

Figure 6.4.4. Force polygon for rear drive.

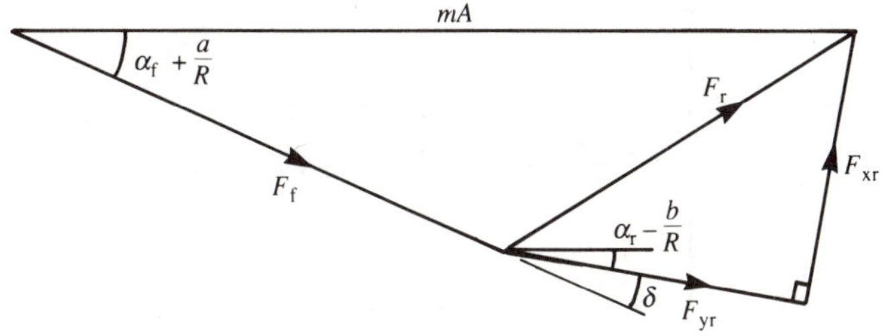

Figure 6.4.5. Force polygon for front drive.

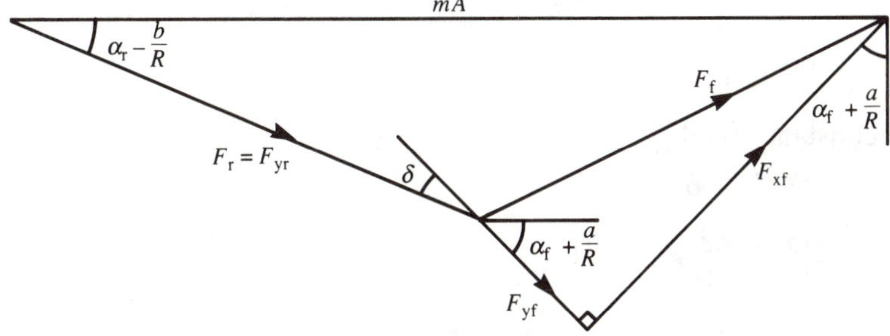

6.5 Linear theory

In the primary handling regime, up to about 3 m/s^2 for a typical car, the steer angle changes linearly with lateral acceleration at a given radius, Figure 6.5.1, and hence

$$\delta_U = kA$$

$$\delta = \frac{l}{R} + kA$$

where k is called the understeer gradient (or understeer coefficient). Physically, k is simply the gradient of the graph of δ against A, Figure 6.5.1. It is expressed in rad/m s^{-2} or deg/m s^{-2} in the S.I. system, although deg/g is convenient and often preferred. A typical value is 5 mrad/m s^{-2} or 3 deg/g. A value of 1.0 deg/g is 0.00178 rad/m s^{-2} or 1.78 mrad/m s^{-2}.

Formally we can define k by differentiating the above equations, giving

$$k \equiv \frac{d\delta_U}{dA} = \left(\frac{d\delta}{dA}\right)_R$$

Figure 6.5.1. Linear vehicle handling ($\delta = l/R + kA$).

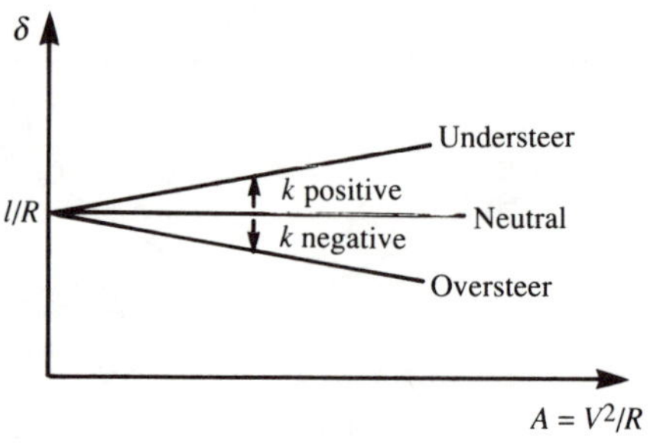

where the subscript R means at constant radius. If R is not constant then the kinematic steer angle also changes with A:

$$\delta_K = \frac{l}{R} = \frac{lA}{V^2}$$

At constant speed:

$$\delta = \delta_K + \delta_U$$

$$\frac{d\delta}{dA} = \frac{d\delta_K}{dA} + \frac{d\delta_U}{dA}$$

$$= \frac{l}{V^2} + k$$

Where the understeer gradient k is positive the total steer angle increases with lateral acceleration, i.e. with increase of speed for a constant-radius test, and the vehicle is described as understeering. For negative k the vehicle is described as oversteering. Zero k is neutral steer. For a real non-linear case a vehicle may have different characteristics at different lateral accelerations.

Substituting $A = V^2/R$ into the above equation for δ,

$$\delta = \frac{l}{R} + k\left(\frac{V^2}{R}\right)$$

$$= \rho\,(l + kV^2)$$

where $\rho = 1/R$ is the path curvature. Thus the path curvature response gain for the linear case is

$$G_\rho \equiv \frac{d\rho}{d\delta} = \frac{\rho}{\delta} = \frac{1}{l + kV^2}$$

$$= \frac{1}{l}\left(\frac{1}{1 + (k/l)V^2}\right)$$

Thus the response gain depends on the speed in a way controlled by k/l, i.e. the understeer gradient per unit of wheelbase. It is convenient to define a characteristic speed

$$V_{ch} = \sqrt{\frac{l}{k}}$$

so that

$$\frac{\rho}{\delta} = \frac{1}{l}\left(\frac{1}{1+(V/V_{ch})^2}\right)$$

If k is negative then V_{ch} would be imaginary. The $\sqrt{(-1)}$ is then discarded and the result is called the critical speed:

$$V_{cr} = \sqrt{\frac{-l}{k}}$$

In the following, k will be assumed positive, as is usual in practice; the results are easily adapted to negative k if required.

The velocity-dependent part of the curvature response is called the response factor:

$$f_R = \frac{1}{1+(k/l)V^2} = \frac{1}{1+(V/V_{ch})^2}$$

Its reciprocal is called the understeer factor U (to be distinguished from the understeer gradient k)

$$U \equiv \frac{1}{f_R} = 1 + \left(\frac{k}{l}\right)V^2 = 1 + \left(\frac{V}{V_{ch}}\right)^2$$

These are sometimes given other names, e.g. f_R may be called the yaw rate factor, but this is misleading because really it is equally applicable to path curvature, yaw rate and lateral acceleration:

$$G_\rho \equiv \frac{\mathrm{d}\rho}{\mathrm{d}\delta} = \frac{\rho}{\delta} = \frac{f_R}{l}$$

$$G_r \equiv V\frac{\mathrm{d}\rho}{\mathrm{d}\delta} = \frac{\rho V}{\delta} = \frac{f_R V}{l} = VG_\rho$$

$$G_A \equiv V^2\frac{\mathrm{d}\rho}{\mathrm{d}\delta} = \frac{\rho V^2}{\delta} = \frac{f_R V^2}{l} = V^2 G_\rho$$

Furthermore, in each case if we divide the response by the response of a neutral-steer vehicle, in every case the ratio is the response factor f_R. Physically, a large response factor means a large lateral response to a given steer input compared with a neutral vehicle for the prevailing conditions of the test.

Figure 6.5.2. Linear vehicle response factor.

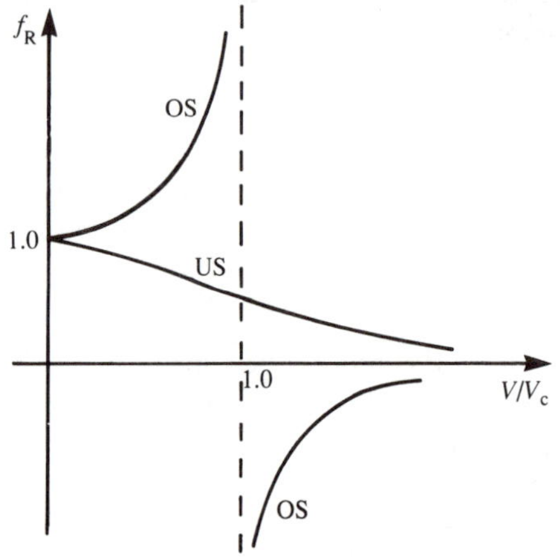

Figure 6.5.2 shows how the response factor varies with V/V_c where V_c represents V_{ch} or V_{cr} as appropriate. For understeer the response factor diminishes with speed, being 0.5 at the characteristic speed. For oversteer, the response increases with speed, going to infinity at the critical speed. Thus as the vehicle approaches the critical speed the steering becomes increasingly sensitive and the directional stability diminishes. Beyond the critical speed, f_R is negative, meaning that for a given response the steering angle must be in the opposite direction to that normally expected. In addition, the vehicle is statically unstable, so with a fixed steering position the vehicle will not maintain a steady path, but will spin out. Such a vehicle can be driven, because the system can be stable when the driver response is included, i.e. open-loop instability but closed-loop stability, but naturally this is difficult and tiring for the driver, so the normal car is designed to have an understeer characteristic.

As a matter of interest, for the simple linear vehicle that we are considering, if the understeer factor $U = 1/f_R$ is plotted against $(V/V_c)^2$ then a simple straight-line plot results.

A typical value of k is 5 mrad/m s^{-2} (2.8 deg/g), which on a wheelbase of 3 m gives a characteristic speed of 25 m/s. The actual value of the understeer gradient depends upon the tyre cornering stiffnesses, the centre-of-mass position, and the suspension geometry and deflection characteristics. This is discussed in detail in Section 6.10. For the linear case the steer angle is

$$\delta = \frac{l}{R} + kA = \frac{l}{R} + (\alpha_f - \alpha_r)$$

and therefore

$$k = \frac{\alpha_f - \alpha_r}{A}$$

The front and rear forces are

$$m_f A = 2\alpha_f C_{\alpha f}$$

$$m_r A = 2\alpha_r C_{\alpha r}$$

where m_f and m_r are the front and rear end-masses:

$$m_f = \frac{bm}{l}$$

$$m_r = \frac{am}{l}$$

Hence $k = \dfrac{m_f}{2C_{\alpha f}} - \dfrac{m_r}{2C_{\alpha r}}$

$$= \frac{m}{2l}\left(\frac{b}{C_{\alpha f}} - \frac{a}{C_{\alpha r}}\right)$$

So for the linear suspensionless vehicle, the understeer gradient can be simply related to the tyre characteristics, the mass and the position of G on the wheelbase.

For real vehicles, beyond the linear handling regime the understeer gradient depends upon the lateral acceleration, as discussed in Sections 6.11 and 6.12.

6.6 Vehicle cornering stiffnesses

For the complete simple model vehicle, the moment and sideforce are related to the vehicle yaw and steer, through coefficients that depend upon the tyre coefficients and the position of the centre of mass on the wheelbase.

Figure 6.6.1 shows the minimal bicycle model vehicle with a steer angle giving a front slip angle (not in steady-state). There are two wheels on the front axle, so the front-axle cornering stiffness is $2C_{\alpha f}$ where $C_{\alpha f}$ is the value for one front wheel. For small angles the resulting vehicle lateral force is

$$F_y = 2\alpha_f C_{\alpha f}$$

As usual, the front axle is distance a in front of the centre of mass G, the rear axle is b behind G, and the wheelbase is

$$l = a + b$$

Figure 6.6.1. Bicycle model with steer angle.

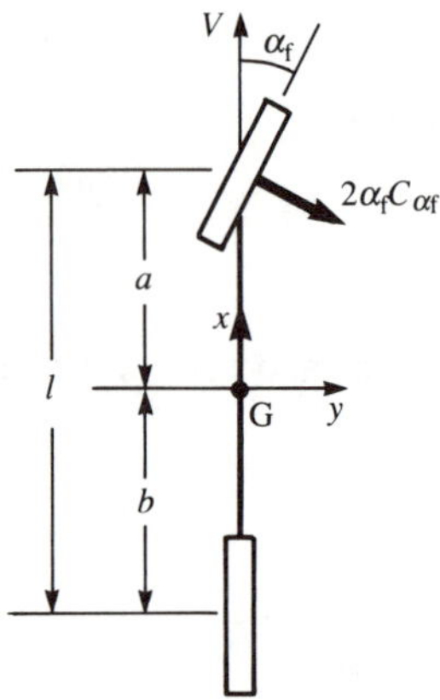

The moment about G of the front-wheel force, clockwise positive, is

$$M_z = 2a\,\alpha_f C_{\alpha f}$$

Figure 6.6.2 shows the vehicle with no steer angle, but at an angular yaw position β. Front and rear slip angles are equal to the yaw. In this case the total lateral force is

$$F_y = 2\beta\, C_{\alpha f} + 2\beta\, C_{\alpha r}$$

so there is a yaw sideforce stiffness

$$\frac{dF}{d\beta} = 2C_{\alpha f} + 2C_{\alpha r}$$

The moment about the centre of mass, clockwise positive, is

$$M_z = 2a\beta\, C_{\alpha f} - 2b\beta\, C_{\alpha r}$$

Thus there is a yaw stiffness

$$\frac{dM}{d\beta} = 2aC_{\alpha f} - 2bC_{\alpha r}$$

Directional yaw static stability requires that this be a restoring moment, i.e. negative (counterclockwise) M for a clockwise β as shown. Thus $dM/d\beta$ must be negative. This requires

$$|bC_{\alpha r}| > |aC_{\alpha f}|$$

Physically, this simply means that for static stability the restoring moment from the rear axle must exceed the disturbing moment from the front axle. If we divide the moment by the total force we get the

Figure 6.6.2. Bicycle model with yaw angle.

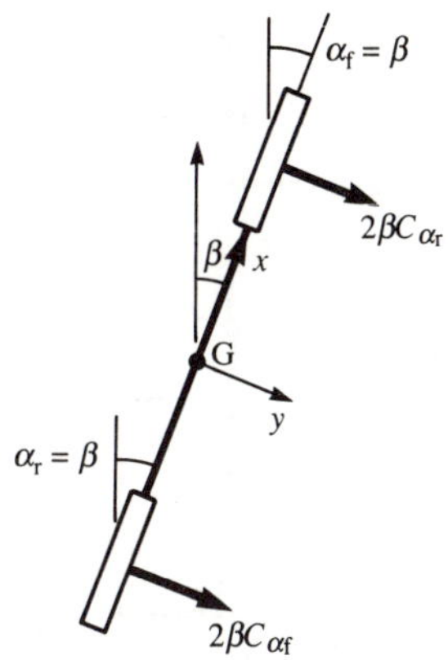

moment arm of the total force. The moment arm x_s as a distance behind G is $-M_z/F_y$:

$$x_s = -\frac{aC_{\alpha f} - bC_{\alpha r}}{C_{\alpha f} + C_{\alpha r}}$$

This is called the static margin, and is alternatively expressed as a fraction of the wheelbase, x_s/l. For static stability the static margin must be positive, i.e. the force must act behind G. The point at which the total tyre force acts is called the neutral steer point, because a small force applied there will cause the vehicle to drift sideways with no change of yaw angle. Actually, because of roll effects there is strictly a neutral steer line in side view, angled back at typically 15° because a force applied high up gives effects such as roll steer, and therefore must generally be more rearwards to give a neutral response. In non-linear analysis the static margin and neutral steer point are defined in terms of increments of force, and the result depends on the trim state, i.e. the reference conditions, but in linear theory the static margin is a constant.

Figure 6.6.3 shows the vehicle with a yaw speed $\dot{\beta} \equiv r$. This gives the velocity diagrams shown, with front and rear slip angles ar/V and br/V (for small r). There is a net lateral force

$$F_y = -\left(\frac{2ar}{V}\right)C_{\alpha f} + \left(\frac{2br}{V}\right)C_{\alpha r}$$

$$= -\frac{2r}{V}(aC_{\alpha f} - bC_{\alpha r})$$

$$\frac{dF_y}{dr} = -\frac{2}{V}(aC_{\alpha f} - bC_{\alpha r})$$

Figure 6.6.3. Bicycle model with yaw speed.

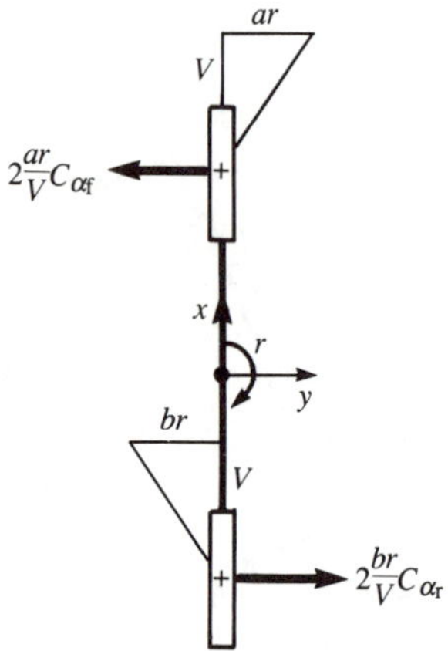

There is a clockwise moment about G of

$$M_z = -\left(\frac{2a^2r}{V}\right)C_{\alpha f} - \left(\frac{2b^2r}{V}\right)C_{\alpha r}$$

$$= -\frac{2r}{V}(a^2C_{\alpha f} + b^2C_{\alpha r})$$

$$\frac{\mathrm{d}M}{\mathrm{d}r} = -\frac{2}{V}(a^2C_{\alpha f} + b^2C_{\alpha r})$$

This moment opposes the yawing velocity and therefore has a damping effect on yawing oscillations, and is known as the yaw damping coefficient. It is inversely proportional to speed, but is always opposed to the yaw velocity, i.e. it is always a positive damping.

In the above expressions, the vehicle characteristics are expressed in terms of the dimensions a and b and the tyre characteristics $C_{\alpha f}$ and $C_{\alpha r}$. The expressions involve the zeroth, first and second moments of the tyre coefficients about the centre of mass, giving three vehicle cornering stiffnesses:

$$C_0 = 2C_{\alpha f} + 2C_{\alpha r}$$

$$C_1 = 2aC_{\alpha f} - 2bC_{\alpha r}$$

$$C_2 = 2a^2C_{\alpha f} + 2b^2C_{\alpha r}$$

Representative values for a medium car would be about 140 kN/rad, –14 kNm/rad and 200 kNm^2/rad respectively.

These complete vehicle cornering stiffnesses are sometimes useful to simplify equations, and occur quite frequently in dynamic analysis. For example, the static margin is $x_s = -C_1/C_0$, and the yaw damping is $-\mathrm{d}M/\mathrm{d}r$ which equals $+C_2/V$. The understeer gradient becomes

$$k = -\frac{mC_1}{4lC_{\alpha f}C_{\alpha r}}$$

Although developed here in terms of a linear vehicle, the coefficients are easily extended to a non-linear vehicle, by considering small, and hence linear, deflections from a trim state.

A serious practical limitation in the use of these coefficients is that although they illuminate understanding of the simple vehicle considered, their neglect of the suspension is not generally acceptable. This can be incorporated by using the cornering compliance concept, Section 6.10.

In stability derivative notation, the sideforce is represented by Y and the moment by N. Derivatives are indicated by a subscript, e.g.:

$$Y_\beta \equiv \frac{\mathrm{d}Y}{\mathrm{d}\beta}$$

In terms of the vehicle coefficients:

$$Y_\beta = C_0$$

$$Y_r = -\frac{C_1}{V}$$

$$N_\beta = C_1$$

$$N_r = -\frac{C_2}{V}$$

Other vehicle cornering stiffnesses that sometimes occur are:

$$C_3 = C_0 a - C_1$$

$$C_4 = C_0 C_2 - C_1^2 \approx C_0 C_2 \approx m C_1 V_{ch}^2$$

$$C_5 = I C_0 + m C_2$$

6.7 Non-linear trim state

The preceding theory showed that the linear regime of handling can be represented by simple equations. This is not true of the more general non-linear regime. The trim state, i.e. an output such as path curvature in terms of an input such as steer angle, is not really amenable to an analytical approach, and for detailed representations computer simulations are invariably used. However, small perturbations about a given trim state, for the study of stability of a given cornering condition, can still be treated as linear. Also, the graphical representation of the relationships between control and response, i.e. of the open-loop transfer function, provides an excellent basis for physical understanding of the trim state and is hardly any more difficult than for the linear case.

There are essentially two controls, the lateral (steering wheel) and the longitudinal (accelerator/brake), and for this reason any output variable will generally be a function of two variables, and hence will be represented by a surface on a three-dimensional graph. For example, vehicle yaw angle as output might be represented as a function of steer angle and accelerator position. However, as described earlier, the forward speed is generally preferred to accelerator position as one control variable. The steering control may be represented by the mean steer angle of the front road wheels δ, the steering wheel angle δ_s, the reference steer angle δ_s/G, the steering-wheel torque T_s, or even the tangential steering force at the wheel rim F_s. Possible steady-state output parameters include anything that varies with trim state, such as path curvature, yaw speed, lateral acceleration, roll angle, attitude (yaw) angle, even pitch angle, and so on. As a result of this large number of variables there are many possible inter-relationships that could be examined, and many different ways of representing a given result: for example at a given speed the path curvature, the yaw speed and the lateral acceleration all contain the same information. However, the principal features are the basic lateral motion parameters, namely attitude angle, roll angle, path curvature, speed and steer angle; and the most basic representation is with attitude angle, roll angle and path curvature each as a function of the speed and steer angle.

In practice a different representation is usually adopted, for good reasons. It is more convenient to think of lateral acceleration as a specified variable, and to have a graph showing the required steer angle, and the consequent attitude angle and roll angle, as in Figure 1.13.1. Another advantage of this representation is that although the driver seeks to control the path radius directly through the steer angle, he does not seek to control, say, the roll angle, i.e. in this sense the roll angle is incidental. However the relationship between roll angle and lateral acceleration is of importance because it may affect the required steer angle through roll steer, and may be used by the driver as feedback. Speed is retained as the second control variable, and this is satisfactory because in general the effect of speed is secondary in the mid-speed range. However, if it is low then for a given acceleration there may be small-radius effects, and if it is high then aerodynamic and tyre speed effects come into play.

For these reasons then, the usual representation of the complete non-linear characteristics of a vehicle is a series of three-dimensional graphs, one for each variable, e.g. steer angle, shown as a function of lateral acceleration and speed. Of course each of these can be shown as a conventional graph against lateral acceleration for a series of specific speed values, or in the most simplified case for a representative single speed. In the usual constant-radius test, the speed varies with lateral acceleration, but for the usual radius (30 to 40 m) the speed effects are small, so this is not a problem.

Figure 6.7.1 shows a possible surface for understeer angle against V and A. We can note that in the V–A plane there is a minimum V for a given A corresponding to a minimum turning radius. There is also a maximum steady-state A for any given V, reducing with V, e.g. because of aerodynamic lift and because of speed sensitivity of tyre friction, and at high speed because of engine power limits. For any particular V there is a curve of δ_U against A, slightly different for each V, the range of V resulting in a complete surface for the understeer angle. For any particular vehicle, similar surfaces exist for the other outputs, e.g. attitude angle. It is only when the speed effect is small that this relatively complex representation can acceptably be reduced to a single curve of δ_U against A.

Figure 6.7.1. Understeer angle versus V and A.

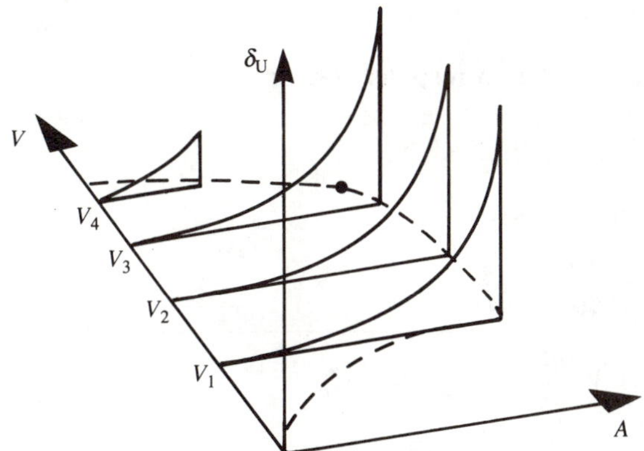

6.8 Non-linear theory

For small perturbations from a trim state, a vehicle will behave in a linear manner, although with characteristic coefficients quite different from the straight-line case. Hence the linear theory developed in earlier sections is valid in principle, provided that the perturbations are now considered from the particular steady-state condition, rather than from the straight running condition.

The definitions of the various control gains still stand:

$$G_\rho \equiv \frac{\mathrm{d}\rho}{\mathrm{d}\delta}$$

$$G_r \equiv \frac{\mathrm{d}r}{\mathrm{d}\delta} = V\,\frac{\mathrm{d}\rho}{\mathrm{d}\delta}$$

$$G_A \equiv \frac{\mathrm{d}A}{\mathrm{d}\delta} = V^2\,\frac{\mathrm{d}\rho}{\mathrm{d}\delta}$$

although it is no longer allowable to substitute the linear relationship $\mathrm{d}\rho/\mathrm{d}\delta = \rho/\delta$. In linear theory we found

$$\delta = \frac{l}{R} + kA \quad \text{(linear)}$$

where $k = \mathrm{d}\delta_U/\mathrm{d}A$ is the understeer gradient; in linear theory this is a constant. For the non-linear case, the concept of the understeer gradient is retained, but it now has a value dependent upon the lateral acceleration. The notion of understeer is considered in more detail in the next section.

For the point-specific understeer gradient it is still allowable to calculate a characteristic or critical speed

$$V_{ch} = \sqrt{\frac{l}{k}}$$

We can also calculate a response factor

$$f_R = \frac{1}{1 + (k/l)V^2}$$

and the various gains

$$G_\rho \equiv \frac{\mathrm{d}\rho}{\mathrm{d}\delta} = \frac{f_R}{l}$$

$$G_r \equiv V\frac{\mathrm{d}\rho}{\mathrm{d}\delta} = \frac{f_R V}{l} = VG_\rho$$

$$G_A \equiv V^2\frac{\mathrm{d}\rho}{\mathrm{d}\delta} = \frac{f_R V^2}{l} = V^2 G_\rho$$

as for the linear case.

The simple expression deduced for a suspensionless vehicle in Section 6.5 for k in terms of the tyre stiffness and centre-of-mass position,

$$k = \frac{m}{2l}\left(\frac{b}{C_{\alpha f}} - \frac{a}{C_{\alpha r}}\right)$$

is still appropriate, although now the tyre stiffness values must be the ones at the active slip angles, not at zero slip as for the straight-running case. In real cases, however, the suspension is not negligible; its effect on k is considered in Section 6.10.

The vehicle cornering stiffnesses C_0, C_1 and C_2 deduced in Section 6.6 remain useful provided that the tyre stiffnesses used are the ones effective at the active slip angles. Again, the suspension effects may be incorporated through the cornering compliance concept.

As an alternative to the understeer gradient, the static stability may still be represented by the static margin

$$x_s = -\frac{C_1}{C_0}$$

and the yaw damping coefficient is

$$-\frac{dM}{dr} = \frac{C_2}{V}$$

For a full representation these would be presented as surface plots against V and A. For the typical vehicle with a completely understeer characteristic, the front-axle sideforce limit is less than the rear, e.g. as in Figure 6.3.2(a); thus the front stiffness coefficient diminishes more rapidly than the rear, and the static margin increases, tending to infinity when the front-axle sideforce is at a maximum. On the other hand, because the tyre stiffnesses reduce at increasing slip, the yaw damping reduces with increased lateral acceleration.

6.9 Understeer and oversteer

Understeer and oversteer are widely used and often abused terms; their definition will be considered in more detail here.

Two tongue-in-cheek qualitative definitions are perhaps worth mentioning. First: 'Understeer is crashing nose first, oversteer is crashing tail first'. This is correct in the sense that if the ultimate grip is less at the front then a crash is liable to occur on full lock with the feeling of going straight on, whereas if the ultimate grip is less at the rear then the vehicle will spin. The second definition is 'Understeer is when the driver is frightened, oversteer is when the passenger is frightened'. This one is more obscure in interpretation. It does reflect the reality that an enthusiastic driver is not necessarily afraid of terminal oversteer, feeling that in such circumstances he can do something about it, i.e. apply opposite lock, whereas in terminal understeer the driver may tend to feel more helpless. This may be true in controlled testing; however in real crisis situations for a given maximum lateral acceleration it is nearly always preferable for the limit to be understeering so that steering control remains progressive rather than reversing into oversteer.

As already explained, the basic quantitative definition of understeer comes from the gradient of the steer angle versus lateral acceleration graph, giving the understeer coefficient or understeer gradient

$$k = \frac{d\delta_U}{dA} = \left(\frac{d\delta}{dA}\right)_R$$

According to the S.A.E. definition, wherever k is positive, the vehicle is described as understeering; if k is negative the vehicle is described as oversteering. If k is zero, the vehicle is neutral steer.

In Figure 6.9.1(a), we see that it is possible for there to be a negative understeer gradient at the same time as a positive understeer angle, or vice versa. It is the gradient that is considered to be decisive, rather than the angle. For $k = 0$, the vehicle is said to be neutral. Figure 6.9.1(a) illustrates these conditions for tests at constant radius.

Because speed is normally considered as one of the control variables, it may be better to plot the results instead at constant speed, Figure 6.9.1(b). However at constant speed the radius changes with lateral acceleration ($A = V^2/R$), so the kinematic steer angle changes too:

$$\delta_K = \frac{l}{R} = \frac{lA}{V^2}$$

$$\frac{d\delta_K}{dA} = \frac{l}{V^2}$$

Hence, for constant speed the kinematic steer angle increases at a rate l/V^2, which is known as the kinematic (S.A.E. Ackermann) steer-angle gradient. For neutral steer, i.e. zero rate of understeer angle change, the total steer angle must also change at this rate; for understeer it will increase more quickly, and for oversteer more slowly. This is illustrated in Figure 6.9.1(b). Note that in this case, in contrast to the constant-radius case of Figure 6.9.1(a), the total steer angle starts at zero, because zero acceleration at a given speed corresponds to infinite radius. For a wheelbase of 3.0 m and speed 20 m/s, the kinematic steer-angle gradient is 7.5 mrad/m s^{-2}, 0.43 deg/m s^{-2} or 4.2 deg/g. In Figures 6.9.1(a) and (b), if we were to plot not the total steer angle but only the understeer angle, then in both cases neutral steer occurs at the horizontal point of the graph, and understeer is any point of positive gradient.

So far it has been assumed that the steer angle referred to is the mean steer angle of the front wheels. However, it is also possible to use the actual steering-wheel angle δ_s, or the reference steer angle δ_{ref}, as

Figure 6.9.1. Steer angle versus lateral acceleration

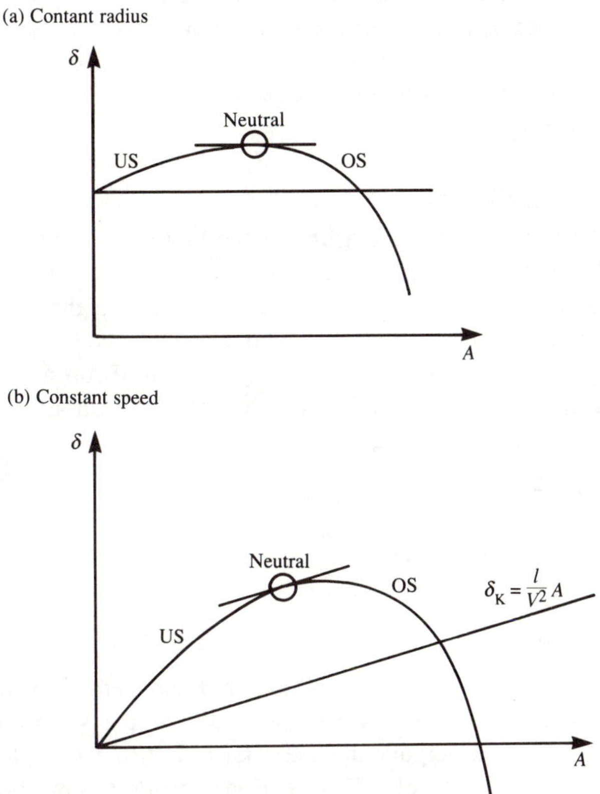

is done in the S.A.E. definitions. Because of steering compliance, there is usually greater understeer for δ_{ref} than for the front wheel angle. It is not fruitful to debate which of these three is the correct understeer value, they are simply parameters of different significance. Steering compliance is sometimes deliberately used to increase the driver's perceived understeer, especially for rear-engined vehicles, when G tends to be more rearwards than is desirable. In some cases the steering compliance seems to improve straight-line stability, in others, possibly those with more compliance, it seems to lead to instability and wandering. The situation is clouded by the fact that rear-heavy cars often have considerable roll understeer, which usually has a significant adverse effect on straight-line stability.

Considering the steer angle as a function of both lateral acceleration and speed, it is apparent that instead of keeping the speed constant and looking at understeer with respect to acceleration, we could instead keep the acceleration constant and consider understeer with respect to speed. In this case, if increased speed requires greater dynamic steer angle, this would be understeer with respect to speed. This sort of understeer has been considered in the research literature but has not found engineering application.

6.10 Primary handling

The primary handling regime is the first stage – that represented adequately by linear relationships. Representation of the vehicle steady-state behaviour in this regime essentially reduces to three numbers – the roll gradient, the attitude gradient and the understeer gradient, measuring the constant of proportionality as each of roll angle, attitude angle and understeer angle varies with lateral acceleration:

$$k_\phi = \frac{\mathrm{d}\phi}{\mathrm{d}A}$$

$$k_\beta = \frac{\mathrm{d}\beta}{\mathrm{d}A}$$

$$k = \frac{\mathrm{d}\delta_\mathrm{U}}{\mathrm{d}A}$$

Evaluation of the roll gradient k_ϕ follows from the methods of Chapter 5. Its value depends basically upon the height of the sprung centre of mass, the roll-centre heights and the total roll stiffness, with a further contribution from axle roll. This section considers how the attitude gradient k_β and the understeer gradient k derive from the vehicle and tyre design features. The attitude gradient is mainly dependent on the rear-tyre cornering stiffness coefficient, although other factors play a part. The understeer gradient depends upon many factors including the position of the centre of mass, front–rear difference of tyre cornering stiffness, roll steer, roll camber, compliance steer, aligning torque compliance steer, and the effect of aligning torque on the vehicle as a rigid body, with further minor effects from such factors as load transfer combined with rolling resistance, static toe and camber, and from axle roll camber. If the differential is not a simple open type, there can be power steer effects because of lateral differences in tractive force acting on the rigid-body vehicle, and from compliance power steer, whether front or rear. If the steering wheel angle or the reference steer angle is being considered, then the steering compliance is also of significance. Aerodynamics may play a part, but this is considered separately in Sections 6.14 and 6.15.

An extremely simple vehicle without suspension was analysed in Sections 6.4 and 6.5, for which the following simple results were found:

$$\delta = \frac{l}{R} + kA$$

$$= \frac{l}{R} + (\alpha_{\mathrm{f}} - \alpha_{\mathrm{r}})$$

$$k = \frac{m}{2l}\left(\frac{b}{C_{\alpha\mathrm{f}}} - \frac{a}{C_{\alpha\mathrm{r}}}\right)$$

A more complex vehicle will now be considered, but remaining in the linear range. Figure 6.10.1 illustrates a 'bicycle model' vehicle with simple suspension, incorporating suspension compliance steer and roll understeer effects of α_{sf} and α_{sr} With these suspension effects

$$\delta = \frac{l}{R} + (\alpha_{\mathrm{f}} - \alpha_{\mathrm{r}}) + (\alpha_{\mathrm{sf}} + \alpha_{\mathrm{sr}})$$

Positive α_{sf} and α_{sr} are defined to call for an increased δ, i.e. an understeer effect. Hence α_{sr} is positive when the rear wheel attempts to increase its cornering force, and α_{sf} is positive when the front wheel tries to decrease its cornering force.

The total steer gradient effect of roll steer and roll camber is

$$k_{\varepsilon} = \varepsilon_{\mathrm{S}} k_{\phi} + \frac{C_{\gamma}}{C_{\alpha}} \varepsilon_{\mathrm{C}} k_{\phi}$$

Compliance steer and camber effects accumulate in a similar but more

Figure 6.10.1. Bicycle model with suspension steer effects.

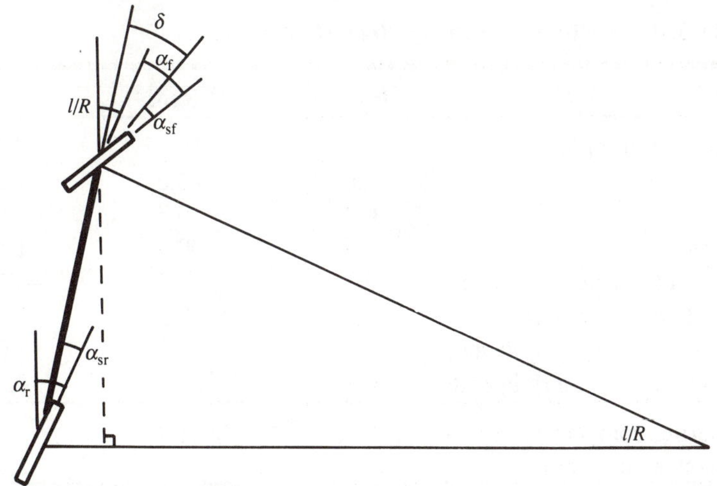

complex way, because there are more terms.

Table 6.10.1 lists representative values of the understeer effects for the principal factors for a free-differential vehicle. The sum of the column for each individual axle is called the cornering compliance. Note that understeer tendency effects are positive and increase the cornering compliance for the front, but in tables of this type are given a negative sign for the rear. The tyre cornering compliance is the angle for which the sideforce would give g (9.81 m/s^2) lateral acceleration, i.e. equal to the ground vertical force, if the tyre remained linear, i.e. with constant cornering stiffness. It is the reciprocal of the cornering stiffness coefficient expressed in g/deg.

For each row, subtracting the rear value from the front one gives the understeer contribution for that feature (column 4). Summing these contributions, or subtracting the total rear compliance from the front one, gives the understeer gradient for this model vehicle. For constant radius conditions, the attitude gradient k_β is equal to the rear cornering compliance. (For smaller radii, the difference of attitude angles at the rear axle and at G, specifically $\delta_{\mathrm{Kr}} = b/R$, is not small, but it is constant.) Thus a table of this kind provides the necessary information to define the parameters of the primary handling regime.

The factors contributing to k_β and k will now be considered in more detail. First consider the simplest possible bicycle model with equal tyre cornering stiffnesses C_α at each end, i.e. $2C_\alpha$ for an axle, and with G at the wheelbase mid-point, i.e. $a = b$.

$$\alpha_{\mathrm{f}} = \alpha_{\mathrm{r}} = \frac{mA}{4C_\alpha}$$

Table 6.10.1. *Example primary steer effects* (deg/g)

Factor	Front	Rear	F – R
Basic tyre compliance	6.0	6.0	0.0
G position	0.6	–0.6	1.2
ΔC_α	–0.3	0.3	–0.6
Roll steer	0.5	–0.5	1.0
Roll camber effect on steer	0.2	0.5	–0.3
Lateral force compliance	0.2	0.2	0.0
Aligning torque compliance	1.1	0.1	1.0
Aligning torque on rigid body	0.1	–0.1	0.2
Cornering compliance	8.4	5.9	
Understeer gradient			2.5

giving

$$k_\beta = \frac{m}{4C_\alpha}$$

$$k = 0$$

With m = 1200 kg and C_α = 500 N/deg, then k_β is 0.6 deg/m s^{-2} or 6 deg/g.

Now consider G to be forward of the mid-point, but with no change to C_α. Considering plan-view moments about the centre of the rear wheel:

$$lC_\alpha \alpha_f = bmA$$

$$\alpha_f = \frac{bmA}{lC_\alpha}$$

$$\alpha_r = \frac{amA}{lC_\alpha}$$

giving $k_\beta = \dfrac{am}{lC_\alpha}$

$$k = \frac{d}{dA}(\alpha_f - \alpha_r) = \frac{m(b-a)}{2lC_\alpha}$$

With $a/l = 0.45$, then k_β = 5.4 deg/g and k = 1.2 deg/g. Thus the forward G has reduced k_β and increased k, Table 6.10.1. This sensitivity to G position is particularly noticeable in practice when a car is loaded at the rear, giving an oversteer increment change. Surprisingly, in practice the opposite sometimes occurs.

In practice a forward G changes the tyre vertical reactions, and also usually demands unequal tyre pressures; these factors change C_α. This depends upon the particular tyres and loads; it may result in virtually no change, or there may be a change almost proportional to the normal force. For $a/l = 0.45$ this may as an example lead to $C_{\alpha f}$ = 525 N/deg and $C_{\alpha r}$ = 475 N/deg. Considering now a model with differing $C_{\alpha f}$ and $C_{\alpha r}$, plan-view moments about the rear wheel give

$$2lC_{\alpha f}\alpha_f = bmA$$

$$\alpha_f = \frac{bmA}{2lC_{\alpha f}}$$

$$\alpha_r = \frac{amA}{2lC_{\alpha r}}$$

This gives

$$k_\beta = \frac{am}{2lC_{\alpha r}}$$

$$k = \frac{m}{2l}\left(\frac{b}{C_{\alpha f}} - \frac{a}{C_{\alpha r}}\right)$$

This result was also derived in Section 6.5. With the values mentioned, $k_\beta = 5.7$ deg/g and $k = 0.60$ deg/g. Thus the changes to C_α compensate to some extent for the forward G position, in proportion to the C_α sensitivity to F_V.

The effect of roll steer depends upon the roll gradient k_ϕ and the front and rear roll understeer coefficients ε_{Sf} and ε_{Sr} The actual steer angle increments are

$$\Delta\alpha_f = \varepsilon_{Sf} k_\phi A$$

$$\Delta\alpha_r = \varepsilon_{Sr} k_\phi A$$

Hence the contributions to the attitude gradient and understeer gradient are

$$\Delta k_\beta = -\varepsilon_{Sr} k_\phi$$

$$\Delta k = (\varepsilon_{Sf} + \varepsilon_{Sr}) k_\phi$$

where ε_{Sf} and ε_{Sr} are defined as positive for an understeer contribution, which means the outer rear wheel tendency to point into the curve, and the outer front tending to point out. Note that positive rear roll understeer reduces the need for body attitude angle, in which case Δk_β is negative. A typical roll steer coefficient is 0.1 with k_ϕ up to 10 deg/g so the roll steer may reduce k_β by say 1 deg/g, and this is often done deliberately. Rear roll understeer is sometimes combined with front roll oversteer. However, significant amounts of roll steer may create unpleasant handling and so are generally avoided, especially as an oversteer effect.

Suspension roll also causes wheel camber relative to the road. The effect depends upon the roll gradient k_ϕ and the roll camber coefficients ε_{Cf} and ε_{Cr},

$$\Delta\gamma_f = \varepsilon_{Cf} k_\phi A$$

$$\Delta\gamma_r = \varepsilon_{Cr} k_\phi A$$

These camber changes give forces depending on the tyre camber stiffnesses C_γ and consequently there are changes of slip angles and steer angles to preserve the correct forces. This gives

$$\Delta k_\beta = -\left(\frac{\varepsilon_{Cr} C_{\gamma r}}{C_{\alpha r}}\right) k_\phi$$

$$\Delta k = \left(\frac{\varepsilon_{Cf} C_{\gamma f}}{C_{\alpha f}} + \frac{\varepsilon_{cr} C_{\gamma r}}{C_{\alpha r}}\right) k_\phi$$

where C_γ/C_α is about 0.04 for radials and 0.15 for bias ply. In the early days of independent suspension, the front roll camber coefficient was 1.0 and for the rear axle it was zero, and because bias tyres were used the camber stiffnesses were large. In this case Δk was typically over 1.0 deg/g. However, nowadays front independent suspensions have smaller roll camber coefficients, and radial tyres have smaller camber stiffnesses, so with a solid rear axle Δk is typically 0.2 deg/g. With some independent rear suspensions, such as the common trailing arm type, the roll camber coefficient is 1.0, and there may be a significant k_β effect. If front and rear suspensions are of the same type, this tends to compensate Δk and make it very small. With a small front roll camber and a large rear one, as on some small front-drive vehicles, the net effect on k may be −0.3 deg/g.

It is difficult to give a useful analytical account of lateral force compliance steer because this depends upon small details such as the stiffness of rubber bushes. Sometimes a compliance steer gradient k_η (deg/m s^{-2}) is used, giving angular deflections $k_{\eta\mathrm{f}}\,A$ and $k_{\eta\mathrm{r}}\,A$. Note that the compliance steer gradient k_η, relating the angle to the lateral acceleration, must be distinguished from the compliance steer coefficients η relating the angle change to a force or moment on the wheel. The latter causes the former. Compliance camber gradient $k_{\eta\mathrm{C}}$ is often neglected. At the front, because of steering compliance, lateral force compliance steer and aligning torque compliance steer are usually significant, and because of relative stiffness at the rear the effect on k_β is small, but on k it is substantial, e.g. 2 deg/g. Nowadays compliance steer effects are closely controlled because of the effect on dynamic handling.

The effect of steering-column compliance, i.e. considering the reference steer angle $\delta_{\mathrm{ref}} = \delta_\mathrm{s}/G$, can be illustrated by considering the kingpin axes to be rigidly mounted. The self-aligning torque is the front sideforce times the pneumatic trail, $mbAt/l$. For an overall steering gear ratio of G, and neglecting friction, the steering-column torque is $mbAt/Gl$; for a column torsional stiffness of k_c the column angular deflection is $mbAt/Glk_\mathrm{c}$ and the associated front-wheel angular deflection is

$$\Delta\alpha_\mathrm{f} = \frac{mbAt}{G^2lk_\mathrm{c}}$$

Hence

$$\Delta k = \frac{mbt}{G^2lk_\mathrm{c}} = \frac{m_\mathrm{f}t}{G^2k_\mathrm{c}}$$

With $G = 16$, $t = 30$ mm and $k_c = 0.8$ N m/deg, the result for Δk is 1.0 deg/g. Actually the effect will be somewhat greater because of the kingpin axis compliance.

The tyre lateral force acts much closer to the kingpin axis than to the line of action of the track-rod, so its effect on steer angles cannot be easily calculated. For a track-rod positioned rearward of the kingpin axes, the wheel angular deflection depends upon the difference between deflections at these two supports. If the track-rod is in front of the kingpin axes then the deflections both contribute to an angular change.

The aligning torque has some effect on the vehicle considered as a rigid body, i.e. separate from the compliance effects; this is because it effectively moves the tyre sideforce backwards, increasing the static margin. For a pneumatic trail t the total plan-view moment is mAt. This requires counteracting sideforces at the axles. By moments

$$F = \frac{mAt}{l}$$

so

$$\Delta\alpha_r = -\frac{mAt}{2lC_{\alpha r}}$$

$$\Delta\alpha_f = +\frac{mAt}{2lC_{\alpha f}}$$

Hence

$$\Delta k_\beta = -\frac{mt}{2lC_{\alpha r}}$$

$$\Delta k = \frac{mt}{2l}\left(\frac{1}{C_{\alpha f}} + \frac{1}{C_{\alpha r}}\right)$$

Using the earlier values, and a trail of 30 mm, gives 0.13 deg/g for $\Delta\beta$ and 0.24 deg/g for Δk.

Considering a four-wheel model instead of the bicycle model, cornering load transfer increases the outer side rolling resistance and reduces the inner one. We consider here only the rolling resistance, not the tyre cornering drag which gives a second-order effect (Section 6.11). The load transfer is $F_T = maH/T$, and with rolling resistance μ_R this gives a plan view moment about G of

$$M = T\mu_R F_T = \mu_R mAH$$

The required rear sideforce is reduced by F and the front increased by F where

$$F = \frac{\mu_R mAH}{l}$$

Hence

$$\Delta\alpha_f = \frac{\mu_R mAH}{2lC_{\alpha f}}$$

$$\Delta\alpha_r = -\frac{\mu_R mAH}{2lC_{\alpha r}}$$

$$\Delta k_\beta = -\frac{\mu_R mH}{2lC_{\alpha r}}$$

$$\Delta k = \frac{\mu_R mH}{2l}\left(\frac{1}{C_{\alpha f}} + \frac{1}{C_{\alpha r}}\right)$$

Using $\mu_r = 0.015$ and $h = 0.6$ m, then Δk_β is 0.04 deg/g and Δk is 0.07 deg/g, so this is a small effect. However, it may be significant for vehicles with high G and large rolling resistance, such as some military vehicles.

Load transfer will interact with the initial running toe and camber angles to give attitude coefficient and understeer coefficient increments. Denoting the straight running rear toe as α_{Rr}, physically a complete load transfer would be equivalent to α_{Rr} change of attitude. Denoting the front and rear distribution fractions of load transfer as d_f and d_r (i.e. $d_f + d_r = 1$), then

$$\Delta\beta = \alpha_{Rr}\left(\frac{mAH}{T}\right)\left(\frac{l}{mgb}\right)d_r$$

$$\Delta k_\beta = -\frac{Hld_r\alpha_{Rr}}{Tag}$$

where positive rear toe-in gives reduced attitude angle.

$$\Delta k = \frac{Hl}{Tg}\left(-\frac{d_f\alpha_{Rf}}{b} + \frac{d_r\alpha_{Rr}}{a}\right)$$

where front toe-in and rear toe-out give an oversteer effect. For example with $d_f = 0.8$, $\alpha_{Rf} = 1°$ and $\alpha_{Rr} = 0$, then Δk is -0.6 deg/g, so this is potentially a significant effect. In practice it is particularly noticeable that cars with light rear loading are sensitive to rear toe.

A similar analysis for camber gives the result

$$\Delta k_\beta = -\frac{Hld_r\gamma_{Rr}(C_\gamma/C_\alpha)}{Tag}$$

Positive rear camber gives slightly increased attitude; negative rear camber and positive front camber give increased understeer.

$$\Delta k = \left(\frac{Hl(C_\gamma/C_\alpha)}{Tg}\right)\left(\frac{d_f\gamma_{Rf}}{b} - \frac{d_r\gamma_{Rr}}{a}\right)$$

Because of the camber/slip stiffness ratio, about 0.04 for radial and 0.15 for bias ply, the camber effects are much less than the slip angle effects, and generally negligible.

Because of load transfer acting on the tyre stiffnesses there is an axle roll angle that results in camber forces. This happens at front and rear, and so has negligible effect on k. For a load transfer $F_T = mAH/T$ and tyre vertical stiffness K_t the axle roll is

$$\phi_A = \frac{2mAH}{K_t T^2}$$

To preserve the correct force

$$\Delta\alpha = \frac{\phi_A C_\gamma}{C_\alpha}$$

$$\Delta k_\beta = \frac{2mH(C_\gamma / C_\alpha)}{K_t T^2}$$

with a typical value of 0.08 deg/g for radial ply and 0.20 deg/g for bias ply, a fairly small effect.

Examining Table 6.10.1 again shows that k_β and k are the result of many factors. The attitude coefficient is dominated by the rear tyre compliance, and although the centre-of-mass position and consequent C_α change largely compensate each other, roll steer and roll camber can be substantial effects, and other minor effects may accumulate in a significant way. For the understeer coefficient, again the centre-of-mass position and C_α largely compensate, but roll steer, roll camber and steering compliance may be major effects. This table gives a fair indication of the main factors for a typical modern vehicle with open differential at low speed. Other differential types, and aerodynamics, may have significant effects; these are discussed in separate sections later.

6.11 Secondary handling

The secondary handling regime is the middle range, for which the non-linearities have become significant, but the vehicle has not yet reached the final, purely friction-limited stage. For a car this range typically covers lateral accelerations of $0.3g$ to $0.6g$. The behaviour is now much more complex than in the linear regime. In general it cannot be represented by simple equations; it can be modelled by detailed numerical simulations, and the result is still simply represented on a graph of roll angle, attitude angle and understeer angle against lateral acceleration, provided that a simple free differential is used and aerodynamic effects and tyre speed effects are neglected.

The roll angle may remain essentially linear. Where it is non-linear, this is usually because of progressive bump stops coming into operation. There may also be non-linear springs or bushes. Many factors may produce non-linearity in the attitude angle and steer angle – essentially any non-linearity in the variables contributing to the coefficients as listed in Table 6.10.1. The main factor is that the tyres now operate on the non-linear part of the sideforce versus slip curve. The second major factor is the lateral load transfer. In addition there are many minor factors. The requirement for tractive force affects the tyre sideforces. The linear roll steer, i.e. a constant roll steer coefficient, may now combine with a non-linear roll angle. On the other hand, second-order roll-steer effects (Section 5.12) tend to cancel out on the axle. Geometry may introduce roll camber non-linearities, or a constant roll camber coefficient may combine with a non-linear roll angle. Compliance effects from rubber bushes will diminish as the bushes stiffen rapidly. The aligning torque compliance will be affected by the diminishing pneumatic trail, i.e. the aligning torque itself becomes extremely non-linear, also diminishing the rigid-body effect. Longitudinal load transfer develops. Jacking causes the non-linear development of camber, which may be especially significant for a high roll centre, e.g. the swing axle.

The first major factor, the attitude angle increment from tyre non-linearity, follows immediately from a comparison of the rear tyre characteristic with the linear model. Representing the deviation from linearity as a power model,

$$\Delta\beta = \Delta\alpha_r = C\left(\frac{F_y}{F_V}\right)^P$$

Neglecting aerodynamic forces we can simply write

$$\Delta\beta = C\,A^P$$

where P has a typical value of 3.0, and C is 0.0065 deg/(m s^{-2})^3 or 6.5 deg/g^3, giving an increment of 1.4° at 0.6g.

The effect of tyre non-linearity on steer angle depends on the difference between the front and rear increments. Typically the front wheels will be more highly loaded and have a smaller maximum lateral force coefficient. This leads to a steer angle deviation from linearity of similar shape to that for individual tyres, but of smaller scale because it is a front-to-rear difference.

The effect of the second major factor, lateral load transfer, is to reduce the cornering force of a pair of wheels, as discussed in Section 2.16. The extra rear slip angle required depends upon the load transfer factor (or vertical force lateral transfer factor) e_V, i.e. the load transfer divided by the initial vertical force F_V for a single wheel

$$e_V = \frac{F_T}{F_{V0}}$$

and on the particular tyre characteristics. This load transfer factor e_V must be distinguished from the front and rear load transfer distribution fractions d_f and d_r. The load transfer will be approximately proportional to the lateral acceleration, giving

$$e_{Vf} = \frac{F_{Tf}}{m_f g/2} = \frac{mAHd_f/T}{mbg/2l} = \frac{2AHd_f l}{bgT}$$

$$e_{Vr} = \frac{2AHd_r l}{agT}$$

where d_f is the front load transfer fraction. The resulting effect on the attitude angle may be modelled by

$$\Delta\beta = C_1 e_{Vr}^P = C_2 A^P$$

where, again, P is typically 3, C_1 is typically 25°, and C_2 is 7 deg/g^3, giving an attitude increment of 1.5° at 0.6g, similar to the direct effect of the tyre F_v–α non-linearity.

The effect of lateral load transfer on steer angle depends on the front-to-rear difference of lateral load transfer. This effect is controllable, for example by roll-centre height adjustment, or more readily by redistribution of the anti-roll bar stiffnesses.

An expression for the effect of longitudinal load transfer on steer angle can be obtained by considering the effect of the new vertical forces on the tyre cornering stiffness. The resulting effect is small, for example 0.4 deg/g^3.

Thus the principal factors controlling the secondary understeer angle contribution are the centre-of-mass position and the distribution of lateral load transfer. In those cases where the tyre characteristics are fundamentally different front and rear, for example because of different tyre sizes as on racing cars and some sports cars, or multiple tyres on trucks, this will also be a major factor.

Tractive forces may be significant in this regime if distributed unevenly side-to-side, Section 6.13, but are not of major importance if a free differential is used. Considering a neutral vehicle, with linear tyre characteristics the tractive force must overcome a total drag of

$$D = \mu_R mg + \tfrac{1}{2}\rho V^2 A_D + \frac{m^2 A^2}{4C_\alpha}$$

At 33 m radius and 6 m/s^2 lateral acceleration the average slip angle is about 5°, and the tractive force is typically 600 N, which if demanded at one end of the vehicle is a tractive coefficient $F_x/F_V \approx 0.1$. The total drag will be a little larger at higher speed, on a greater radius, because of the greater aerodynamic drag. It is sometimes said that the total tyre drag is greater on a small radius because of the greater steer angle. This is false – it is only the slip angles that contribute to the tyre drag, not the kinematic steer angle. This may be seen by taking moments about the centre of path curvature, or by considering an energy analysis.

The tractive force, required to maintain speed, affects the tyre F_y–α relationship; in some cases there may be a small increase in F_y, but generally, and always for large traction, F_y is reduced and greater slip angle is required. Thus a front-drive vehicle has an increase of steer angle, and a rear drive one has a decrease; four-wheel-drive may remain broadly unaffected if the front-to-rear torque distribution is appropriate. For a traction coefficient of 0.1 on a normal road surface, the effect is in any case rather small, but may become significant at high speed or on low-friction surfaces.

6.12 Final handling

The final handling regime is the last 25% or so of the handling curve. For passenger cars this corresponds to lateral accelerations from 0.6g (6 m/s^2) up to the maximum value of about 0.8g. The ultimate value is generally less for commercial vehicles, e.g. 0.5g, because of poorer suspension systems, harder-wearing lower-friction tyre materials, higher tyre–road contact pressures, roll-over potential because of high centre of mass, and so on. It may exceed 1.0g for good sports cars with wide high-grip tyres, and may exceed even 3g for some racing cars with ultra high-grip tyres and aerodynamic downforce.

As far as the handling diagram is concerned, the basic question is whether the vehicle has final understeer or final oversteer, and this depends upon which end of the vehicle has the greater ultimate lateral acceleration capability. Neglecting the tractive force requirement, this simply depends upon the graph of lateral force over end-mass for the two axles. Referring back to Figure 6.3.2(a), which is representative for a typical modern car with final understeer, the maximum lateral acceleration is achieved at the peak of the curve for the front axle. At this acceleration, the rear axle achieves the required sideforce at a smaller slip angle. The attitude angle at the limit therefore follows from this rear slip angle plus the roll steer and compliance steer of the rear suspension. The steer-angle curve as a function of acceleration, curve A of Figure 6.3.4, increases extremely steeply near to the limit, and may curve back on itself if the tyre characteristic is peaked, or if account is taken of the tractive requirement for overcoming the tyre drag component at high slip angles.

For a final oversteer vehicle, the front suspension has the greater lateral acceleration capability, i.e. the greater sideforce to end-mass ratio, and the curve labels of Figure 6.3.2(a) are reversed. In the limit the attitude angle grows extremely rapidly, calling for large negative increments of steer angle; the understeer gradient is negative, and the understeer angle and possibly even the total steer angle will become negative, curve B of Figure 6.3.4. Because of this need for steering reversal, final oversteer is generally considered bad; final understeer has the advantage of more progressive and consistent control behaviour.

The vehicle final handling behaviour can be represented fairly well by just two variables, the front and rear maximum lateral accelerations, A_f and A_r. However, it is more useful to think in terms of the actual vehicle maximum lateral acceleration A_M (which is the lesser of A_f and A_r) and a final handling balance parameter. This is equivalent to using the attitude gradient and the understeer gradient, instead of the front and rear compliances, for primary handling. This final handling balance parameter should measure the commitment of the vehicle to terminal understeer or oversteer. It would be possible to use A_r minus A_f as a final understeer margin, but this can be non-dimensionalised to give a final understeer number:

$$N_U = \frac{A_r}{A_f} - 1$$

Hence a final understeer vehicle, for which $A_r > A_f$, has positive N_U; final oversteer has negative N_U, and in principle a final neutral vehicle has $N_U = 0$. The typical modern car has a final understeer number of about 0.2, usually somewhat less for rear-drive and more for front-drive. The design factors and operating conditions that contribute to the final understeer number will be considered in this and subsequent sections.

Table 6.12.1 indicates example main contributions to the final understeer number. The changes are considered relative to a vehicle with $a = b$ and equal tyres all round.

The principal factors affecting the ultimate ability of each axle are the centre-of-mass position, the tyre friction coefficient sensitivity to F_V, the lateral load transfer distribution, the longitudinal load transfer, and the tractive force requirements. Roll steer and compliance steer still have

Table 6.12.1. *Example final understeer number effects (front drive),* N_U

Factor	$\Delta A_f/A_0$	$\Delta A_r/A_0$	ΔN_U
G at 45% ($a = 0.45l$)	–0.100	+0.100	+0.200
Corresponding N_f, N_r (constant μ)	+0.100	–0.100	–0.200
Friction sensitivity with G posn ($p = -0.15$)	–0.015	+0.015	+0.030
Lateral load transfer 70%/30%	–0.059	–0.011	+0.048
Longitudinal load transfer	–0.029	+0.029	+0.058
Traction (tyre drag only, front drive)	–0.040	0	+0.040
Steer angle (6°)	–0.044	0	+0.044
Rolling resistance	–0.008	+0.006	+0.014
Totals (front drive)	–0.195	+0.039	+0.234

some effect on attitude angle and steer angle, but do not significantly influence the ultimate steady-state cornering ability, other than through the longitudinal load transfer. Camber angles have some effect on maximum cornering force of tyres, but this is not well documented.

A forward centre of mass increases the front end-mass, and increases the axle normal reaction, but it increases the maximum front sideforce capability in a smaller proportion because the maximum force coefficient reduces. Hence a forward G gives a final understeer tendency. Consider an initially neutral vehicle with G at 50% and four equal tyres, and representing this condition with the subscript 0, to give for example a maximum cornering coefficient μ_{C0}. Now move G forward by x. The front axle reaction is

$$N_f = mg\left(\frac{1}{2} + \frac{x}{l}\right)$$

$$\frac{N_f}{N_{f0}} = 1 + \frac{2x}{l}$$

The maximum sideforce coefficient can be represented as being proportional to the vertical force to the power p, where p is found experimentally to be typically –0.15 for passenger car tyres and about –0.23 for racing tyres. Representing the maximum lateral force coefficient by μ_C,

$$\frac{A_\mathrm{f}}{A_0} = \frac{\mu_\mathrm{Cf}}{\mu_\mathrm{Cf0}} = \left(\frac{N_\mathrm{f}}{N_\mathrm{f0}}\right)^p = \left(1 + \frac{2x}{l}\right)^p \approx 1 + \frac{2xp}{l}$$

$$\frac{\Delta A_\mathrm{f}}{A_0} = +\frac{2xp}{l}$$

$$\frac{\Delta A_\mathrm{r}}{A_0} = -\frac{2xp}{l}$$

For a most extreme likely mass distribution of 70/30, i.e. $x = 0.2l$, this gives $A_\mathrm{f}/A_0 = 0.94$, $A_\mathrm{r}/A_0 = 1.06$, and $A_\mathrm{r}/A_\mathrm{f} \approx 1.13$. Thus the maximum sideforce coefficient of the complete vehicle is 6% less than for an even balance, and there is a strong final understeer tendency with an N_U contribution of +0.13. To a good approximation these effects are proportional to x. For negative x (rearwards G), the maximum lateral acceleration also deteriorates, but is now limited at the rear so there is a final oversteer tendency. The maximum capability occurs at a central G; it is notable that this is not a curved optimum but a peaked one because of the sharp transition from front limitation to rear limitation. If the tyres are different front and rear, the optimum will no longer be a central G, but the sensitivity for deviation from the optimum position will be similar.

The influence of lateral load transfer may be examined by again beginning with an $a = b$ symmetrical vehicle, with lateral load transfer distribution factors d_f and d_r. The individual front-wheel standing reaction without load transfer is

$$N_\mathrm{f0} = \tfrac{1}{2} m_\mathrm{f} g = \frac{mbg}{2l}$$

The front load transfer is

$$F_\mathrm{Tf} = \frac{mAH}{T} d_\mathrm{f}$$

giving reaction forces, front inner and front outer, of

$$N_\mathrm{fi} = N_\mathrm{f0}\left(1 - \frac{2AHld_\mathrm{f}}{Tbg}\right)$$

$$N_{\text{fo}} = N_{\text{f0}}\left(1 + \frac{2AHld_{\text{f}}}{Tbg}\right)$$

Similar equations may be written for the rear. The axle maximum front cornering force will be

$$\begin{aligned} F_{\text{yf}} &= \mu_{\text{Cfi}} N_{\text{fi}} + \mu_{\text{Cfo}} N_{\text{fo}} \\ &= \mu_{\text{C0}} N_{\text{f0}} \left(\frac{N_{\text{fi}}}{N_{\text{f0}}}\right)^{1+p} + \mu_{\text{C0}} N_{\text{f0}} \left(\frac{N_{\text{fo}}}{N_{\text{f0}}}\right)^{1+p} \end{aligned}$$

The axle maximum lateral acceleration is then given by

$$A_{\text{f}} = \frac{F_{\text{yf}}}{m_{\text{f}}} = \frac{F_{\text{yf}}\, g}{2N_{\text{f0}}}$$

giving

$$\frac{A_{\text{f}}}{\mu_{\text{C0}}g} = \tfrac{1}{2}\left(\frac{N_{\text{fi}}}{N_{\text{f0}}}\right)^{1+p} + \tfrac{1}{2}\left(\frac{N_{\text{fo}}}{N_{\text{f0}}}\right)^{1+p}$$

Substituting, and using the approximate binomial expansion to the second order, now gives for the acceleration change, with $A_0 = \mu_{\text{C0}}\, g$,

$$\frac{\Delta A_{\text{f}}}{A_0} = 2p\,(1+p)\left(\frac{AHLd_{\text{f}}}{Tbg}\right)^2$$

For the rear, a similar expression may be obtained, but with d_{r} and a. With $a = l/2$

$$\frac{\Delta A_{\text{f}}}{A_0} = 8p\,(1+p)\left(\frac{AHd_{\text{f}}}{Tg}\right)^2$$

For representative car values this gives a deterioration of maximum lateral force coefficient of about 4% from lateral load transfer. The effect on the final understeer number is

$$\Delta N_{\text{U}} = 8p\,(1+p)\left(\frac{AH}{Tg}\right)^2 (d_{\text{f}} - d_{\text{r}})$$

For a representative car this is $+0.15(d_{\text{f}} - d_{\text{r}})$.

Thus putting 70% of the load transfer distribution at the front will give a final understeer number contribution $\Delta N_{\text{U}} = 0.06$, another strong final understeer tendency. The load transfer distribution is easily amenable to tuning by the anti-roll bars, and provides an important design variable in this respect.

Perhaps surprisingly, longitudinal load transfer may have a significant effect on N_{U}. Because of the attitude angle, there is an acceleration component $A \sin\beta$ giving a longitudinal load transfer

$$F_{\text{TX}} = \frac{mAH \sin\beta}{l}$$

This influences the axle reactions according to

$$\frac{N_f}{N_{f0}} = 1 - \frac{AH\sin\beta}{bg}$$

$$\frac{N_r}{N_{r0}} = 1 + \frac{AH\sin\beta}{ag}$$

With friction sensitivity p,

$$\frac{\Delta A_f}{A_{f0}} \approx -\frac{(1+p)\,AH\sin\beta}{bg}$$

$$\frac{\Delta A_r}{A_{r0}} \approx +\frac{(1+p)\,AH\sin\beta}{ag}$$

$$\Delta N_U = (1+p)\left(\frac{AH\sin\beta}{g}\right)\left(\frac{1}{a}+\frac{1}{b}\right)$$

For an example vehicle with $\beta = 6°$ the fractional acceleration changes at each axle are of magnitude 0.029. The total ΔN_U of 0.058 is a large effect.

Traction requirements may have a significant influence on final handling. In this regime slip angles are relatively large, say 10°, and the associated tyre cornering drag is substantial:

$$D_T \approx mA\sin\alpha$$

and the traction coefficient for the driven tyres for two-wheel drive is

$$f_T = \frac{F_x}{F_V} \approx \frac{D}{mg/2} \approx \frac{2A\sin\alpha}{g} \approx 0.28$$

At the basic test radius of 33 m, 0.8g corresponds to 16 m/s, so the aerodynamic drag is relatively small – about 110 N against 2 kN tyre cornering drag. Even at 100 m radius and 0.8g the aerodynamic drag is only about 350 N, although the force diagram, e.g. as Figure 6.4.2, can easily be amended to include aerodynamic drag and sideforce, and rolling resistance, if required. The effect of the traction coefficient depends in detail on the tyre (Section 2.18), but as a simple approximation the friction ellipse model may be used. A traction coefficient of 0.28 will reduce $F_{y\,max}$ by about 4%. This is therefore a significant effect, encouraging final understeer for front-drive and final oversteer for rear-drive.

Because of the steer angle there is a negative contribution to attitude angle at G, giving a forward load transfer. There is also a reduction of the moment arm of the front sideforce, with a further reduction because of front lateral load transfer. This is discussed in detail in Section 6.16. The outcome for notional infinite radius and 6° final understeer is a ΔN_U

of 0.037. At 33 m radius this increases by 0.010.

As a result of lateral load transfer there is a rolling resistance understeer moment about G, neglecting the steer angle, of

$$M = \mu_R F_T T = \mu_R mAH$$

This requires counteracting front and rear sideforces, and hence

$$\frac{\Delta A_f}{A_0} = \frac{\mu_R mH}{m_f L} = \frac{\mu_R H}{b}$$

$$\frac{\Delta A_r}{A_0} = \frac{\mu_R H}{a}$$

$$\Delta N_U = \mu_R H \left(\frac{1}{a} + \frac{1}{b}\right)$$

$$\approx \frac{4\mu_R H}{l}$$

with a typical value of 0.014.

The maximum lateral force of a tyre depends upon the camber angle because this controls the presentation of the footprint to the road. Broadly, the best camber angle will be the one that gives the greatest contact area. However this is not simply zero camber, because the tyre distorts with side load. Depending on the nature of the distortions, which depends amongst other things on the rim width, the optimum camber may be a few degrees positive or negative. For conventional passenger car tyres the peak is not highly sensitive, although camber angle settings can affect the maximum lateral acceleration by 5% or more, and for the extremely wide tyres used in racing camber may be critical. A rigid axle introduces small camber angles equal to the axle roll angle, which is 1 to 2 deg/g. Independent suspensions vary considerably. Simple trailing arms give camber equal to roll. Parallel wishbones may give camber equal to roll, or be arranged at the other extreme to give opposite camber. Thus the suspension camber may be significant in influencing the final cornering limit and the final understeer number, but it is very difficult to make any useful general statements or to quantify the effects in the absence of specific tyre and suspension data.

On the basis of force diagrams such as Figure 6.4.2, arguments are sometimes presented to compare the merits of front and rear drive in terms of maximum cornering acceleration. Such analyses do not seem to be borne out by experiment, the most favourable end for traction really depending on many details. Also, from a practical point of view a change of drive implies changes of mass positions, so it is not necessarily useful to compare drives with other things being equal.

Rear-drive (described by one rally driver as "handling a man can understand") has some controllability advantages at the limit. Four-wheel-drive has had only limited success in racing because of controllability problems ("like trying to write your signature with someone jogging your elbow"). The advantage of four-wheel-drive lies in its superior traction on poor surfaces, and hence its success in rallying.

6.13 Differentials

In this section, the effect of differentials on the vehicle as a rigid body will be treated. For a non-free differential, there may be different tractive forces on the two sides of the vehicle, giving an understeer or oversteer moment.

With zero net traction, a solid differential will have an inner wheel slip $T/2R$ and outer wheel slip $-T/2R$ (Section 1.9). For a tyre longitudinal stiffness C_x (N/unit slip) the forward tractive forces are, for small slip, i.e. $R > 15$ m,

$$F_{xi} = \frac{C_x T}{2R}$$

$$F_{xo} = -\frac{C_x T}{2R}$$

giving an understeer moment

$$M = \tfrac{1}{2}(F_{xi} - F_{xo})T$$

$$= \frac{C_x T^2}{2R}$$

This moment requires counteracting front and rear slip angle increments to restore steady-state. The required axle sideforce increments are M/l. For the linear case

$$\Delta\alpha_f = -\frac{M}{2C_{\alpha f}l}$$

$$\Delta\alpha_r = +\frac{M}{2C_{\alpha r}l}$$

$$\Delta\delta = \Delta\alpha_f - \Delta\alpha_r = -\frac{M}{2l}\left(\frac{1}{C_{\alpha f}} + \frac{1}{C_{\alpha r}}\right)$$

$$= \frac{C_x T^2}{4lR}\left(\frac{1}{C_{\alpha f}} + \frac{1}{C_{\alpha r}}\right)$$

which is typically about $33/R$ degrees. This is valid for the linear tyre and lateral acceleration region, i.e. $R > 15$ m and $A < 0.3g$. At the basic test radius of 33 m, $\Delta\delta$ is typically 1°, independent of lateral acceleration.

In the steady state it is necessary only to overcome the vehicle resistance, so the tractive force is generally small and the steer effect fairly small. In transient conditions, such as acceleration out of a corner, there may be strong steer effects from non-free differentials, which vary considerably in their torque distribution characteristics. A simple locking differential may apply all its torque to the inner wheel because of the small angular speed of this wheel. On the other hand an 'intelligent' differential seeking to avoid wheelspin might allocate torque in proportion to the tyre normal force. The handling contribution of these differentials will be analysed approximately here for a simple $a = b$ vehicle with equal tyre characteristics all round.

The total vehicle drag including tyres is D. The simple over-run ratcheting differential, assumed to be applying all its thrust on the inner wheel, gives a tractive understeer moment

$$M = \tfrac{1}{2}DT$$

and hence, arising from the tractive moment, there is a compensating sideforce at each axle of

$$F = \frac{M}{l} = \frac{DT}{2l}$$

$$\Delta\alpha_{\mathrm{r}} = -\frac{DT}{4lC_\alpha}$$

$$\Delta\alpha_{\mathrm{f}} = \frac{DT}{4lC_\alpha}$$

$$\Delta\delta = \frac{DT}{2lC_\alpha}$$

In the linear regime at moderate speeds the drag is mainly tyre drag:

$$D \approx D_{\mathrm{T}} \approx mA\sin\alpha$$

$$\approx \frac{m^2A^2}{4C_\alpha}$$

$$\Delta\delta = \frac{m^2TA^2}{8LC_\alpha^2}$$

$$\approx (93\times 10^{-6})A^2 \qquad (\mathrm{rad}/(\mathrm{m\ s^{-2}})^2)$$

$$= 0.6A^2 \qquad (\mathrm{deg}/g^2)$$

Hence this type of differential has no effect on the understeer coefficient k, and has a small secondary understeer effect in A^2.

Considering now final handling, the tractive moment influences the axle maximum lateral accelerations:

$$\Delta A_f = -\frac{DT/2l}{m/2} = -\frac{AT\sin\alpha}{l}$$

$$\Delta A_r = \frac{AT\sin\alpha}{l}$$

The change in final understeer number is

$$\begin{aligned}\Delta N_U &= \frac{\Delta A_r}{\mu_{C0}\,g} - \frac{\Delta A_f}{\mu_{C0}\,g}\\ &= \frac{2AT\sin\alpha}{l\mu_{C0}\,g}\\ &\approx \frac{2T\sin\alpha}{l}\end{aligned}$$

with a value of typically +0.16. Thus there is a very strong final understeer tendency when the traction is on the inner wheel only.

Considering now the 'intelligent' differential applying torque proportional to tyre vertical force, with load transfer factor d at the powered end, i.e. d_f or d_r as appropriate, the axle load transfer has inner and outer vertical forces such that

$$\frac{N_o}{N_i} = \frac{1+CA}{1-CA}$$

where $C \approx 4Hd/Tg$. The inner and outer tractive forces F_{xo} and F_{xi} give

$$F_{xo} + F_{xi} = D$$

and the differential property is

$$\frac{F_{xo}}{F_{xi}} = \frac{N_o}{N_i}$$

The result, with some manipulation, is a tractive understeer moment

$$\begin{aligned}M &= -\tfrac{1}{2}DTCA\\ &= -\frac{2mHd\sin\alpha}{g}A^2\end{aligned}$$

In the primary handling regime, $\alpha \approx mA/4C_\alpha$ and the effect on steer angle is

$$\begin{aligned}\Delta\delta &= \frac{M}{lC_\alpha}\\ &\approx -\frac{m^2HdA^3}{2gLC_\alpha^2}\end{aligned}$$

which is about $-0.50A^3$ (deg/g^3). Thus there is no primary effect, and a small secondary oversteer effect in A^3.

For final handling, we still have

$$M = -\left(\frac{2mHd\sin\alpha}{g}\right)A^2$$

$$\Delta A_\mathrm{f} = \frac{M/l}{m/2} = -\left(\frac{4Hd\sin\alpha}{gL}\right)A^2$$

and $\Delta A_\mathrm{r} = -\Delta A_\mathrm{f} \approx 0.45\ \mathrm{m/s^2}$.

$$\Delta N_\mathrm{U} = \frac{\Delta A_\mathrm{r}}{\mu_\mathrm{CO}\,g} - \frac{\Delta A_\mathrm{f}}{\mu_\mathrm{CO}\,g}$$

$$= -\frac{8Hd\sin\alpha\ A^2}{\mu_\mathrm{CO}\,g^2 l}$$

which is typically -0.12. Thus there is a very strong final oversteer contribution, in this case with the traction mainly on the outer wheel, not quite as great as for the over-run ratcheting differential understeer because of incomplete load transfer.

6.14 Aerodynamics – Primary

This section discusses the effect of aerodynamics on wind-free steady-state primary handling. As discussed in Chapter 3, the total aerodynamic force on the vehicle may be represented by lift, drag and sideforce coefficients, and pitching, yawing and rolling moment coefficients.

Table 6.14.1 gives example effects quantified for a representative car at 50 m/s, i.e. a top speed, where aerodynamic effects will be at their most extreme.

Table 6.14.1. *Example primary aerodynamic effects* (deg/g at 50 m/s)

Component			Δk	Δk_β
1	Lift – front	$C_{\mathrm{Lf}} = 0.1$	+0.22	0.00
2	Lift – rear	$C_{\mathrm{Lr}} = 0.1$	−0.14	+0.14
3	Drag (front drive)	$C_\mathrm{D} = 0.32$	+0.10	0
4	Sideforce	$C_\mathrm{S}' = 0.04$/deg	+0.06	−0.32
5	Yaw	$C_\mathrm{Y}' = 0.01$/deg	−0.32	−0.28
Totals			−0.08	−0.46

At the legal speed limits of most countries these effects will be smaller, but of course it is a manufacturer's responsibility to ensure that a vehicle behaves properly at all speeds of which it is capable. For an aerodynamically bad vehicle the net effect can be significant and bad. For a good one the effects can be controlled and balanced so that there is only a small and possibly even favourable effect. To summarise broadly, the sideforce contributes directly to lateral acceleration, the yaw moment is usually destabilising, the lift and pitch affect the tyre characteristics, drag affects the tyres through traction requirements, and the roll moment gives some load transfer and possible roll-steer effects, usually small.

To analyse in more detail, lift and pitching will be taken in terms of front and rear lift. The front lift is

$$L_{\mathrm{f}} = \tfrac{1}{2}\rho V^2 S C_{\mathrm{Lf}}$$

which changes the tyre normal reactions and hence the cornering stiffness according to

$$\frac{C_\alpha}{C_{\alpha 0}} = \left(\frac{m_{\mathrm{f}}g - L_{\mathrm{f}}}{m_{\mathrm{f}}g}\right)^f$$

$$= \left(1 - \frac{\tfrac{1}{2}\rho V^2 S C_{\mathrm{Lf}}\, l}{mbg}\right)^f$$

which by approximate expansion gives

$$\frac{C_\alpha}{C_{\alpha 0}} \approx 1 - \frac{f\rho V^2 S C_{\mathrm{Lf}} l}{2mbg}$$

where $m_{\mathrm{f}} = mb/l$ is the front end-mass and f is an empirical constant for the tyre. The change in front cornering compliance is

$$\frac{\Delta D_{\mathrm{f}}}{D_{\mathrm{f}}} \approx -\frac{\Delta C_\alpha}{C_{\alpha 0}} = \frac{f\rho V^2 S C_{\mathrm{Lf}} l}{2mbg}$$

with an equivalent expression for the rear. The proportional change in the attitude gradient is

$$\frac{\Delta k_\beta}{k_\beta} = -\frac{\Delta C_{\alpha\mathrm{r}}}{C_{\alpha 0}} = \frac{f\rho V^2 S C_{\mathrm{Lr}} l}{2mag}$$

With $f = 0.5$ this evaluates to $0.22C_{\mathrm{Lr}}$, which could be significant for a bad vehicle, but with $C_{\mathrm{Lr}} = 0.1$ and $k_\beta = 6$ deg/g the result is a negligible 0.14 deg/g change. The change of understeer gradient is

$$\Delta k = \Delta D_{\mathrm{f}} - \Delta D_{\mathrm{r}}$$

$$\approx \frac{\rho V^2 S l f}{2mg}\left(\frac{C_{\mathrm{Lf}} D_{\mathrm{f}}}{b} - \frac{C_{\mathrm{Lr}} D_{\mathrm{r}}}{a}\right)$$

This is approximately $8C_P$ (deg/g) which could be significant for a bad vehicle.

The above expressions are for a vehicle with the same tyre C_α to F_V sensitivity f value front and rear, but may easily be extended to different values. Typically with a forward G position f_f becomes smaller and f_r larger, making C_{Lr} more important. Even if f_f is close to zero, C_{Lf} may remain important because of attitude changes and consequent coefficient changes, especially to drag which is sensitive to pitch angle.

The aerodynamic drag is

$$D = \tfrac{1}{2}\rho V^2 S C_D$$

calling for a traction coefficient

$$f_T = \frac{D}{F_{V0}}$$

where F_{V0} is the vertical reaction of the driven wheels, e.g. for rear drive and no rear lift,

$$F_{V0} = m_r g = \frac{mag}{l}$$

To gain some insight into the likely scale of the consequences of the tractive force, we can use the cornering force ellipse model in the form

$$\frac{C_\alpha}{C_{\alpha 0}} = \sqrt{1 - \left(\frac{f_T}{\mu_L}\right)^2}$$

By approximate expansion

$$\frac{\Delta C_\alpha}{C_{\alpha 0}} = -\frac{1}{2}\left(\frac{f_T}{\mu_L}\right)^2$$

$$= -\frac{1}{8}\left(\frac{\rho V^2 S C_D l}{\mu_L mag}\right)^2$$

For rear drive at 50 m/s a representative result is –0.10 deg/g on both k_β and k. For front drive k_β is unchanged, and Δk is positive. These values are not large. On the other hand, on a poor friction surface the effects can be substantial.

The aerodynamic sideforce as a function of aerodynamic yaw angle β_{Ae} is

$$F_S = \tfrac{1}{2}\rho V^2 S C'_S \beta_{Ae}$$

where

$$C'_S = -\frac{dC_S}{d\beta_{Ae}} \approx 0.04/\text{deg}$$

This contributes a lateral acceleration

$$A'_S = \frac{F'_S}{m} = \frac{\rho V^2 S C'_S}{2m} \quad (\text{m s}^{-2}/\text{deg})$$

By definition of the standard aerodynamic axes, the sideforce acts at the centre of the wheelbase, i.e. there is an equal effective sideforce at each axle. The counteracting tyre slip angles are therefore

$$\Delta\alpha_f = \frac{F_S}{4C_{\alpha f}}$$

$$\Delta\alpha_r = \frac{F_S}{4C_{\alpha r}}$$

Hence

$$\Delta k_\beta = -\frac{\rho V^2 S C'_S k_\beta}{8C_{\alpha r}}$$

$$\Delta k = -\tfrac{1}{8}\rho V^2 S C'_S k_\beta \left(\frac{1}{C_{\alpha f}} - \frac{1}{C_{\alpha r}}\right)$$

Representative values are –0.32 deg/g for Δk_β, and +0.06 deg/g for Δk.

Turning now to the yaw moment, as a function of yaw angle this is

$$M = \tfrac{1}{2}\rho V^2 SLC'_Y \beta_{Ae}$$

where

$$C'_Y = \frac{dC_Y}{d\beta_{Ae}} \approx 0.01\ /\text{deg}$$

The moment requires counteracting forces at the axles, giving

$$\Delta\alpha_r = \frac{M/l}{2C_{\alpha r}}$$

$$\Delta\alpha_f = -\frac{M/l}{2C_{\alpha f}}$$

Using $\beta_{Ae} \approx \alpha_r$, this results in

$$\frac{\Delta k_\beta}{k_\beta} = -\frac{\Delta\alpha_r}{\alpha_r} \approx -\frac{\Delta\alpha_r}{\beta} = -\frac{M\,\Delta\alpha_r}{2lC_{\alpha r}\beta}$$

$$= -\frac{\rho V^2 S C'_Y}{4C_{\alpha r}}$$

$$\Delta\delta = \Delta\alpha_f - \Delta\alpha_r$$

$$= -\tfrac{1}{4}\rho V^2 S C'_Y \beta \left(\frac{1}{C_{\alpha f}} + \frac{1}{C_{\alpha r}}\right)$$

Using $\beta_{Ae} = \beta = k_\beta A$

$$\Delta k = \frac{\Delta\delta}{A} = -\tfrac{1}{4}\rho V^2 S C'_Y k_\beta \left(\frac{1}{C_{\alpha f}} + \frac{1}{C_{\alpha r}}\right)$$

For the example vehicle, Δk_β is –0.28 deg/g and Δk is –0.32 deg/g, which are small effects. However, for a vehicle with an unfavourably-shaped body and rearward G the effect on k may be much larger.

Detailed equations can also be worked out for the effect of the aerodynamic roll moment on lateral load transfer and roll steer; in general for a car such effects are small, although they may be noticeable for high-sided vehicles.

Reconsidering Table 6.14.1, even at 50 m/s the total effects on primary handling are not great, although the effect on understeer coefficient will depend very much upon the aerodynamic characteristics of the particular vehicle.

6.15 Aerodynamics – Final

This section discusses the influence of aerodynamics on final steady-state handling, i.e. on the maximum lateral acceleration A_M and on the final understeer number N_U. Table 6.15.1 indicates example effects quantified for a representative car at 50 m/s.

The effects will vary considerably between vehicles. Actually, because of tyre cornering drag the maximum steady-state speed at which the lateral acceleration is limited by handling rather than simply speed is usually only about 0.6 times the straight-line maximum speed, and hence typically 30 m/s, corresponding to maximum lateral acceleration at about 110 m radius. However a vehicle can be turned into a severe corner at its maximum straight-line speed, and although

Table 6.15.1. *Example final aerodynamic effects* (50 m/s)

Component		ΔA_M m/s²	ΔN_U —
1 Lift	$C_L = 0.1$	–0.160	–0.010
2 Pitch	$C_P = 0.1$	–0.320	+0.080
3 Drag	$C_D = 0.32$	0	–0.060
4 Sideforce	$C'_S = 0.04$/deg	+0.550	+0.030
5 Yaw	$C'_Y = 0.01$/deg	+0.270	–0.070
Totals		+0.340	–0.030

this strictly gives a transient condition it is convenient to adopt it here as the most severe aerodynamic condition. At 30 m/s or legal limit speeds the effects will be substantially smaller.

The basic effect of lift or downforce on the limiting cornering ability may be investigated as follows. For a lift coefficient C_L the total tyre vertical reaction is

$$N = mg - \tfrac{1}{2}\rho V^2 S C_L$$

For simplicity, considering first that the maximum cornering force coefficient μ_C is independent of N, the maximum speed is given by

$$mA = \frac{mV^2}{R} = \mu_C N$$

For a given radius the maximum speed is then

$$V = \sqrt{\frac{\mu_C mgR}{m + \frac{1}{2}\mu_C \rho RSC_L}}$$

Without aerodynamics, the maximum lateral acceleration is $\mu_C g$; by comparison

$$\frac{A}{\mu_C g} = \frac{1}{1 + \mu_C \rho RSC/2m}$$

Lift reduces the maximum lateral acceleration; but for downforce, which is negative C_L, the maximum A goes to infinity for a radius exceeding

$$R^* = -\frac{2m}{\mu_C \rho S C_L}$$

This radius was about 180 m for extreme ground-effect Formula 1 racing cars.

Figure 6.15.1 (curve for $p = 0$) shows how the maximum speed for a corner varies with R/R^* for this simple model, where V_0 is the maximum speed with no downforce.

The above analysis neglects the deterioration of maximum sideforce coefficient with increasing vertical force. In this case, with the usual friction power sensitivity p, and coefficient μ_{C0} when there is zero aerodynamic lift,

$$mA = \frac{mV^2}{R} = \mu_C N = \mu_{C0}\left(\frac{N}{mg}\right)^p N$$

$$\frac{V^2}{\mu_{C0} Rg} = \left(\frac{N}{mg}\right)^{1+p}$$

At a given speed, the minimum radius is increased by lift and reduced

Figure 6.15.1. Downforce effect on maximum cornering speed.

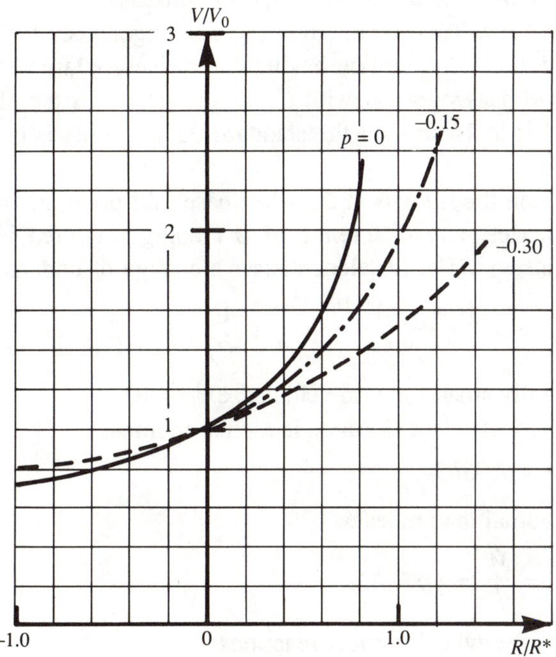

by downforce. For zero lift, $R_0 = V^2/\mu_{C0}g$, so

$$\frac{R}{R_0} = \left(1 - \frac{\rho V^2 S C_L}{2mg}\right)^{-(1+p)}$$

$$\frac{R}{R^*} = \frac{V^2}{\mu_{C0} R^* g}\left(1 - \frac{V^2}{\mu_{C0} R^* g}\right)^{-(1+p)}$$

Therefore

$$\frac{R}{R^*} = \left(\frac{V}{V^*}\right)^2 \left[1 - \left(\frac{V}{V^*}\right)^2\right]^{-(1+p)}$$

where

$$V^* = \sqrt{\mu_{C0} R^* g} = \sqrt{\frac{2mg}{\rho S C_L}}$$

which we may call the critical aerodynamic speed. Physically it is the (normally unachievable) speed when lift equals weight. For downforce it is imaginary, but introducing a negative sign it is the speed when downforce equals weight. For Formula 1 racing cars it is about 50 m/s. For a representative passenger car with $C_L = 0.2$ it is 240 m/s. These equations allow us to plot on Figure 6.15.1 how the maximum speed varies with R/R^* for a realistic friction sensitivity of −0.15, and an extreme of −0.30. The maximum speed no longer goes to infinity, the cornering friction deterioration having a surprisingly large effect. For a representative passenger car with $C_L = 0.1$, at 50 m/s the effect on A_M is −0.16 m/s^2; at 16 m/s on the standard 33 m radius it is only −0.02 m/s^2.

By definition the lift acts at the wheelbase mid-point, giving equal lift at the two axles, any difference of lift being reflected in the pitch coefficient instead. The lift alone causes a change of understeer number

$$\Delta N_U = -\frac{(1+p)\rho V^2 S l C_L}{4mg}\left(\frac{1}{a}+\frac{1}{b}\right)$$

This is generally small for road vehicles, e.g. −0.01.

For a pitch coefficient C_P, there is a pitch moment

$$M = \tfrac{1}{2}\rho V^2 S l C_P$$

and a longitudinal load transfer

$$F_{TX} = \frac{M}{l} = \tfrac{1}{2}\rho V^2 S C_P$$

This changes the wheel vertical reactions:

$$\frac{F_{Vf}}{F_{Vf0}} = 1 - \frac{\rho V^2 S l C_P}{2mbg}$$

The maximum lateral acceleration at the front is then given by

$$\frac{A_f}{A_{f0}} = \left[1 - \left(\frac{\rho V^2 S l C_P}{2mbg}\right)\right]^{1+p}$$

$$\approx 1 - \frac{(1+p)\rho V^2 S l C_P}{2mbg}$$

$$\frac{\Delta A_f}{A_{f0}} \approx -(1+p)\frac{\rho V^2 S l C_P}{2mbg}$$

$$\frac{\Delta A_r}{A_{r0}} \approx +(1+p)\frac{\rho V^2 S l C_P}{2mag}$$

For a final understeer vehicle it will be A_f that controls A_M, giving a reduction of 0.32 m/s^2 for the example case.

$$\Delta N_U = \frac{\Delta A_r}{A_{r0}} - \frac{\Delta A_f}{A_{f0}}$$

$$= \frac{(1+p)\rho V^2 S l C_P}{2mg}\left(\frac{1}{a}+\frac{1}{b}\right)$$

This is about 0.08 for $C_P = 0.1$, and hence aerodynamic pitch may exert a substantial effect on final balance.

The aerodynamic drag combines with the tyre drag to require traction. The total drag force is

$$\begin{aligned} D &= D_A + D_T + D_R \\ &\approx \tfrac{1}{2}\rho V^2 S C_D + mA \sin\alpha + \mu_R mg \end{aligned}$$

The tractive coefficient, assuming rear drive, is

$$f_T = \frac{D}{m_r g}$$

Using the friction ellipse model, the limiting lateral acceleration is given by

$$\frac{A_r}{A_{r0}} = \sqrt{1 - \left(\frac{f_T}{\mu_L}\right)^2}$$

Rather unrealistically, assuming that sufficient engine power is available to sustain steady-state limit cornering at 50 m/s, with a mean α of 10°, this evaluates to $\Delta A_r = 0.32$ m/s^2 for the tyres alone, and $\Delta A_r = 0.76$ m/s^2 for the total drag, i.e. an increment of 0.44 m/s^2 for the aerodynamic drag when added to the existing tyre drag. For rear drive and terminal understeer this will have little effect on A_M, and give N_U an increment of –0.06. For final understeer front drive, A_M will be reduced and the final understeer number increment will be positive, i.e. greater understeer.

Considering now the aerodynamic sideforce, and taking $\beta_{Ae} = \beta$, the sideforce is F_S giving an extra lateral acceleration

$$\Delta A_M = \frac{F_S}{m} = \frac{\rho V^2 S C'_S \beta}{2m}$$

For $C'_S = 0.04$/deg and a final β of 6° this gives $\Delta A_M \approx 0.55$ m/s^2 at 50 m/s, a useful increase of about 7%.

By definition of the standard aerodynamic axes at the wheelbase mid-point, the sideforces are equally distributed at the front and rear axles. Hence the change of N_U is given by

$$\Delta N_U = \frac{F_S\, l}{2mA_f}\left(\frac{1}{a} - \frac{1}{b}\right)$$

having a value of typically +0.03 at 50 m/s for a centre of mass 5% forward of the mid-point.

Considering now the yaw moment, for positive yaw coefficient this is a turn-in moment, i.e. an oversteer tendency. The moment changes the required steady-state force at each axle by M/l, giving a limit acceleration increment

$$\Delta A_f = +\frac{M}{lm_f} = \frac{\rho V^2 S l C'_Y \beta}{2mb}$$

$$\Delta A_r = -\frac{M}{lm_r} = -\frac{\rho V^2 S l C'_Y \beta}{2ma}$$

This effect assists the front, and so for terminal understeer there is an improvement in A_M of typically about 0.3 m/s^2 (3%) for $C'_Y = 0.01$/deg and $\beta = 6°$.

$$\Delta N_U = \frac{\Delta A_r - \Delta A_f}{A_f}$$

$$\approx -\frac{\rho V^2 S l C'_Y \beta}{2mA_M}\left(\frac{1}{a} + \frac{1}{b}\right)$$

This evaluates to about –0.07 for the example case, illustrating that there may be a substantial oversteer effect from yaw.

Considering broadly the above developments, and Table 6.15.1, it is apparent that at high speed the aerodynamic effects on maximum lateral acceleration may be substantial. The total effect on understeer number for this example is small, but the effect will vary considerably from one car to another because of different aerodynamic coefficients, especially for yaw, lift and pitch. At the basic test radius of 33 m the limiting speed is only about 16 m/s, giving effects only about one-tenth of those of Table 6.15.1; in this case the aerodynamic contribution would usually be negligible, although it is possible that a combination of a strong nose down pitch (negative C_P) and positive yaw could become relevant. In practice, the minimisation of drag does not severely interfere with the trimming of pitch or control of lift, although low-drag vehicles tend to have high yaw coefficients. The oversteer tendency of yaw can be offset by some pitch-up, as can rear-drive effects. On the whole, it is desirable for a vehicle to have greater primary and final understeer at high speed and so some net aerodynamic understeer is preferable. Within legal speed limits such effects remain fairly modest, and unsteady-state problems tend to be more pressing, e.g. wind gust response.

In some of the fields of competition vehicles, aerodynamic effects have become of major importance, because of bodywork optimisation and higher speeds, as discussed in Section 3.6. Downforce from wings and underbody flow is used to greatly enhance maximum lateral acceleration and traction. To maintain good handling, the position of the centre of pressure on the wheelbase, i.e. the front and rear distribution of downforce, must be carefully controlled by adjustment of underbody shape, or by trimming the wing incidences or wing flaps, or even by adjusting the wheelbase by moving the front wheels forwards or backwards. The optimum position for the total downforce is not simply at the centre of mass, but usually further forward; possibly this is because the smaller front tyres have a maximum sideforce coefficient that diminishes with load more rapidly than the larger rear tyres.

For wings, the induced drag increases as the square of the downforce coefficient, so the downforce/drag ratio of a wing deteriorates rapidly as greater downforce is demanded, although because of the irreducible zero-lift drag it may improve for the complete vehicle. Hence the downforce must be optimised, not simply maximised. This optimum depends upon the particular circuit, with high mean speeds or long straights tending to favour less downforce and less drag. Of course it is desirable to have minimum drag for given downforce, and venturi underbodies with sealed skirts are much superior to separate wings in this respect.

6.16 Path radius

Turning radius affects the understeer gradient at small radii. Because the front wheels are (usually) steered, considering a simple bicycle model the moment arm to G for the front tyre sideforce is reduced from a to

$$a' = a \cos \delta_K$$

neglecting the small understeer angle. Hence to preserve moment equilibrium about G, a larger front sideforce is needed than would otherwise be the case. Compared with the large radius turn, the increase of front slip angle is given by

$$\frac{\Delta\alpha_f}{\alpha_f} = 1 - \cos \delta_K$$

$$\approx \tfrac{1}{2}\left(\frac{l}{R}\right)^2$$

The effect on steer angle is

$$\Delta\delta = \Delta\alpha_f = \tfrac{1}{2}\left(\frac{l}{R}\right)^2 \alpha_f$$

$$\Delta k = \frac{\Delta\delta}{A} = \tfrac{1}{2}\left(\frac{l}{R}\right)^2 D_f$$

This is insignificant at 33 m, but may be as high as 2 deg/g at 5 m radius.

For a vehicle with parallel steering (zero Ackermann factor), in a corner the front wheels each have an effective toe-in, relative to the direction of travel for low speed, of

$$\alpha_i = \frac{T\sin\delta}{2R} \approx \frac{Tl}{2R^2}$$

The sideforce is generally produced more by the outer wheel. Defining a sideforce transfer factor e_S, where $e_S = 0$ represents equal sideforces, and $e_S = 1$ represents all force from the outer wheel, then a changed steer angle is required:

$$\Delta\delta = -e_S\alpha_i$$

giving

$$\Delta k = -\frac{de_S}{dA}\frac{Tl}{2R^2}$$

This is small for 33 m radius, but may be –4 deg/g at 5 m radius.

Considering now the effect of radius on final handling, the turning radius affects the understeer number in two main ways: the steering results in a negative attitude angle (front steering) and also in a reduced moment arm for the tyre forces on the steered wheels. In addition there may be effects because of cambering of the wheels at large steer angles, but this is hard to quantify because of lack of tyre data.

The attitude angles at the centre of mass G and at the rear axle differ by the rear kinematic steer angle

$$\delta_{Kr} = \frac{b}{R}$$

This reduces the attitude angle at G caused by the rear slip angle, which has already been dealt with in Section 6.12. It gives a longitudinal load transfer

$$F_{TX} = -\frac{mAH\delta_{Kr}}{l}$$

Adapting the equation derived for slip attitude effects (Section 6.12),

$$\Delta N_U = -(1+p)\left(\frac{AH\delta_{Kr}}{g}\right)\left(\frac{1}{a}+\frac{1}{b}\right)$$

$$= -(1+p)\left(\frac{AH}{Rg}\right)\left(\frac{b}{a}+1\right)$$

which is an oversteer effect.

Figure 6.16.1 shows a four-wheeled vehicle with steer and slip angles and traction at the rear. The front vertical force lateral transfer factor

Figure 6.16.1. Reduced moment arm of steered-wheel forces.

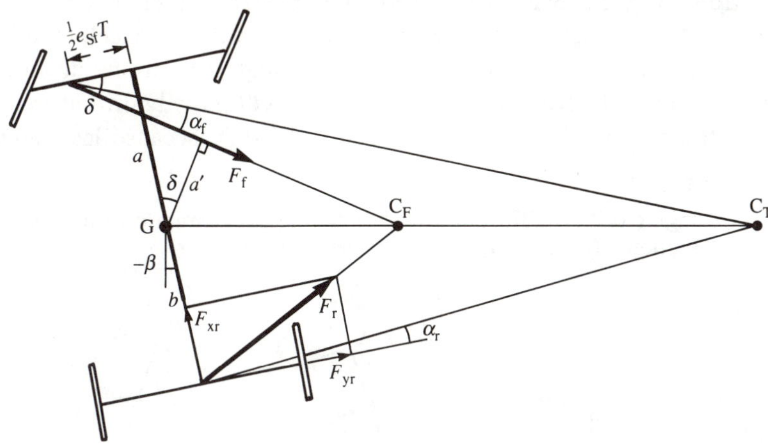

e_{Vf} results in a front sideforce transfer factor e_{Sf}. The parameter e_{Sf} is zero for equal sideforces and unity for all sideforce on the outer wheel. The moment arm of the front sideforce is

$$a' = a\cos\delta - \tfrac{1}{2}e_{\mathrm{Sf}}T\sin\delta$$

This causes

$$\Delta N_{\mathrm{U}} = \frac{a-a'}{a'} = \frac{a}{a'} - 1$$

$$= \left(\cos\delta - \frac{e_{\mathrm{Sf}}T}{2a}\sin\delta\right)^{-1} - 1$$

The total radius effect including longitudinal load transfer is therefore

$$\Delta N_{\mathrm{U}} = \left(\cos\delta - \frac{e_{\mathrm{Sf}}T}{2a}\sin\delta\right)^{-1} - 1 - (1+p)\left(\frac{AH}{Rg}\right)\left(\frac{b}{a}+1\right)$$

At a small radius the reduced moment arm effect predominates, and there is a strong understeer tendency.

6.17 Banking

Road slopes can be divided into two main types: longitudinal and lateral. Lateral slopes are commonly known as banking, or just as camber in the case of modest slopes used for water drainage. On the straight, banking results in a sideforce component $mg\sin\theta$ where θ is the road angle. The tyre vertical reaction becomes $mg\cos\theta$ which can be approximated as mg for practical camber cases up to 8°. For a primary neutral vehicle, the lateral force at G can be opposed by equal slip angles at the two ends, so the vehicle will drift sideways without rotation, Figure 6.17.1(a). With no control steer change, an understeering vehicle will tend to generate less lateral acceleration at the front, and so will rotate away from the sideforce, turning down the slope, Figure 6.17.1(b). An oversteer vehicle will initially drift away but will ultimately turn up the slope, Figure 6.17.1(c), because less lateral acceleration is generated at the rear.

Figure 6.17.1. Effect of sideforce at G (e.g. on entry to road camber section): (a) neutral, (b) understeer, (c) oversteer.

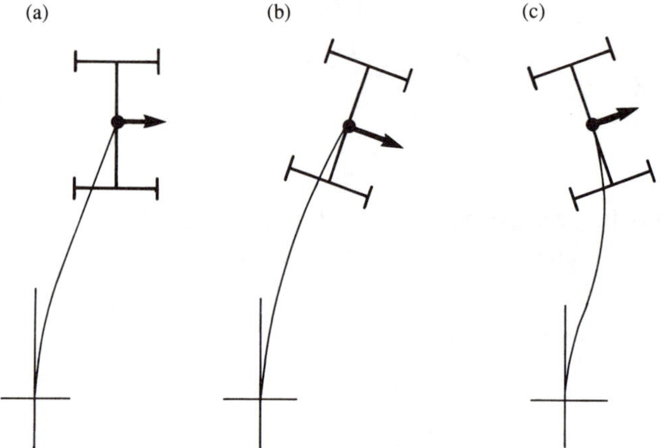

To maintain a straight path in the steady state, the sideforce acting at G calls for slip angles, but no kinematic steer angle:

$$\alpha_f = \frac{mgb}{2lC_{\alpha f}} \sin\theta$$

$$\alpha_r = \frac{mga}{2lC_{\alpha r}} \sin\theta$$

Allowing for suspension effects using cornering compliance, the attitude angle is

$$\beta = D_r g \sin\theta$$

and the steer angle required is

$$\delta = (D_f - D_r)\, g \sin\theta = kg \sin\theta$$

where k is the primary understeer gradient. Thus a neutral steer vehicle will be in equilibrium at an attitude angle but zero steer angle. A primary understeer vehicle will need to be steered up the slope, and an oversteer vehicle down it. At the fairly large value of $k = 0.5$ deg/m s^{-2} (5 deg/g) and a side slope of 5°, then $\delta = 0.44°$, or about 8° at the steering wheel, and $\beta \approx 0.5°$.

A straight road of banking θ will require a lateral friction coefficient of $\tan\theta$ merely to maintain a straight path. This may be difficult to achieve in icy conditions.

A small amount of banking is also common on corners, and this may be favourable or unfavourable. More extreme banking, sometimes called super-elevation and with angles as great as 40°, Figure 6.17.2, is found on some racing circuits and on high-speed test circuits. In such cases the road cross-section is usually curved with greater angles high up, so that the driver can choose the desired banking angle.

Figure 6.17.2. Free-body diagram for a car on a super-elevated corner.

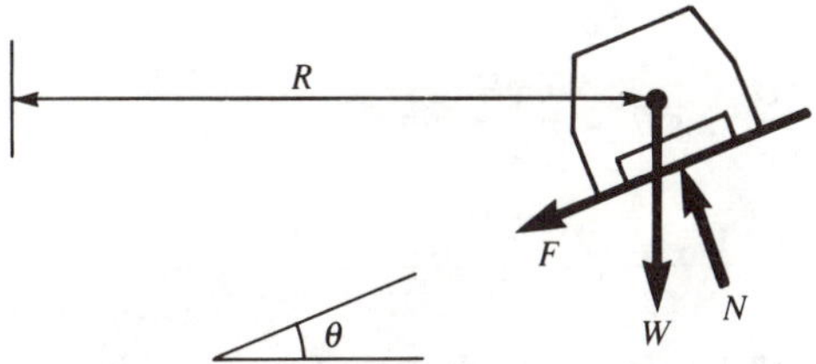

In Figure 6.17.2, a vehicle is cornering with lateral acceleration $A = V^2/R$ at constant height on a banked road. The tyre forces required are

$$F = mA\cos\theta - mg\sin\theta$$

$$N = mA\sin\theta + mg\cos\theta$$

Setting F to zero, no cornering tyre force will be required if the slope angle is

$$\theta_0 = \arctan\frac{V^2}{Rg}$$

and this condition may actually be achieved on super-elevated circuits. In this case the normal reaction is greatly increased, to $mg/\cos\theta$. This is liable to bring a conventional vehicle down onto its bump stops. Vehicles designed for such conditions generally have very stiff or rising-rate suspensions, and appropriately uprated tyres.

With a maximum cornering coefficient μ_C, the maximum speed on banking is given by

$$V^2 = \mu_C gR\,\frac{1 + \dfrac{\tan\theta}{\mu_C}}{1 - \mu_C\tan\theta}$$

compared with $V^2 = \mu_C gR$ on the level.

all cambers can be significant with low friction. On ice, maximum cornering force coefficient μ_C as 0.1, with 5° ie speed is increased 38% for favourable camber, and reduced adverse camber. At $\mu_C = 1$, the changes are +9% and –9%.

t-to-rear differences of maximum lateral force friction sensitivity p would imply some final handling variation on banking, but this is normally unimportant because the required sideforce coefficients are generally small. However, there may still be significant changes of primary handling.

6.18 Hills

On longitudinal slopes, uphill at an angle of θ, Figure 6.18.1, at steady speed, there is a front-to-rear load transfer. The axle reactions are

$$N_f = \frac{mgb}{l}\cos\theta - \frac{mgH}{l}\sin\theta$$

$$N_r = \frac{mga}{l}\cos\theta + \frac{mgH}{l}\sin\theta$$

Figure 6.18.1. Free-body diagram for a car driving uphill.

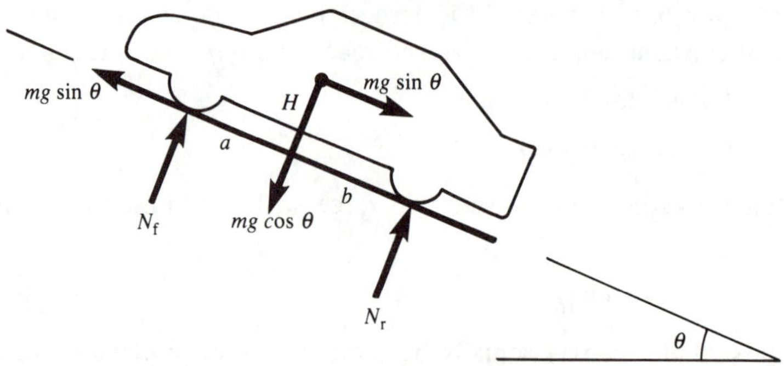

When driving uphill, the load transfer to the rear leads to increased primary understeer, and reduced attitude coefficient, but these are not large effects; for example at 1 in 5 (12°) uphill Δk_β is 0.25 deg/g and Δk is 0.5 deg/g, and the opposite (negative values) downhill. The effect on final handling is more important:

$$\frac{A_f}{A_{f0}} = \frac{\mu_{Cf} N_f}{\mu_{C0} N_{f0}} = \left(\frac{N_f}{N_{f0}}\right)^{1+p}$$

$$= \left(1 - \frac{H}{b}\tan\theta\right)^{1+p}$$

$$\approx 1 - (1+p)\frac{H}{b}\tan\theta$$

$$\frac{A_r}{A_{r0}} \approx 1 + (1+p)\frac{H}{a}\tan\theta$$

For a vehicle which is initially neutral on level ground, for small θ,

$$\Delta N_U = \frac{A_r}{A_f} - 1$$

$$\approx \left(1 + (1+p)\frac{H}{a}\theta\right)\left(1 - (1+p)\frac{H}{b}\theta\right)^{-1} - 1$$

$$\approx (1+p)H\theta\left(\frac{1}{a} + \frac{1}{b}\right)$$

For $a \approx b \approx l/2$

$$\Delta N_U = 2(1+p)\frac{H\theta}{l}$$

$$\frac{dN_U}{d\theta} = 2(1+p)\frac{H}{l}$$

This is about 0.34/rad or 0.006/deg. For 1 in 5 uphill (12°) ΔN_U is +0.07, a substantial final understeer effect. Downhill ΔN_U is –0.07 which is a readily noticeable effect, and may in some cases change final understeer into final oversteer.

Examining now vertical curves, these are not strictly steady-state, but may be analysed approximately as such. In the vehicle-fixed coordinate system the vehicle has a vertical downward compensation (pseudo) force

$$F_V = \frac{mV^2}{R_V}$$

where R_V is the vertical radius of curvature; F_V is positive downwards for a positive R_V which therefore corresponds to a centre of curvature above the road, i.e. a trough. Thus tyre reaction forces are increased in a trough and reduced over a crest. The main practical consequences are a deterioration of braking and cornering capability over a crest.

6.19 Loading

Adding loads to a vehicle, such as passengers, luggage or fuel, increases the total mass and moves the centre of mass. Usually this movement is rearward and upward; in some cases it may be forwards, as in the case of rear engine and front boot (trunk). A higher centre of mass means more load transfer and roll. A larger weight increases the suspension deflection, affecting camber and load transfer. Front-to-rear movement of the centre of mass affects the relative tyre characteristics. This can have complex effects on the handling, and in general it is necessary to analyse the loaded case quite independently of the unloaded case. Of course this is especially true of trucks and commercial vehicles where the load is a large proportion of the unladen weight.

For cars the total maximum load is likely to be about 40% of the kerb weight. The driver and front passenger are close to the initial centre of mass and so cause little G movement. The rear passengers are usually in front of the rear axle, and for a rear boot the load is behind the rear axle. Hence a simple indication of the effect of loading on a typical car may be obtained by studying the case of a load m_{ar} added over the rear axle. In any case, an added load may be analysed as load increments over the two axles.

At the rear the vertical reaction changes in the ratio

$$\begin{aligned}\frac{F_V}{F_{V0}} &= \frac{m_r + m_{ar}}{m_r}\\ &= 1 + \frac{m_{ar}\,l}{ma}\end{aligned}$$

Denoting the tyre cornering coefficient sensitivity to F_V as f then

$$\frac{C_\alpha}{C_{\alpha 0}} = \left(1 + \frac{m_{ar}l}{ma}\right)^f$$

Neglecting suspension, the rear cornering compliance is then given by

$$\begin{aligned}\frac{D_r}{D_{r0}} &\approx \frac{(m_r + m_{ar})/C_\alpha}{m_r/C_{\alpha 0}}\\ &= \left(1 + \frac{m_{ar}\,l}{ma}\right)^{1-f}\end{aligned}$$

With $m_{ar} = 0.2$ m ($0.4m_r$) and $f = 0.5$, then ΔD_r is 1.1 deg/g. This is an increase of the attitude gradient, and there is also an equal primary oversteer tendency effect, i.e. Δk is -1.1 deg/g.

The effect on final handling depends mainly on the reduced rear lateral acceleration limit because of the tyre maximum cornering coefficient sensitivity p. At the rear

$$\frac{\mu_C F_V}{\mu_{C0} F_{V0}} = \left(1 + \frac{m_{ar}\,l}{ma}\right)^{1+p}$$

$$\frac{\mu_C F_V}{m_r + m_{ar}}\;\frac{m_r}{\mu_{C0} F_{V0}} = \left(1 + \frac{m_{ar}\,l}{ma}\right)^p$$

$$\Delta N_U \approx \frac{p m_{ar} l}{m a}$$

Because p is negative this is an oversteer effect. The equivalent expression for a load increment at the front is

$$\Delta N_U = -\frac{p m_{af} l}{m a}$$

Taking $m_{ar}/m = 0.2$ and $p = -0.15$, this gives $\Delta N_U = -0.06$, a substantial final oversteer tendency. This neglects other possible factors, such as greater rear load transfer because of lowering of the rear suspension towards the bump stops.

The above simple analysis shows how rear loading can result in substantial primary and final oversteer tendencies; these are readily observable in practice. It is common practice for vehicle operating manuals to call for increased rear tyre pressures when operating with full load. This limits tyre deflections, but also generally increases C_α helping to offset the primary oversteer effect, and generally also improves the tyre maximum cornering force coefficient with the increased load, helping to offset the final oversteer tendency.

In the case of racing cars with substantial aerodynamic downforce, there may be great sensitivity to changes of mass and to centre-of-mass position. Considering first the simplest possible model vehicle (Section 6.15) with no aerodynamics, a mass increase will increase the tyre normal forces in proportion, and the maximum lateral acceleration will deteriorate according to the reduction in maximum cornering force coefficient, i.e. it will be directly dependent on the friction sensitivity p. On the other hand, for a vehicle with very large aerodynamic downforce, the maximum lateral acceleration will be inversely proportional to the mass, because the normal forces will not be affected proportionally by a weight change.

The effect of movement of the centre of mass on the final balance of a racing car follows a similar pattern. With no aerodynamics, moving the centre of mass forward will reduce the front cornering friction and increase it at the rear, according to the sensitivity p. However with strong downforce, a movement of the centre of mass will have much more dramatic effects because the normal forces and maximum cornering forces will hardly be affected. This is one reason for the location of fuel tanks close to the centre of mass on downforce racing vehicles, to avoid trim changes as fuel is consumed.

6.20 Wind

In straight running a vehicle may be subject to a sidewind, giving a sideforce and a yaw moment. To maintain the required course an attitude angle and steer angle will be required.

Given the wind speed and direction and the vehicle speed and direction, the relative wind speed V_r and aerodynamic attitude angle β_{Ae} can be calculated (Section 3.4). The resulting dynamic air pressure is

$$q = \tfrac{1}{2}\rho V_r^2$$

and the sideforce is

$$F_S = q\, C_S\, S$$

and the yaw moment is

$$M = q\, C_Y\, Sl$$

To maintain the required straight course the forces needed at the front and rear axles are

$$F_f = q\,(C_S + C_Y)\, S$$

$$F_r = q\,(C_S - C_Y)\, S$$

Although there is generally some small aerodynamic roll, the vehicle does not roll as for cornering, so the resulting attitude angle does not depend significantly upon the roll steer. On the other hand, suspension compliance may have an effect. But neglecting suspension effects,

$$\Delta\alpha_f = \frac{F_f}{2C_{\alpha f}} \quad \text{and} \quad \Delta\alpha_r = \frac{F_r}{2C_{\alpha r}}$$

Hence the attitude angle is

$$\alpha_r = \frac{q(C_S - C_Y)S}{2C_{\alpha r}}$$

and the steer angle required towards the wind is

$$\delta = \frac{q(C_S + C_Y)S}{2C_{\alpha f}} - \frac{q(C_S - C_Y)S}{2C_{\alpha r}}$$

From this, the crosswind steer correction coefficient $d\delta/dw$ may be found.

For a pure sidewind the steer angle is, for the linear case, proportional to $\beta_{Ae}V_r^2$. For a given wind speed, the steer angle increases with vehicle speed. Taking a case with a sidewind speed of 10 m/s and S = 1.8 m^2, a vehicle ground speed of 50 m/s gives β_{Ae} = 11.3°, and typically C_S = 0.4 and C_Y = 0.1, giving F_r = 857 N, α_r = 0.86°, F_f = 1428 N, α_f = 1.43° and δ = 0.57°. The actual steering-wheel deflection will be some 10°. Hence the steady-state effects of wind are not very great. On the other hand, wind gusting is an important unsteady effect.

6.21 Testing

In principle the main result of testing is the path curvature $\rho = 1/R$ as a function of the control inputs δ and V. In practice the required steer input δ is usually plotted as a function of lateral acceleration A, as in Figure 1.13.1 for example. The steer angle is measured at the steering wheel and in some cases also at the road wheels. The lateral acceleration may be found by accelerometer, or from $A = \Omega V$ with a rate gyro measuring the angular speed, or from $A = V^2/R$.

There are three basic forms of the test. These are constant radius, constant speed and constant steer angle, although of course for a full picture of the handling characteristic it is necessary to use a succession of different constant values.

The most common test is the constant radius one, where the speed is varied and the associated required steer angle measured, plus other variables such as attitude angle. The radius chosen is likely to be around 30 m depending on available facilities, but may be 100 m or more if it desired to include aerodynamic effects, if the vehicle has sufficient power, and if a suitable test facility is available. At 33 m the steering kinematic steer angle is only about 5°, thus avoiding extreme steering geometry effects, whilst for A = 8 m/s^2 the speed is 16 m/s, so aerodynamics are usually negligible; hence 33 m radius gives a good reference characteristic. In practice, for a constant-radius test the driver is required to follow a paint line whilst at the same time keeping the control positions constant; this may be difficult near to limit conditions because of reduced or exaggerated steering response.

In the constant-speed type of test the vehicle is usually tested at various specific radii, and the steer angle, attitude angle, etc. observed. When the steer angle is plotted against lateral acceleration at constant speed, rather than at constant radius, then the kinematic steer angle varies. In this case the understeer coefficient is no longer $d\delta/dA$, but it is still $d\delta_U/dA$. This is discussed in Section 6.9. An advantage of the constant-speed test is that this is a more realistic representation in so far as aerodynamic effects are concerned. The constant speed is no easier for the driver because the differing lateral accelerations give different tyre drags and require different throttle settings.

In the third type of steady-state test, the steer angle is held constant, and a sequence of steady speeds used. This test is the easiest to perform as far as driver skill is concerned, because both controls are fixed. It is possible to use a varying speed, but the variation must be fairly slow or else the tractive forces will be sufficiently far from equilibrium to influence the results.

In practice the constant-radius test is most commonly used. For large radii a sufficiently large test-pad may not be available, and curved tracks may be used, such as the Dunlop–M.I.R.A. handling circuit with its variety of radii of sufficient length to allow steady state to be achieved.

Some description of instrumentation is given in Section 1.13, along with a broader description of testing.

The international standard for steady-state testing is I.S.O. 4138 *Road Vehicles – Steady State Circular Test Procedure*. This requires a constant-radius test, with no particular radius specified, other than a minimum of 30 m. The track gradient may not exceed 2%, and the windspeed not exceed 7 m/s. Tyres should preferably be conditioned by normal use for 150–200 km, and warmed-up at the time of test by driving 500 m at a lateral acceleration of 3 m/s^2. Data should be taken at increments not exceeding 0.5 m/s^2, and averaged over 3 s, and the path should be maintained within 0.3 m. The recommended form of presentation of results is separate graphs of steering-wheel angle, sideslip angle, roll angle and steering-wheel torque, each against lateral acceleration with right turn considered positive lateral acceleration and left turn negative.

The following derived parameters and notation are defined: steering-wheel angle gradient $\mathrm{d}\delta_s/\mathrm{d}A$, sideslip angle gradient $\mathrm{d}\beta/\mathrm{d}A$, roll angle gradient $\mathrm{d}\phi/\mathrm{d}A$, steering-wheel torque gradient $\mathrm{d}T_s/\mathrm{d}A$, and steering-wheel angle/sideslip gradient $\mathrm{d}\delta_s/\mathrm{d}\beta$.

Various normalised parameters are also defined, including the understeer sideslip gradient $\delta_s/(G\,\mathrm{d}\beta/\mathrm{d}A)$, the steer coefficient $(\mathrm{d}\delta_s/\mathrm{d}A)/Gl$, and the directional coefficient $(\mathrm{d}\delta_s/\mathrm{d}\beta)/Gl$.

Testing can also be performed by operating the vehicle on a complete chassis dynamometer. One such testing system has been built, and is described by Odier (1972).

6.22 Moment method

Figure 6.22.1 shows an example plot of total moment coefficient against lateral force coefficient, the vehicle behaviour being represented by a carpet plot for various values of steering-wheel angle and attitude angle. Here the vehicle is considered to be travelling in a straight line with no lateral or rotational acceleration, i.e. the force and moment are considered balanced by applied external forces. This can be achieved

Figure 6.22.1. Moment-method carpet plot.

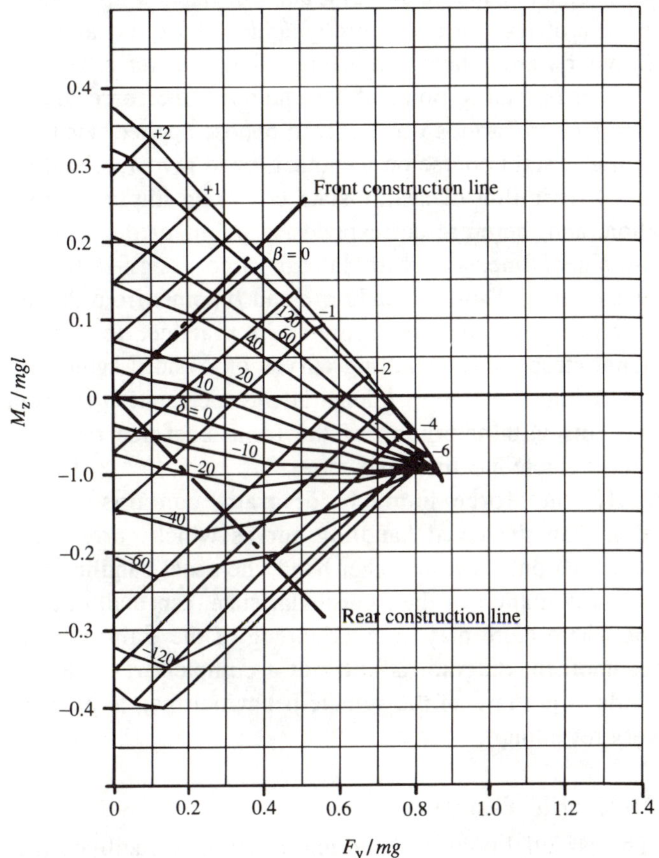

experimentally, but in practice such plots are produced by computer simulation. Also shown are front and rear 'construction lines'. The front construction line is the locus of points with lateral force at the front axle only. The $\beta = 0$ line diverges from this because of effects such as roll steer. These lines have gradient a/l and $-b/l$, or just a and $-b$ if the plot is of force F_y and moment M_z instead of coefficients F_y/mg and M_z/mgl.

Considering any particular point, this corresponds to some specific value of moment and force which may be read from the axes. Constructing a parallelogram to the construction lines shows the corresponding front and rear lateral forces. The mesh lines of the carpet plot will show the particular δ_s and β that will give the specified force and moment; usually the values are unique except close to limit conditions. The steer angle derived from this diagram is the understeer angle only; the kinematic steer angle must be considered separately. Actually the exact shape of the diagram will depend upon an assumed speed because of the effect on aerodynamics and tyres, and of tractive forces. In other words, there are different diagrams for different speeds.

For a given operating point of the diagram, the force and moment may be assigned in various ways, e.g. to oppose applied wind effects, or to maintain a straight course on a camber, or to give accelerations.

The usual definition of steady-state cornering implies zero angular acceleration, and therefore corresponds to points of the mesh lying on the zero-moment line, and with lateral force assigned to producing lateral acceleration. Thus by reading δ and β points from that axis it is possible, although somewhat roughly in practice, to construct the conventional steady-state handling curves of δ and β against A. From Figure 6.22.1, the maximum trimmed lateral acceleration is about $0.73g$, and this occurs with saturation of the front sideforce rather than the rear, i.e. this vehicle has final understeer.

Evidently the force–moment diagram contains much more information than the usual handling curves which correspond to the force axis points only. On the other hand, the usual handling curves are a clearer representation of their particular state, especially near to limit conditions where there may be overlapping of the δ lines of the mesh. The force–moment diagram is still not a complete representation, but does provide a 'portrait' of the vehicle behaviour that, to an experienced eye, is very revealing.

6.23 Desirable results

The essential point of handling theory is to facilitate the design and production of better handling vehicles. An important question, then, is: What is the optimum form of the handling diagram curves? One difficulty here, as discussed in Section 1.12, is that small design changes that have little effect on the theoretical behaviour of the vehicle may be strongly noticeable to the driver, and make the difference between good and bad driver-feel, especially dynamically. Furthermore, because handling is so subjective, there are differences of opinion regarding the best behaviour. There is therefore scope for disagreement with the following comments. Also, of course, handling cannot in practice be optimised without consideration of other factors such as ride.

For the steering angle versus lateral acceleration curve, the initial slope, i.e. the initial understeer gradient, should be small, say between zero and 2 deg/g at the road wheels. American practice is to use higher values, e.g. 5 deg/g. The graph should then curve smoothly through gradually increasing understeer in the mid-range into final understeer, with a final understeer number of about 0.20. The argument for final understeer is that steering control should be progressive, and not reverse as it does if there is final oversteer. As far as the maximum lateral acceleration is concerned, the greater the better. If a vehicle with final oversteer is altered at the front to reduce the front acceleration ability then the final controllability is improved, but only at the expense of the overall maximum lateral acceleration, so the result is not necessarily a safer vehicle.

For the attitude angle graph, the initial slope, i.e. the attitude gradient or coefficient, should be as small as possible; this mainly depends upon the rear tyre cornering stiffness. The roll curve is not critical. However all three curves should be smooth with no irregularities or sharp changes of gradient. Even if power steering is used, it is possibly best to have a steering-torque gradient of about 20 Nm/g, rather than extremely light steering (position control only). Driver preferences depend to a large extent upon previous experience. The actual value of the steering torque gradient is perhaps not as important as the fact that there should be a gradient. Traditional U.S. practice, still seen on some European vehicles for the U.S.A. market, is to have a steep gradient for lateral acceleration up to 0.1g, and then zero gradient. Typical European practice is to mimic a manual system, but at lower force levels.

The handling of European passenger cars has steadily improved over the last thirty years, and is now generally very good indeed. On the other hand, commercial vehicles tend to be rather poor with low limit accelerations, and final oversteer. Basically this is because of the low performance limit at the rear because of high loads, and hard tyre materials to give good mileage. There are also roll-over limitations because of the high centre of mass. From a handling point of view, the best solution is a sufficient number of rear wheels and good enough tyres to give a rear limit performance close to the roll limit, with a slightly inferior front-end performance to give final understeer without roll-over. This is, however, complicated by the proneness of commercial vehicles to dynamic roll-over in swerving.

As far as the process of tuning the suspension is concerned, this is basically a matter of achieving the desired understeer gradient and final understeer number. The parameters that the designer can usefully influence have been discussed in some detail in Sections 6.10 and 6.12.

6.24 Problems

Q 6.1.1 Discuss the relative merits of various definitions of steady-state vehicle motion.

Q 6.2.1 Discuss the relative merits of various choices of control parameters.

Q 6.2.2 List, define and explain the steady-state vehicle motion parameters, and any simple relationships between them.

Q 6.2.3 Give the principal alternative units used for the vehicle controls, responses and gains.

Q 6.3.1 Explain, with graphs, how the handling characteristic curves of a simple vehicle can be deduced from the tyre force curves.

Q 6.3.2 Define and explain the relationship between steer angle δ, kinematic steer angle, geometric steer angle, Ackermann angle, dynamic steer angle and understeer angle.

Q 6.3.3 Describe the possible shapes of the steer angle curve, as related to the shape and relative position of the tyre force curves.

Q 6.3.4 Sketch handling steer curves for a typical modern car and truck, and explain their merits and drawbacks.

Q 6.3.5 Sketch steer curves for vehicles with primary secondary and final steer as follows: (a) under, over, over; (b) under, under, over; (c) over, over, under; (d) over, under, under; (e) under, over, under.

Q 6.4.1 Sketch plan-view free-body diagrams for a 'bicycle model' vehicle in cornering without traction, and in steady cornering for front-, rear- and four-wheel drive. Explain them.

Q 6.4.2 Sketch simple plan-view force polygons for front- and rear-drive bicycle model vehicles. Extend them to include rolling resistance and aerodynamic forces.

Q 6.4.3 Draw the plan-view force diagram (as Figure 6.4.2) for rear drive at small A and small R.

Q 6.4.4 Draw the vehicle with forces (as Figure 6.4.1) for four-wheel drive, for the case with parallel front and rear forces. Discuss alternatives.

Q 6.5.1 Define and explain a linear handling vehicle. What practical conditions are covered?

Q 6.5.2 Define and explain characteristic speed and critical speed for the linear case. Derive equations for them.

Q 6.5.3 For the linear case, define and derive the response factor, also giving graphs, and derive expressions for curvature gain, yaw gain and acceleration gain.

Q 6.5.4 Discuss the problems that would be experienced in driving a linear oversteer vehicle at various speeds in a straight line.

Q 6.5.5 Derive an expression for the understeer gradient for a suspensionless bicycle model vehicle.

Q 6.6.1 Derive expressions for the yaw stiffness and static margin for a simple vehicle. Explain the neutral steer point and line.

Q 6.6.2 Derive an equation for k/x_s for a simple suspensionless vehicle.

Q 6.6.3 Explain yaw damping. Give a diagram and derive a suitable equation for a simple linear vehicle.

Q 6.6.4 Obtain an expression for $d\delta/d\beta$ for a simple suspensionless vehicle for steady cornering.

Q 6.7.1 Sketch three-dimensional graphs of understeer angle against speed and lateral acceleration, for a vehicle with primary understeer and final oversteer at low speed and strong aerodynamic understeer effects.

Q 6.8.1 For the non-linear case, explain understeer angle and understeer gradient/coefficient.

Q 6.8.2 Explain how the idea of characteristic or critical speed can be applied to a non-linear vehicle in a given trim state.

Q 6.8.3 Plot graphs showing how static margin and yaw damping typically depend upon speed and lateral acceleration.

Q 6.9.1 Explain the difference between primary and final understeer.

Q 6.9.2 Describe the concept of understeer gradient for the non-linear case.

Q 6.9.3 Derive and explain the kinematic steer-angle gradient.

Q 6.9.4 Giving example graphs, show how the required steer angle is likely to vary with lateral acceleration at constant radius, and also at constant speed, for cars and for trucks. Explain which regions of the curves represent understeer and oversteer.

Q 6.9.5 Explain the difference between understeer angle and understeer gradient. Explain how a vehicle with positive understeer angle can be 'oversteer'.

Q 6.10.1 Explain, with a diagram, the difference that suspension makes to the linear equation for steer angle.

Q 6.10.2 Give example values of the primary understeer effects, in a table, for a typical modern small front-drive front-wishbone rear plain trailing-arm car, and discuss it.

Q 6.10.3 Explain, with equations and graphs, the effect of G position on k and k_β for a vehicle with the same C_α for all tyres, and then also for the case when C_α varies with load. Give example values.

Q 6.10.4 Explain the effect of suspension roll steer on primary handling, with equations, and give extreme and representative example values.

Q 6.10.5 Explain the effect of suspension camber on primary handling, with equations and give extreme and representative example values.

Q 6.10.6 Explain the effect of steering-column compliance on primary understeer.

Q 6.10.7 Explain the effect of tyre pneumatic trail on primary handling.

Q 6.10.8 Explain the effect of rolling resistance on primary handling.

Q 6.10.9 Cars with lightly-loaded rear axles are sensitive to rear toe angle. Explain the effect of front and rear toe on primary handling.

Q 6.10.10 Give an overview, in 1000 words, of the main factors affecting the primary understeer of a modern car.

Q 6.10.11 For an $a = b$ suspensionless vehicle with tyre $C_S = 10$/rad, what are the values of k and k_β?

Q 6.10.12 For a suspensionless vehicle with $l = 2.8$ m, $a = 1.3$ m, $C_{\alpha f} =$ 520 N/deg, $C_{\alpha r} = 480$ N/deg, $k_f = 9$ deg/g and $m = 1600$ kg, evaluate k and k_β.

Q 6.10.13 For the vehicle of Q 6.10.12, with front and rear roll understeer coefficients of –0.1 and +0.3, and a roll gradient of 10 deg/g, find k and k_β.

Q 6.10.14 For the vehicle of Q 6.10.12, with no roll steer, but with front and rear roll camber coefficients of 0.8 and 0.1 and tyre $C_\gamma / C_\alpha = 0.15$, find k and k_β.

Q 6.10.15 For the vehicle of Q 6.10.12, with pneumatic trail of 40 mm, $G = 16$, and a steering column compliance of 4 N m/deg, compare understeer with respect to δ and δ_{ref}.

Q 6.10.16 For the vehicle of Q 6.10.15, evaluate the pneumatic trail on the rigid-body effect on k and k_β.

Q 6.10.17 For the vehicle of Q 6.10.12, with a rolling resistance of 0.03 and a G height of 1.1 m, what is the effect on k and k_β?

Q 6.10.18 For steering backlash $\pm\delta_b$ at the front wheels, obtain a simple expression for the range of lateral acceleration in terms of k and V. Evaluate this for 0.1° and $k = 0$, at 50 m/s. Find the time to veer 2 m.

Q 6.10.19 Explain the relationship between the compliance steer gradients $k_{\eta S}$ and $k_{\eta A}$ and the compliance steer coefficients η_S and η_A.

Q 6.10.20 Explain how the compliance camber gradient arises from compliance camber coefficients.

Q 6.10.21 Explain the effect of compliance camber on the total compliance steer gradient k_η.

Q 6.10.22 Explain how the roll steer gradient k_ε is influenced by front and rear roll steer and the roll angle gradient, on a rigid axle.

Q 6.10.23 Explain how the roll steer gradient k_ε is influenced by bump steer and bump camber on an all-independent suspension vehicle.

Q 6.10.24 Explain the total roll steer gradient k_ε and how it arises from roll steer and roll camber coefficients on a rigid axle.

Q 6.11.1 Give an overview of the main factors influencing secondary handling.

Q 6.11.2 Derive an equation representing the effect of longitudinal load transfer on steer angle for moderate A.

Q 6.12.1 Define and explain the term 'final handling regime'.

Q 6.12.2 Define and explain final understeer number. List the main factors controlling its value.

Q 6.12.3 Give a table showing typical values of N_U contributions to a modern car. Discuss it.

Q 6.12.4 Describe the likely process in a crash because of trying to round a corner too quickly, contrasting final understeer and final oversteer. Compare their merits.

Q 6.12.5 Explain how the effect of G position and tyre friction effects influence N_U and A_M. Give an example graph of A_M against G position.

Q 6.12.6 Explain how the distribution of roll stiffness affects N_U.

Q 6.12.7 Compared with $a = b$ at which A_M is $0.8g$, if G is moved forward 120 mm on a wheelbase of 2.8 m, what would be the likely quantitative effect on A_f, A_r, A_M and N_U?

Q 6.12.8 With $p = -0.15$, $A = 0.7g$, $H = 0.8$ m, $T = 1.5$ m, $a = 1.25$ m and $l =$ 2.7 m, what will be the effect on N_U of changing the front load transfer fraction d_f from 0.6 to 0.75?

Q 6.12.9 Explain how longitudinal load transfer arises in cornering, and how it affects N_U. Give example values.

Q 6.12.10 For the vehicle of Q 6.12.8, with a final attitude angle of 5.7° what will be the effect of longitudinal load transfer on N_U?

Q 6.12.11 Discuss the effect of traction on final handling.

Q 6.12.12 Discuss the effect of camber on tyre forces and final handling.

Q 6.12.13 Explain the effect on primary, secondary and final handling of adding a stiff front anti-roll bar for vehicles with no roll steer and with (a) considerable front roll camber, none at rear, (b) considerable rear roll camber, none at the front.

Q 6.13.1 Explain the effect of an over-run ratcheting differential on primary and final handling.

Q 6.13.2 Explain the effect on handling of a differential that apportions torque in proportion to tyre normal force. Give example values.

Q 6.14.1 Give an overview of the effect of aerodynamics on primary handling.

Q 6.14.2 A rear-drive vehicle at 40 m/s has $S = 2$ m^2, $a = 1.26$ m, $l = 2.8$ m, $m = 1500$ kg, tyre stiffness sensitivity $f = 0.5$, $D_f = 12$ deg/g, $D_r = 8$ deg/g, $C_{Lf} = 0.1$, $C_{Lr} = -0.2$, $C_D = 0.5$, $C'_S = 0.06$/deg, $C'_Y = 0.02$/deg, and $\mu_L = 0.7$. Evaluate the various aerodynamic effects on k.

Q 6.14.3 Give a detailed algebraic development of equations representing the effect of aerodynamic lift (or drag, etc.) on primary handling.

Q 6.15.1 Give a table listing the typical contributions to A_M and N_U for a modern car at maximum speed, and discuss it.

Q 6.15.2 For the vehicle of Q 6.14.2, with a final attitude angle of 7°, a maximum lateral accceleration of 7.5 m/s^2, and a tyre friction sensitivity of –0.20, evaluate the effect of aerodynamic yaw moment on N_U.

Q 6.15.3 Show that for a simple vehicle with lift and tyre friction sensitivity, the minimum cornering radius is given by

$$\frac{R}{R_0} = \left(1 - \frac{\rho V^2 SC}{2mg}\right)^{-(1+p)}$$

where R_0 is the radius value achievable at zero lift. Explain the idea of a critical aerodynamic speed.

Q 6.15.4 Explain Figure 6.15.1 in qualitative terms.

Q 6.15.5 Explain why, for a racing car with aerodynamic downforce, it becomes even more advantageous to reduce the mass.

Q 6.16.1 Describe the effect of path radius on the understeer coefficient and number.

Q 6.17.1 A straight-running vehicle passes from a level road onto a cambered one. Explain its subsequent path if no control is applied, for understeer, neutral and oversteer cases.

Q 6.17.2 For the vehicle of Q 6.14.2, on a side slope of 6° what will be the required δ and β for straight running?

Q 6.17.3 Explain the principle of super-elevation, with equations, including one for maximum cornering speed. A vehicle of mass 1800 kg has speed 22 m/s at cornering radius 40 m on banking angle 10°. Find the minimum cornering friction coefficient, and the angle for zero lateral force.

Q 6.18.1 Explain longitudinal load transfer on hills, and the consequences for final handling A_M and N_U.

Q 6.18.2 A vehicle with $a = 1.3$ m and $l = 2.8$ m has G height 0.9 m, and $N_U = 0.07$ on level ground. Estimate N_U on a downhill slope of 10°.

Q 6.18.3 A vehicle has a speed of 42 m/s over a crest of radius 300 m. Estimate the proportional deterioration in A_M, compared with the level, assuming constant tyre cornering friction.

Q 6.19.1 Explain with equations the influence of adding a load over a rear axle, on primary and final handling.

Q 6.19.2 A vehicle initially with $a/l = 0.46$, $m = 1900$ kg, tyre cornering stiffness sensitivity 0.4, $D_f = 10$ deg/g and $D_r = 6$ deg/g has a mass of 300 kg added over the rear axle. Estimate the initial and final k and k_β before and after adding the load.

Q 6.20.1 Explain the effect of side wind on steer and attitude angle for straight running, with equations.

Q 6.20.2 A vehicle has ground speed 40 m/s, frontal area 2.2 m^2, $C'_S =$ 0.07/deg, $C'_Y = 0.02$/deg, $C_\alpha = 500$ N/deg, $G = 20$ and is in a sidewind of 8 m/s. Find the effective forces at front and rear axles, and the change of slip angles and steering-wheel angle to maintain a straight course.

Q 6.20.3 For a vehicle attempting a steady circular path with modest lateral acceleration, describe the action of the steering wheel required as a result of a steady sidewind. Obtain an expression for an estimate of the amplitude of wheel motion.

Q 6.20.4 Derive an equation for the crosswind steer correction coefficient $\mathrm{d}\delta/\mathrm{d}w$.

Q 6.21.1 Describe the various principal types of steady handling test.

Q 6.21.2 Describe the I.S.O. recommended procedure for a steady-state handling test.

Q 6.21.3 What factors influence the choice of radius for a handling test?

Q 6.22.1 Sketch a representative moment coefficient versus force coefficient plot for a typical car. Describe and explain its main features.

Q 6.22.2 Read the moment-method references, and write a 2000 word review of the method.

Q 6.23.1 Describe the desirable form of the standard handling curves, and discuss the reasons.

Q 6.24.1 For one of the vehicles of Appendix B, apply the theory of each section of this chapter to analyse the steady-state handling.

6.25 Bibliography

Frankly, it is not easy to give a useful bibliography of books for handling theory because the subject has largely been restricted to research papers. One of the few books to tackle the subject at all, and from a quite different perspective from that presented here, is *Road Vehicle Dynamics* by Ellis (1989), a development of the previous book *Vehicle Dynamics* (Ellis, 1969). Wong (1978) has a chapter on handling characteristics; see also Steeds (1960). Bastow (1980) has an example numerical calculation of the basic handling curves.

For the practical side of suspension modifications to achieve desired handling, Smith (1978) and Puhn (1981) are worth examination.

For terminology, the primary reference is S.A.E. J670e *Vehicle Dynamics Terminology*, but the I.S.O. will probably soon produce a standard; this will certainly differ from J670e and will be preferred in Europe. See also Hales (1965).

From the research literature, the following can be singled out as particularly relevant or interesting, either expanding slightly on the material here, or giving an alternative perspective: Ellis (1963), Radt & Pacejka (1965), Milner (1967), Bundorf (1967), Grylls (1972), Pacejka (1973), Topping (1974), Bundorf & Leffert (1976), I.S.O. (1982), Dixon (1987b) and (1988).

The moment method is fully described in Milliken *et al.* (1976), Rice & Milliken (1980), and Milliken & Rice (1983). A complete car chassis handling dynamometer is described by Odier (1972).

The influence of aerodynamics on the performance of racing cars is examined in Wright (1983) and Dominy & Dominy (1984).

7

Unsteady-state handling

7.1 Introduction

This chapter deals with the vehicle in unsteady state. This includes theory of response to varying control inputs or to disturbances, and covers the natural transient motions after a disturbance. Theory of stability of motion is included here, because this is basically the study of the transient motion following a notional small disturbance. The theory of vehicle transient behaviour is complex, and can be highly mathematical; therefore this chapter is not comprehensive, but serves as an introduction and overview.

Control disturbances may be due to motion of the steering or due to tractive or braking forces. The most basic disturbance is the step steer input; a practical approximation to this is the ramp-step input. Non-zero longitudinal acceleration is also unsteady, with possible effects from tractive forces unequally distributed side-to-side by non-free differentials, from steer compliance effects, and from tractive force effect on tyre characteristics. Of more critical practical importance is longitudinal deceleration combined with cornering, because this is likely to involve large longitudinal forces, and occurs in accident avoidance.

External disturbances arise from wind and road. The classic wind disturbance occurs when suddenly moving into or out of a crosswind; this is a problem of practical importance because it occurs when overtaking or being overtaken on motorways. Road disturbances include anything other than a smooth flat uniform surface, e.g. moving onto a camber, road roughness, or a change of friction coefficient.

The theory of motion stability may be investigated for specific initial conditions by computer, even for large disturbances in the non-linear regime. However, it is still valuable to have an understanding of the results of the linear mathematical theory, first given by Rocard (1946) for a simple vehicle without suspension or load transfer (two degrees of freedom, yaw and sideslip), and extensively explored by Segel (1957a and 1957b) for three degrees of freedom including roll. Because of the widespread use of computing nowadays, no attempt will be made here to give a full mathematical justification of the linear theory; rather the emphasis is on physical understanding of the results. The chapter begins with the rather unphysical single-degree-of-freedom models of a vehicle, because these, although of little value themselves, throw light on the more complex models.

7.2 1-d.o.f. vibration

Physical interpretation of the equations of motion of the vehicle is enhanced by an understanding of basic stability and vibration theory for a single-degree-of-freedom (1-d.o.f.) system, which will therefore be briefly reviewed here. There is an inertia M, stiffness K and damping coefficient C, Figure 7.2.1. When displaced by x there is a stiffness force Kx. If this is a restoring force the system is said to be statically stable; if the stiffness force tends to move the mass further away from its equilibrium position then it is statically unstable. The damper exerts a force $C\dot{x}$; if this opposes the motion then the damper will remove energy from any motion – this is dynamic stability. If the speed-dependent force acts in the same direction as the motion then it will add energy, giving dynamic instability. The motion of the object after a disturbance will depend upon the type of stability, Figure 7.2.2.

For static instability, regardless of dynamic stability, there is a continuous divergence from the equilibrium position, although this occurs more rapidly if there is also dynamic instability. Static stability coupled with dynamic instability gives an oscillation of growing amplitude. Static stability coupled with dynamic stability gives an oscillation of reducing amplitude; if the damping is sufficient then the oscillation is suppressed and there is simply a smooth return to the equilibrium position.

Taking K and C positive for static and dynamic stability, i.e. opposing displacement and velocity, Figure 7.2.1(b), the equation of motion is

$$\Sigma F_x = -Kx - C\dot{x} = M\ddot{x}$$

$$M\ddot{x} + C\dot{x} + Kx = 0$$

$$\ddot{x} + \frac{C}{M}\dot{x} + \frac{K}{M}x = 0$$

Using the operational notation D for d/dt, or assuming a solution of the form e^{Dt} and substituting, this becomes

Figure 7.2.1. Single-degree-of-freedom system: (a) system diagram, (b) free-body diagram of mass.

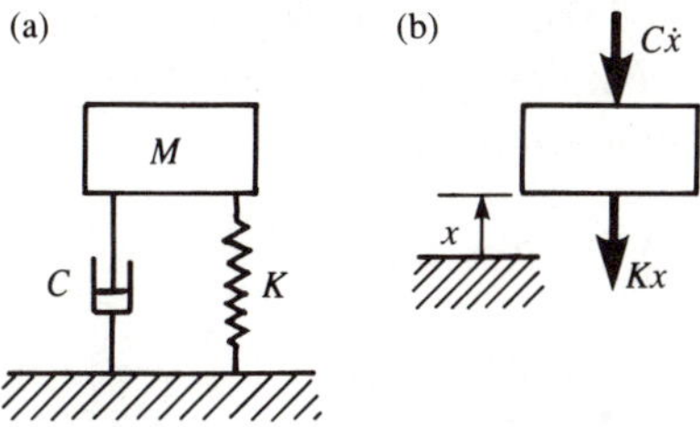

Figure 7.2.2. Possible motions after displacement.

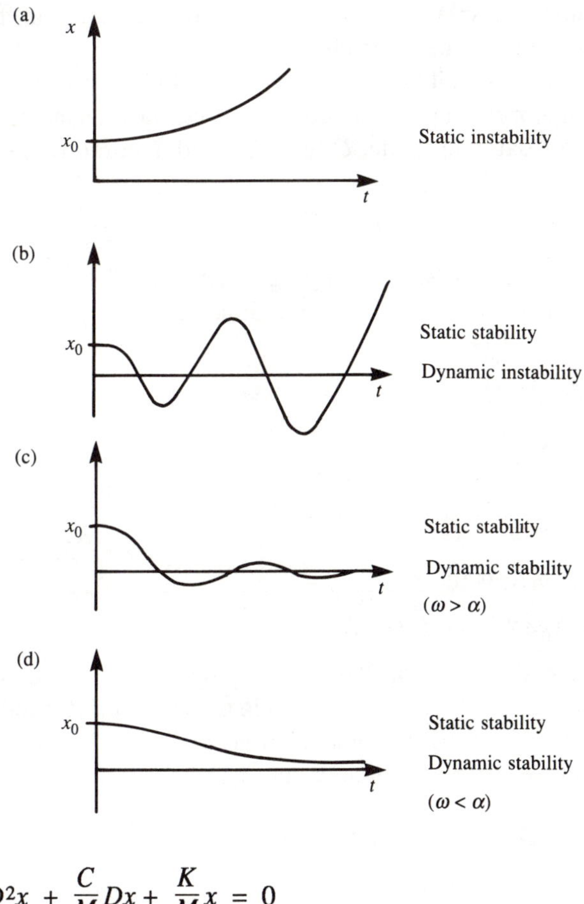

$$D^2x + \frac{C}{M}Dx + \frac{K}{M}x = 0$$

Dividing by x then gives the characteristic equation:

$$D^2 + \frac{C}{M}D + \frac{K}{M} = 0$$

The physical nature of the solution depends upon whether D is real (non-oscillatory) or complex (damped oscillatory); D is found by the usual quadratic equation solution:

$$D = \frac{-b \pm \sqrt{b^2 - 4ac}}{2a}$$

$$= -\frac{C}{2M} \pm \sqrt{\frac{C^2}{4M^2} - \frac{K}{M}}$$

$$= -\alpha \pm \sqrt{\alpha^2 - \omega^2}$$

where α (units of s^{-1}) is called the damping factor and ω is the undamped natural frequency (rad/s).

The mathematical solution will be complex (which physically means damped oscillatory) if $\omega > \alpha$, in which case we have undamped natural frequency ω, damping ratio ζ and damped natural frequency ω_d according to:

$$\ddot{x} + 2\zeta\omega\dot{x} + \omega^2 x = 0$$

Note the distinction between damping coefficient C (Ns/m), damping factor α (s^{-1}) and damping ratio ζ (non-dimensional).

$$\omega = \sqrt{\frac{K}{M}}$$

$$\zeta = \frac{\alpha}{\omega} = \frac{C}{2M\omega} = \frac{C}{2\sqrt{MK}}$$

$$\omega_d = \sqrt{\omega^2 - \alpha^2} = \omega\sqrt{1 - \zeta^2}$$

The actual displacement is

$$x = X_0 e^{-\alpha t} \sin(\omega_d t + \phi)$$

where α and ω depend upon the system properties, and the amplitude X_0 and phase angle ϕ depend upon the initial conditions of x and $\dot{x}$.

If $\omega < \alpha$, there will be two real solutions to D, and a non-oscillatory response with two time constants, τ_1 and τ_2:

$$-\frac{1}{\tau_1} = D_1 = -\alpha + \sqrt{\alpha^2 - \omega^2}$$

$$-\frac{1}{\tau_2} = D_2 = -\alpha - \sqrt{\alpha^2 - \omega^2}$$

The actual displacement is then

$$x = X_1 e^{-t/\tau_1} + X_2 e^{-t/\tau_2}$$

where τ_1 and τ_2 depend on the system properties and X_1 and X_2 depend upon the initial conditions of x and $\dot{x}$.

The system considered so far has had no additional external forces, so the behaviour is called the natural or free response.

If the system of Figure 7.2.1 is subject to an external driving force $F(t)$ then the equation of motion becomes

$$\ddot{x} + \frac{C}{M}\dot{x} + \frac{K}{M}x = \frac{F(t)}{M}$$

The solution to this is called the forced response. It is mathematically the sum of the complementary function and the particular integral. Physically, it is the sum of the corresponding transient and steady-state solutions. The complementary function represents the transient, and is the natural response with arbitrary constants depending on initial conditions, as before. The particular integral represents the 'steady-state' forced response once the starting transient has been damped out. The usual method of solution for the particular integral is to try a solution of the same form as $F(t)$. For example, for $F(t) = F_1 \sin \omega_f t$ where ω_f is the forcing frequency in rad/s and F_1 is the amplitude of the disturbing force, try

$$x = X \sin(\omega_f t + \phi)$$

where X is the displacement amplitude and ϕ is the phase angle of the response. Differentiating, and substituting in the equation of motion, the coefficient of the sine term must be equal for both sides. Assuming an oscillatory natural solution

$$\ddot{x} + 2\zeta\omega\,\dot{x} + \omega^2 x = 0$$

the response amplitude is given by

$$X = \frac{F_1/K}{\sqrt{[1 - (\omega_f/\omega)^2]^2 + 2\zeta(\omega_f/\omega)^2}}$$

If the force F_1 were applied steadily to stiffness K then the displacement would be F_1/K. The dynamic effect may be seen by comparing the displacement amplitude X with F_1/K. Figure 7.2.3 shows how the magnification ratio $X/(F_1/K)$ depends upon the frequency ratio ω_f/ω for various values of damping ratio ζ. Particularly notable is the resonant response for ω_f close to the natural frequency ω. For road vehicles the yaw damping ratio is generally in the range 0.2 to 1.0, and the yaw natural frequency is typically about 6 rad/s (1 Hz), so there may be observable resonance in yaw behaviour.

Figure 7.2.3. Magnification ratio versus frequency ratio.

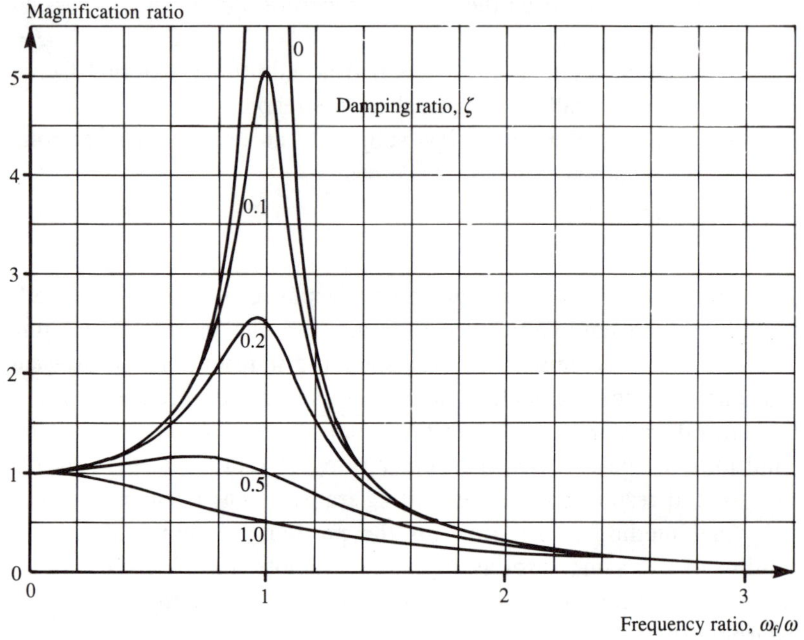

7.3 1-d.o.f. sideslip

In the one-degree-of-freedom sideslip model the vehicle is considered incapable of yaw; this is of course unrealistic and is investigated here only in order to throw light on the more complex two- and three-degrees-of-freedom models considered later. Figure 7.3.1 shows the bicycle model vehicle with sideforce $F(t)$ positioned to give zero yaw, with resultant lateral velocity component $\dot{y}$ giving attitude angle β. It is apparent from the figure that a steady $F(t)$ will cause the development of a steady $\dot{y}$ such that β gives adequate tyre forces to oppose $F(t)$. Also, because there is a damping force but no stiffness force, there will not be a natural frequency.

$$\alpha_f = \alpha_r = \beta = \frac{\dot{y}}{V}$$

$$\Sigma F_y = F(t) - 2C_{\alpha f}\alpha_f - 2C_{\alpha r}\alpha_r = m\dot{y}$$

$$m\ddot{y} + \frac{C_0}{V}\dot{y} = F(t)$$

where C_0 is the zeroth moment vehicle cornering stiffness:

$$C_0 = 2C_{\alpha f} + 2C_{\alpha r}$$

Figure 7.3.1. Single-degree-of-freedom sideslip model.

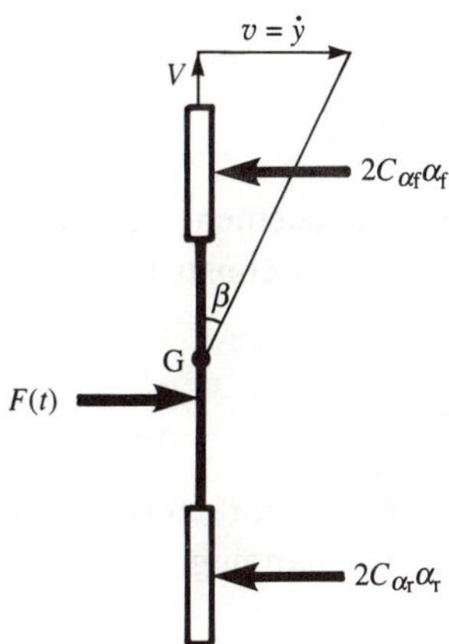

The characteristic equation of the free motion is

$$D^2 + \frac{C_0}{mV} D = 0$$

$$-\frac{1}{\tau} = D = -\frac{C_0}{mV}$$

$$\tau = \frac{mV}{C_0}$$

This confirms that the natural response is exponential rather than oscillatory. A typical value of τ is 0.2 s at 20 m/s. The solution for y is

$$y = Y\, e^{-t/\tau}$$

where Y depends on the initial conditions. The forced response to a step input force F is

$$\dot{y} = \frac{FV}{C_0}(1 - e^{-t/\tau})$$

The initial lateral acceleration is

$$\ddot{y}_0 = -\frac{FV}{\tau C_0} = \frac{F}{m}$$

as would be expected. The final drift speed is

$$\dot{y}_\infty = \frac{FV}{C_0}$$

which is proportional to V, because this gives the necessary value of $\beta = F/C_0$.

The lateral displacement response to the step force input is found by integrating $\dot{y}$, and evaluating the constant of integration from the initial condition $y = 0$, to give

$$y = \frac{FV}{C_0}[t - \tau(1 - e^{-t/\tau})]$$

In response to a forced displacement held constant at y there is no restoring force. In response to a constant displacement velocity $\dot{y}$ there is an opposing damping force

$$F = C_0\beta = \frac{C_0\dot{y}}{V}$$

Thus the sideslip motion is characterised by no stiffness and no oscillation, but strong positive damping.

7.4 1-d.o.f. yaw

In the one-degree-of-freedom yaw model the vehicle is considered pinned at G; it can yaw but not sideslip. Again, this is unrealistic, but the results help to illuminate the more complex models.

Figure 7.4.1. Single-degree-of-freedom yaw model.

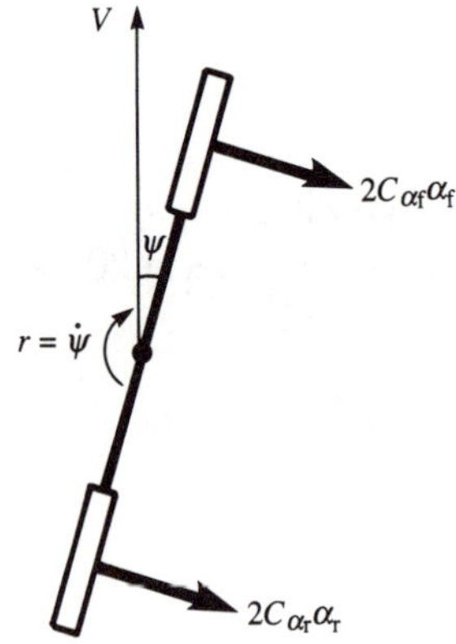

From Figure 7.4.1 the slip angles are

$$\alpha_f = \psi - \frac{a\dot{\psi}}{V}$$

$$\alpha_r = \psi + \frac{b\dot{\psi}}{V}$$

where ψ is the heading angle and $\dot{\psi} \equiv r$ is the yaw angular speed. The equation of motion in yaw is

$$\Sigma M = 2a\, C_{\alpha f}\, \alpha_f - 2b\, C_{\alpha r}\, \alpha_r = I\ddot{\psi}$$

$$2aC_{\alpha f}\left(\psi - \frac{a\dot{\psi}}{V}\right) - 2bC_{\alpha r}\left(\psi + \frac{b\dot{\psi}}{V}\right) = I\ddot{\psi}$$

$$I\ddot{\psi} + \left(\frac{2a^2C_{\alpha f} + 2b^2C_{\alpha r}}{V}\right)\dot{\psi} - (2aC_{\alpha f} - 2bC_{\alpha r})\,\psi = 0$$

$$\ddot{\psi} + \frac{C_2}{IV}\dot{\psi} - \frac{C_1}{I}\psi = 0$$

where

$$C_1 = 2a\,C_{\alpha f} - 2b\,C_{\alpha r}$$

$$C_2 = 2a^2\,C_{\alpha f} - 2b^2\,C_{\alpha r}$$

are the first and second moment vehicle cornering stiffnesses.

By comparison with the standard one-degree-of-freedom vibration equation of Section 7.2, which was

$$\ddot{x} + \frac{C}{M}\dot{x} + \frac{K}{M}x = 0$$

the yaw stiffness is $-C_1$ and the yaw damping coefficient is C_2/V. For static stability the yaw stiffness should be restoring, i.e. C_1 should be negative. Since, from Chapter 6, the understeer gradient is

$$k = -\frac{mC_1}{4lC_{\alpha f}C_{\alpha r}}$$

static stability requires positive k, i.e. understeer. Dynamic stability requires $C_2/I > 0$, and hence $C_2 > 0$, which will always be the case; i.e. because $C_2 > 0$ there is always positive damping.

If the vehicle is oversteer then the motion will be a non-oscillatory divergence from the equilibrium position. For an understeer vehicle the motion will be a return to the equilibrium position. Whether it is oscillatory or not depends on the characteristic equation:

$$D^2 + \frac{C_2}{IV}D - \frac{C_1}{I} = 0$$

$$D = -\frac{C_2}{2IV} \pm \sqrt{\left(\frac{C_2}{2IV}\right)^2 + \frac{C_1}{I}}$$

The motion will be oscillatory if D is mathematically complex, i.e. if

$$\left(\frac{C_2}{2IV}\right)^2 + \frac{C_1}{I} < 0$$

$$V^2 > \frac{-C_2^2}{4IC_1}$$

where the right-hand side is positive because C_1 is negative for understeer. Hence there is a transition speed

$$V_t = \sqrt{\frac{-C_2^2}{4IC_1}}$$

below which the response is overdamped. Above this speed there is a natural frequency and damping ratio that may be found by comparing the characteristic equation with this form of the standard single-degree-of-freedom vibration equation:

$$\ddot{x} + 2\zeta\omega\dot{x} + \omega^2 x = 0$$

Hence

$$\omega = \sqrt{\frac{C_1}{I}}$$

which is typically about 10 rad/s (1.6 Hz). The damping ratio is

$$\zeta = \frac{C_2}{2IV\omega} = \frac{C_2}{V\sqrt{4IC_1}} = \frac{V_t}{V}$$

where V_t is the transition speed given by the equation earlier. Hence the damping ratio is inversely proportional to speed. More understeer (greater magnitude of C_1) will increase the natural frequency, and reduce V_t, i.e. reduce the damping.

In accordance with standard vibration theory, the damped natural frequency in rad/s is

$$\omega_d = \omega\sqrt{1-\zeta^2}$$

This is zero at the transition speed.

Thus the single-degree-of-freedom yaw motion is characterised by continuous divergence for oversteer; for understeer it is convergent and possibly oscillatory. There is both stiffness and damping.

7.5 2-d.o.f. model (vehicle-fixed axes)

The equations of motion for the vehicle may be expressed in coordinates fixed to the Earth (*XYZ*) or coordinates fixed to the vehicle (*xyz*); vehicle-fixed axes will be used in this section. First it is necessary to find the vehicle lateral acceleration in terms of the absolute motion. Figure 7.5.1 shows the vehicle in axes *XYZ*, moving substantially in the *X* direction, with angles shown exaggerated. The path angle from the *X*-axis is ν (nu), the heading angle is ψ and the attitude angle is β. The total speed V tangent to the path may be resolved into V_{ax} and $V_{ay} \equiv v$ as used in this section, or into V_X and $V_Y \equiv u$ as in the next section.

Figure 7.5.1. Angles and velocity components. (Note distinction between u and v, V_{ax} and V_X, and ν and β.)

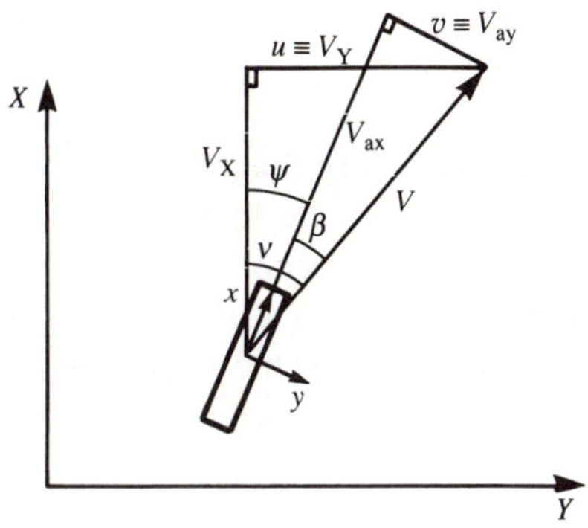

$$V_{ax} = V \cos\beta \approx V$$

$$V_X = V \cos\nu \approx V$$

$$V_{ay} \equiv v \approx \beta V$$

$$V_Y \equiv u \approx \nu V$$

Care is needed here to distinguish angle ν (Greek nu) from speed component v (vee).

Here V_{ax} is the component, in the x direction, of the absolute velocity, in contrast to V_x which would be the component of velocity in xyz, which is zero because xyz are the vehicle-fixed axes. The vehicle has yaw angular velocity $r = \dot{\psi}$ where ψ is the heading angle.

Figure 7.5.2 shows the vehicle position and orientation at time t and $t + \mathrm{d}t$. The relative rotation is $r\,\mathrm{d}t$.

Figure 7.5.2. Velocity components at path points.

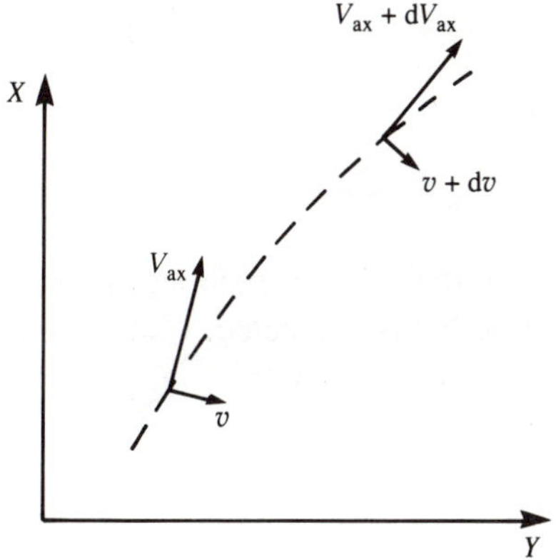

The absolute acceleration in the y direction is

$$A_{ay}\,dt = dv\cos(r\,dt) + V\sin(r\,dt)$$

$$\approx dv + Vr\,dt$$

$$A_{ay} = \dot{v} + Vr$$

Figure 7.5.3 shows the free-body diagram in vehicle-fixed axes, for a bicycle model vehicle, with roll and load transfer being neglected. The vehicle is considered in the accelerating vehicle-fixed axes, so appropriate compensation force and moment (mA_{ay} and $I\dot{r}$) are included (as explained in Chapter 1). The vehicle has been subjected to a small disturbance, now having a lateral velocity component v. As a result there is an attitude angle

$$\beta = \frac{v}{V_{ax}} \approx \frac{v}{V}$$

Figure 7.5.3. Vehicle free-body diagram in vehicle-fixed axes.

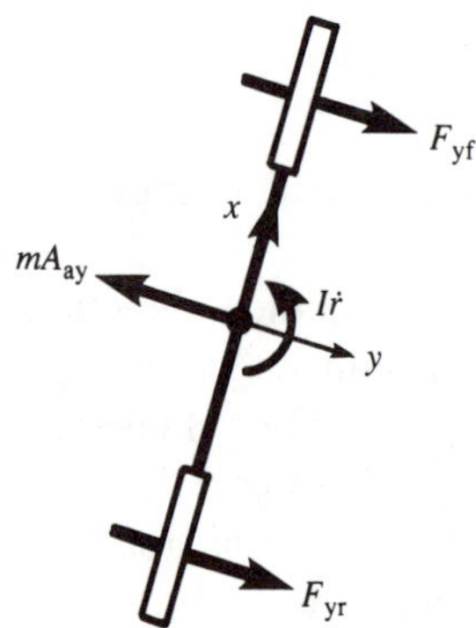

Allowing for the attitude angle and the yaw rotation speed r, the slip angles are

$$\alpha_f = \frac{v}{V} + \frac{ar}{V}$$

$$\alpha_r = \frac{v}{V} - \frac{br}{V}$$

This is a fixed control analysis; in practice there would still be some control compliance, but this is neglected. Using the linear approximation there are corresponding tyre forces of

$$F_{yf} = 2\,C_{\alpha f}\,\alpha_f$$

$$F_{yr} = 2\,C_{\alpha r}\,\alpha_r$$

Bearing in mind that the acceleration in the vehicle-fixed axes is zero by definition, the equations of lateral motion for the vehicle of Figure 7.5.3 are

$$\Sigma F_y = F_{yf} + F_{yr} - mA_{ay} = 0$$

$$\Sigma M = aF_{yf} - bF_{yr} - I\dot{r} = 0$$

Substituting in the above two equations and collecting terms gives

$$m\dot{v} + \left(\frac{2C_{\alpha f} + 2C_{\alpha r}}{V}\right)v + \left(mV + \frac{2aC_{\alpha f} - 2bC_{\alpha r}}{V}\right)r = 0$$

$$\left(\frac{2aC_{\alpha f} - 2bC_{\alpha r}}{V}\right)v + I\dot{r} + \left(\frac{2a^2C_{\alpha f} + 2b^2C_{\alpha r}}{V}\right)r = 0$$

These equations may be simplified by using the vehicle cornering stiffness constants C_0, C_1 and C_2 (Section 6.6), e.g. $C_0 = 2C_{\alpha f} + 2C_{\alpha r}$. Also introducing D to represent differentiation with respect to time, the force and moment equations can be represented more concisely as:

$$\left(mD + \frac{C_0}{V}\right)v + \left(mV + \frac{C_1}{V}\right)r = 0$$

$$\left(\frac{C_1}{V}\right)v + \left(ID + \frac{C_2}{V}\right)r = 0$$

These equations are a representation of the natural motion of the vehicle observed in the vehicle-fixed axes. If there are external forces additionally present, e.g. aerodynamic or steering forces, then these would appear as time-dependent forcing functions on the right-hand side.

Actually these equations can be expressed not only in v and r as above; r may be replaced by $D\psi$. However it is not useful to replace v by Dy because the cumulative displacement in the varying y direction has no practical significance. A useful alternative form may be found by using the attitude angle $\beta = v/V$, hence expressing the equations in β and r, or in β and ψ.

The characteristic equation for the above pair of simultaneous equations is found by multiplying the first by $(ID + C_2/V)$ and the second by $(mV + C_1/V)$ and subtracting, giving:

$$\left(aD + \frac{C_0}{V}\right)\left(ID + \frac{C_2}{V}\right) - \left(mV + \frac{C_1}{V}\right)\left(\frac{C_1}{V}\right) = 0$$

Expanding this gives

$$mID^2 + \left(\frac{mC_2}{V} + \frac{IC_0}{V}\right)D + \left(\frac{C_0C_2}{V^2} - mC_1 - \frac{C_1^2}{V^2}\right) = 0$$

This is the characteristic equation. It is the condition for there to be a non-zero solution to the simultaneous differential equations of motion, and therefore tells us the nature of the free response, and in particular tells us the vehicle's natural frequency and damping (Section 7.7).

7.6 2-d.o.f. model (Earth-fixed axes)

If it is desired to compare theory with data measured by vehicle-fixed instrumentation, e.g. lateral acceleration, then the formulation of the previous section is suitable. However, if it is desired to analyse the vehicle motion as seen in Earth-fixed coordinates, e.g. to find the amplitude of path oscillations in response to an oscillatory steer input, then the equations of motion should be expressed in Earth-fixed axes.

The slip angles in Earth-fixed axes are:

$$\alpha_\mathrm{f} = \frac{u}{V} + \frac{ar}{V} - \psi$$

$$\alpha_\mathrm{r} = \frac{u}{V} - \frac{br}{V} - \psi$$

where $r = \dot{\psi}$ is the yaw angular speed. Compared with the slip angle expressions of the last section, the use of u instead of v results in the appearance of ψ in the expressions.

Figure 7.6.1, shows the free-body diagram in the Earth-fixed axes XYZ. These are inertial axes, so compensation forces are not required. From the free-body diagram, the equations of motion in Earth-fixed axes are:

Figure 7.6.1. Vehicle free-body diagram in Earth-fixed axes.

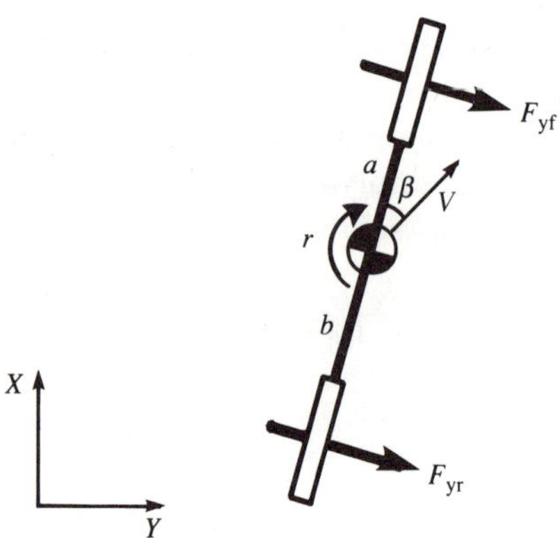

$$\Sigma F_Y = F_{yf} + F_{yr} = mA_Y$$

$$\Sigma M = aF_{yf} - bF_{yr} = I\dot{r}$$

Substituting, as in the last section, collecting terms and using the D operator notation gives:

$$\left(mD + \frac{C_0}{V}\right)u + \left(\frac{C_1}{V}D - C_0\right)\psi = 0$$

$$\left(\frac{C_1}{V}\right)u + \left(ID^2 + \frac{C_2}{V}D - C_1\right)\psi = 0$$

This may also be expressed in other forms, e.g. by using $u = DX$, or using $u = vV$. Using the latter substitution gives:

$$(mVD + C_0)\, v + \left(\frac{C_1}{V}D - C_0\right)\psi = 0$$

$$(C_1)v + \left(ID^2 + \frac{C_2}{V}D - C_1\right)\psi = 0$$

The above equations can be contrasted with the equivalent formulation, in β and ψ, for vehicle-fixed axes from the last section, which were

$$(mDV + C_0)\,\beta + \left(mVD + \frac{C_1}{V}D\right)\psi = 0$$

(vehicle-fixed axes)

$$(C_1)\,\beta + \left(ID^2 + \frac{C_2}{V}D\right)\psi = 0$$

Elimination of v or ψ from the Earth-fixed axes formulation, after cancellation of terms and dividing by VD, gives the following characteristic equation:

$$mI\,D^2 + \left(\frac{mC_2}{V} + \frac{IC_0}{V}\right)D + \left(\frac{C_0C_2}{V^2} - mC_1 - \frac{C_1^2}{V^2}\right) = 0$$

Despite the different starting equations for the Earth-fixed axes, this is the same characteristic equation as was found for the vehicle-fixed axes; physically this is correct because the natural frequency and damping ratio must be the same in both coordinate systems.

If there are additional forces, e.g. aerodynamic or front steer angle, these will appear on the right-hand side of the equations of motion, as functions of time.

Eliminating ψ from the various sets of equations will give different results for u and v, as it should, because the different coefficients of ψ for the two sets of axes will affect the right-hand side differently.

7.7 2-d.o.f. free response

The nature of the free response, i.e. the stability, natural frequency and damping ratio, depends upon the characteristic equation as found for both Earth-fixed and vehicle-fixed axes:

$$mID^2 + \left(\frac{mC_2}{V} + \frac{IC_0}{V}\right)D + \left(\frac{C_0C_2}{V^2} - mC_1 - \frac{C_1^2}{V^2}\right) = 0$$

The conditions for stability of motion can be seen by comparing the characteristic equation with the single-degree-of-freedom vibration equation in the following form, from Section 7.2, where the spring stiffness K is positive for a restoring force (opposing displacement) and C is positive for a positive damping force (opposing velocity):

$$\ddot{x} + \frac{C}{M}\dot{x} + \frac{K}{M}x = 0$$

In the characteristic equation, the inertia term mI is always positive. Dynamic stability requires that the damping coefficient C be positive, i.e. that

$$\frac{mC_2}{V} + \frac{IC_0}{V} > 0$$

This will always be true, so if there is instability it must be static, i.e. due to a negative stiffness. For static stability $K > 0$, so

$$\frac{C_0C_2}{V^2} - mC_1 - \frac{C_1^2}{V^2} > 0$$

From Chapter 6, the understeer gradient is

$$k = -\frac{mC_1}{4lC_{\alpha f}C_{\alpha r}}$$

so for an understeer vehicle C_1 is negative, and, from the previous equation, static stability is assured for all speeds. For an oversteer vehicle (positive C_1) static stability requires

$$V^2 < \frac{C_0C_2}{mC_1} - \frac{C_1}{m}$$

Substituting and simplifying gives

$$V^2 < \frac{4l^2C_{\alpha f}C_{\alpha r}}{mC_1}$$

Hence an oversteer vehicle will be stable up to a critical speed V_{cr} beyond which it will be unstable, where

$$V_{cr} = \sqrt{\frac{4l^2 C_{\alpha f} C_{\alpha r}}{mC_1}} = \sqrt{\frac{1}{-k}}$$

$$= \sqrt{\frac{C_0 C_2 - C_1^2}{mC_1}}$$

For understeer, negative C_1, the characteristic speed is

$$V_{ch} = \sqrt{\frac{-(C_0 C_2 - C_1^2)}{mC_1}}$$

The undamped natural frequency of the free motion may be found by comparing the characteristic equation with the single-degree-of-freedom vibration equation in the following form, where ω is the undamped natural frequency in rad/s, and ζ is the damping ratio:

$$\ddot{x} + 2\zeta\omega\,\dot{x} + \omega^2 x = 0$$

Hence the undamped natural frequency in rad/s is given by

$$\omega^2 = \frac{\dfrac{C_0 C_2}{V^2} - mC_1 - \dfrac{C_1^2}{V^2}}{mI}$$

At high speeds, ω tends to the value $\sqrt{-C_1/I}$ which is the value for the single-degree-of-freedom pure yaw model. Hence the sideslip tends to increase the undamped natural frequency, especially at low speed.

Using

$$V_{ch}^2 = -\frac{C_0 C_2 - C_1^2}{mC_1}$$

then

$$\omega^2 = -\frac{C_1}{I}\left[1 + \left(\frac{V_{ch}}{V}\right)^2\right]$$

Expressing ω at infinite speed as ω_∞,

$$\omega_\infty^2 = -\frac{C_1}{I}$$

$$\frac{\omega}{\omega_\infty} = \sqrt{1 + \left(\frac{V_{ch}}{V}\right)^2}$$

At $V = V_{ch}$, $\omega/\omega_\infty = \sqrt{2}$, Figure 7.7.1.

For a neutral steer vehicle, $C_1 = 0$ and $k = 0$, so $\omega_\infty = 0$ and

$$\omega = \sqrt{\frac{C_0 C_2/mI}{V}}$$

tending to zero at infinite speed.

Figure 7.7.1. Undamped yaw natural frequency versus speed.

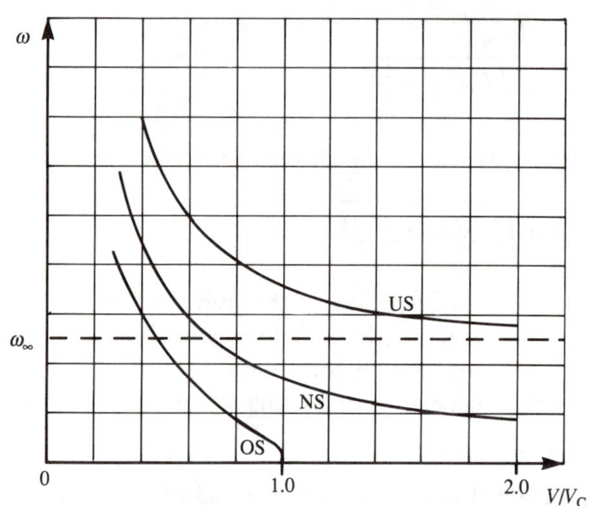

For an oversteer vehicle,

$$\omega^2 = \frac{C_1}{I}\left[\left(\frac{V_{cr}}{V}\right)^2 - 1\right]$$

and the undamped natural frequency ω goes to zero, Figure 7.7.1, at

$$V = \sqrt{\frac{C_0 C_2 - C_1^2}{mC_1}} = V_{cr}$$

Above this speed the response is non-oscillatory, and there is no natural frequency, and no real solution for ω.

The damping coefficient in the characteristic equation is $(mC_2 + IC_0)/V$; this is the sum of the damping coefficients from pure sideslip and pure yaw, and these motions contribute approximately equally to the damping ($a \approx b \approx k_z$, so $mC_2 \approx IC_0$). The damping ratio of the free motion may be found by comparing the characteristic equation with the single-degree-of-freedom vibration equation in the following form, where ζ is the damping ratio:

$$\ddot{x} + 2\zeta\omega\,\dot{x} + \omega^2 x = 0$$

Hence the damping ratio is given by

$$\zeta = \frac{mC_2 + IC_0}{2mIV\omega}$$

For understeer,

$$\omega^2 = -\frac{C_1}{I}\left[1 + \left(\frac{V_{ch}}{V}\right)^2\right]$$

so

$$\zeta = \frac{mC_2 + IC_0}{2mI\sqrt{\frac{-C_1}{I}}\sqrt{V^2 + V_{ch}^2}}$$

$$= \frac{V_d}{\sqrt{V^2 + V_{ch}^2}}$$

where V_d is the characteristic damping speed:

$$V_d = \frac{mC_2 + IC_0}{\sqrt{-4m^2C_1I}}$$

$$= \frac{C_2 + k_z^2 C_0}{\sqrt{-4IC_1}}$$

where k_z is the yaw radius of gyration. Hence

$$\zeta = \frac{V_d/V_{ch}}{\sqrt{1 + \left(\frac{V}{V_{ch}}\right)^2}}$$

$$= \frac{\zeta_0}{\sqrt{1 + \left(\frac{V}{V_{ch}}\right)^2}}$$

where ζ_0 is the damping ratio at zero speed.

$$\zeta_0 = \frac{V_d}{V_{ch}}$$

$$= \frac{C_2 + k_z^2 C_0}{\sqrt{4k_z^2(C_0C_2 - C_1^2)}}$$

For the neutral steer case, V_{ch} is infinite, and C_1 is zero, so the damping ratio is constant for all speeds, at

$$\zeta = \frac{C_2 + k_z^2 C_0}{\sqrt{4k_z^2 C_0 C_2}}$$

For the oversteer case, an analysis similar to the understeer case may be applied, giving a characteristic damping speed

$$V_d = \frac{C_2 + k_z^2 C_0}{\sqrt{4IC_1}}$$

and a damping ratio

$$\zeta = \frac{\zeta_0}{\sqrt{1 - \left(\frac{V}{V_{cr}}\right)^2}}$$

The zero-speed damping ratio is

$$\zeta_0 = \frac{V_d}{V_{cr}}$$

$$= \frac{C_2 + k_z^2 C_0}{\sqrt{4k_z^2(C_0 C_2 - C_1^2)}}$$

as before.

The possible values of ζ_0 may be investigated as follows. In practice

$$C_1^2 \ll C_0 C_2$$

and

$$a \approx b \approx k_z$$

so

$$C_2 \approx k_z^2 C_0$$

giving

$$\zeta_0 \approx 1.0$$

Actually, it is possible to place a definite limit on ζ_0 in the following way:

$$\zeta_0 = \frac{C_2 + k_z^2 C_0}{\sqrt{4k_z^2 (C_0 C_2 - C_1^2)}}$$

Hence

$$\zeta_0 > \frac{C_2 + k_z^2 C_0}{\sqrt{k_z^2 C_0 C_2}} = \frac{\frac{1}{2}(C_2 + k_z^2 C_0)}{\sqrt{C_2 k_z^2 C_0}}$$

This is the ratio of the arithmetic mean of C_2 and $k_z^2 C_0$ to their geometric mean, which must be greater than or equal to 1.0, so

$$\zeta_0 \geqslant 1.0$$

Thus $\zeta_0 < 1$ is excluded from physical solutions.

Figure 7.7.2 shows the variation of ζ_0 with radius of gyration k_z for an example case of various C_1 values, for $a = b = 1.5$ m, $C_0 = 120$ kN/rad, and $C_2 = 270$ kNm²/rad. Again, this indicates that, for the two-degrees-of-freedom model, values of ζ_0 are a little in excess of 1.0, and not below 1.0.

Figure 7.7.3 shows how ζ varies with V/V_c for understeer, neutral and oversteer cases. For understeer, $\zeta = \zeta_0/\sqrt{2}$ at the characteristic speed, and $\zeta_0/\sqrt{5}$ at $2V_{ch}$. For oversteer, ζ goes to infinity at V_{cr}.

From a practical point of view, it is notable that at high speed the damping ratio becomes rather small if there is a large understeer coefficient. To avoid excessive overshoot, V_{ch} should be not less than $\frac{1}{2}V_{max}$, i.e. $\zeta > 0.45$. On a wheelbase of 2.8 m with a maximum speed of 50 m/s, this implies that the characteristic speed should be at least 25 m/s, and therefore, using

$$V_{ch} = \sqrt{\frac{l}{k}}$$

that the understeer gradient k should not exceed 0.0045 rad/m s^{-2} (2.5 deg/g).

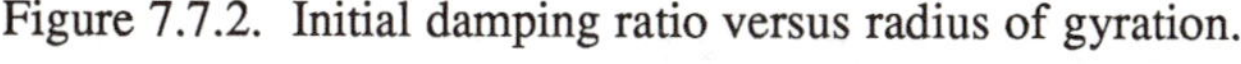

Figure 7.7.2. Initial damping ratio versus radius of gyration.

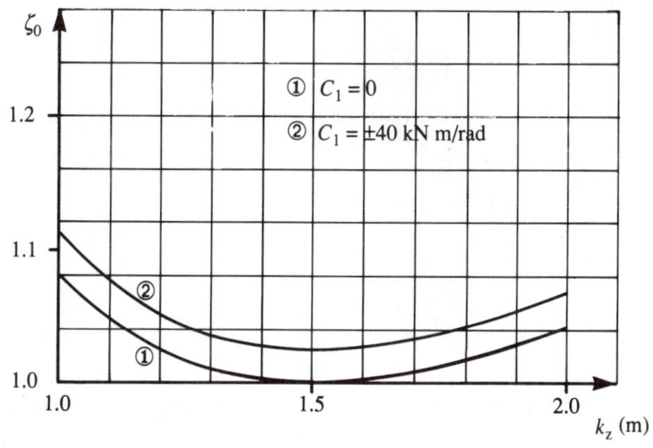

Figure 7.7.3. Damping ratio versus speed.

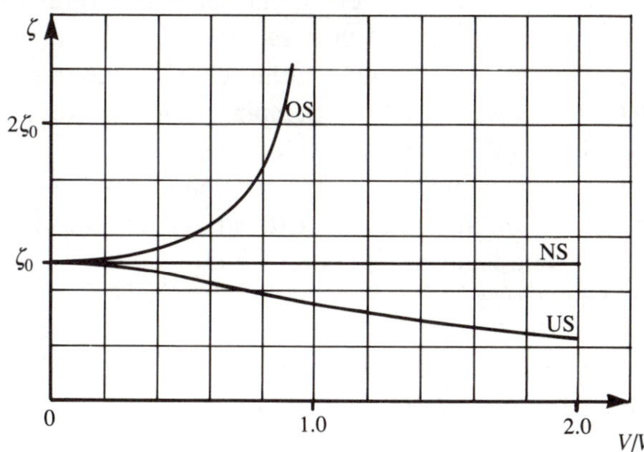

The damped natural frequency, which is the frequency actually observed, is

$$\omega_d = \omega\sqrt{1-\zeta^2}$$

The damping ratio is reasonably high except for extreme understeer vehicles, so the damped natural frequency is significantly lower than the undamped one.

For the understeer case

$$\frac{\omega_d^2}{\omega^2} = 1-\zeta^2$$

$$= 1 - \frac{\zeta_0^2}{1+\left(\dfrac{V}{V_{ch}}\right)^2}$$

$$= \frac{1+\left(\dfrac{V}{V_{ch}}\right)^2 - \zeta_0^2}{1+\left(\dfrac{V}{V_{ch}}\right)^2}$$

Also

$$\frac{\omega^2}{\omega_\infty^2} = 1+\left(\frac{V_{ch}}{V}\right)^2$$

giving

$$\frac{\omega_d^2}{\omega_\infty^2} = 1-\left(\frac{V_{ch}}{V}\right)^2(\zeta_0^2-1)$$

where $\zeta_0 \gtrless 1$. Hence it is necessary to distinguish two cases, in principle, according to the value of ζ_0, Figure 7.7.4. For the limiting case of $\zeta_0 = 1$, ω_d is constant and equal to ω_∞. For $\zeta_0 < 1$, when V/V_{ch} is less than $\sqrt{(\zeta_0^2-1)}$ then $\zeta > 1$ and there is no damped natural frequency. For V/V_{ch} greater than $\sqrt{(\zeta_0^2-1)}$ then ω_d is real, and increases up to ω_∞. This increase occurs because although the undamped natural frequency decreases with speed, the damping ratio also decreases with speed, and the latter is the dominant effect.

For a neutral-steer vehicle, the damping ratio always equals ζ_0, which is greater than or equal to 1.0, so the response is always overdamped, and there is no real solution for ω_d.

For the oversteer case,

$$\frac{\omega_d^2}{\omega^2} = 1 - \frac{\zeta_0^2}{1-\left(\dfrac{V}{V_{cr}}\right)^2}$$

Figure 7.7.4. Damped natural frequency versus speed (understeer).

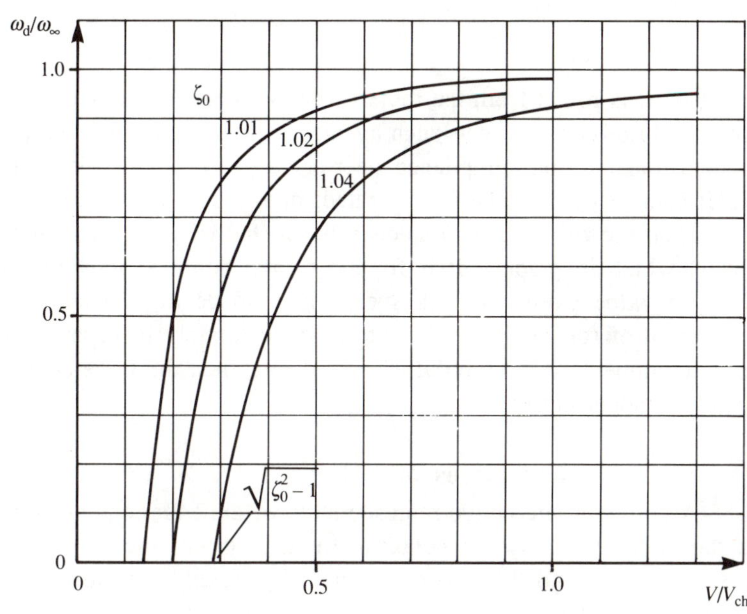

There is an oscillatory solution for $V < V_{cr}$, with

$$\omega^2 = \frac{C_1}{I}\left[\left(\frac{V_{cr}}{V}\right)^2 - 1\right]$$

giving

$$\frac{\omega_d^2}{C_1/I} = -1 - \left(\frac{V_{cr}}{V}\right)^2 (\zeta_0^2 - 1)$$

Thus, since $\zeta_0 \geqslant 1$, there is no speed with real ω_d, and all responses are overdamped. Above V_{cr}, of course, the response is divergent.

7.8 Improved 2-d.o.f. model

The accuracy and utility of the various equations involving C_1 found for the two-degrees-of-freedom model may be enhanced by using an effective C_1 deduced from a known understeer gradient or static margin, i.e. instead of the simplistic

$$C_1 = 2aC_{\alpha f} - 2bC_{\alpha r}$$

we may adapt the simple equation for understeer from Chapter 6:

$$k = -\frac{mC_1}{4lC_{\alpha f}C_{\alpha r}}$$

to give

$$C_1 = -\frac{4klC_{\alpha f}C_{\alpha r}}{m}$$

or

$$C_1 = -x_s C_0$$

where k or x_s is found from the more extensive steady-state analysis of Chapter 6. In this way, effects such as pneumatic trail, rolling resistance with load transfer, and compliance steer can be incorporated. Roll steer and roll camber can also be incorporated; in this case the model really requires that the roll natural frequency is substantially higher than that for yaw, which in practice it is for most cars, being about twice the undamped value even for high speed. If it is desired to study the development of roll angle itself, or to more accurately incorporate roll steer and camber and the inertial effects of roll, then the three-degrees-of-freedom model must be used.

7.9 Stability derivatives

A common alternative representation of the equations of motion uses the stability derivative notation. In this system the sideforce is represented by Y and the yawing moment by N, and the variation of these with respect to a motion parameter is shown by a subscript; for example the rate of change of sideforce with attitude angle is

$$Y_\beta \equiv \frac{\mathrm{d}Y}{\mathrm{d}\beta}$$

Hence the equations of motion may be expressed in the vehicle-fixed axes by

$$\Sigma Y = Y_\beta \beta + Y_r r + Y - mV(\dot{\beta} + r) = 0$$

$$\Sigma N = N_\beta \beta + N_r r + N - I\dot{r} = 0$$

where Y and N are additional forces and moments, either external (e.g. aerodynamic) or steering ($Y_\delta \delta$), and $\dot{\beta} = \dot{v}/V$. As discussed in Section 6.6, the actual derivatives for the two-degrees-of-freedom model are

$$Y_\beta = -(2C_{\alpha f} + 2C_{\alpha r}) = -C_0$$

$$Y_r = -\frac{1}{V}(2aC_{\alpha f} + 2bC_{\alpha f}) = -\frac{C_1}{V}$$

$$N_\beta = -(2aC_{\alpha f} - 2bC_{\alpha r}) = -C_1$$

$$N_r = -\frac{1}{V}(2a^2C_{\alpha f} + 2b^2C_{\alpha r}) = -\frac{C_2}{V}$$

With this notation, the simultaneous differential equations of free motion in the vehicle-fixed axes become

$$(mVD - Y_\beta)\,\beta + (mV - Y_r)\,r = 0$$
$$(-N_\beta)\,\beta + (ID - N_r)\,r = 0$$

and in the Earth-fixed axes they become

$$\left(mD - \frac{Y_\beta}{V}\right)u \quad + \quad ((-Y_r)D + Y_\beta)\,\psi \;=\; 0$$
$$\left(\frac{-N_\beta}{V}\right)u \;+\; (ID^2 - N_r D + N_\beta)\,\psi \;=\; 0$$

The characteristic equation for both cases becomes

$$mI\,D^2 - \left(N_r m + \frac{IY_\beta}{V}\right)D + \left(\frac{Y_\beta N_r}{V} + mN_\beta - \frac{Y_r N_\beta}{V}\right) = 0$$

For the two-degrees-of-freedom model there is no great advantage in the use of stability derivatives; in fact rather the contrary because Y_r and N_r vary with speed and hence obscure the effect of speed on the equations. However the stability derivative notation is useful for the more complex three-degrees-of-freedom model because it is more concise.

7.10 3-d.o.f. model

As discussed in Section 7.8, some of the effects of roll, for example roll steer, can be included indirectly in the two-degrees-of-freedom model. However a fuller representation requires that roll is included as an explicit variable, usually denoted by ϕ for the roll angle and p for the roll angular speed. This is known as the three-degrees-of-freedom model (sideslip, yaw and roll), and was first systematically investigated by Segel.

The vehicle is considered as two masses – the rolling sprung mass m_S and the unsprung mass m_U. The sprung mass is considered to roll about a longitudinal axis parallel to the ground, at the same height as the 'real' roll axis where it meets the vertical z-axis. The origin of the vehicle-fixed xyz axes is not taken at the centre of mass, but on the roll axis below the centre of mass. The basic equations of motion for sideforce, yaw moment and roll moment are

$$\Sigma Y = m(\dot{v} + Vr) + m_S\, d\,\dot{p}$$
$$\Sigma N = I_z \dot{r} + I_{xz}\dot{p}$$
$$\Sigma L = I_x \dot{p} + m_S\, d\,(\dot{v} + Vr) + I_{xz}\,\dot{r}$$

where I_x is the second moment of sprung mass about the x-axis; because this does not pass through G, this includes an $m_S d^2$ term, where d is the height of G above the roll axis.

The external forces and moments acting on the vehicle, in stability derivative notation, are

$$\Sigma Y = Y_\beta \beta + Y_r r + Y_\phi \phi + Y$$

$$\Sigma N = N_\beta \beta + N_r r + N_\phi \phi + N$$

$$\Sigma L = L_p p + L_\phi \phi + L$$

where some small terms, e.g. Y_p, are neglected, and Y, N and L are additional applied force and moment, e.g. aerodynamic or steering ($Y_\delta \delta$). The actual values of the stability derivatives depend upon the design parameters of the vehicle, but are generally more complex than for the simple two-degrees-of-freedom model, being

$$Y_\beta = -(2C_{\alpha f} + 2C_{\alpha r})$$

$$Y_r = -\frac{2aC_{\alpha f} - 2bC_{\alpha r}}{V}$$

$$Y_\phi = -(2\varepsilon_{Sf} C_{\alpha f} - 2\varepsilon_{Sr} C_{\alpha r}) - (2\varepsilon_{Cf} c C_{\alpha f} - 2\varepsilon_{Cr} c C_{\alpha r})$$

where ε_S is the roll steer coefficient, ε_C is the roll camber coefficient and c is the tyre camber/slip stiffness ratio C_γ / C_α.

$$N_\beta = -(2aC_{\alpha f} - 2bC_{\alpha r}) + \frac{dM_{Zf}}{d\alpha_f} + \frac{dM_{Zr}}{d\alpha_r} + \mu_R (2C_{\alpha f} h_f + 2C_{\alpha r} h_r)$$

where M_Z is the tyre self-aligning moment, μ_R is the rolling resistance coefficient and h is the height of the roll axis above the ground.

$$N_r = -\frac{2a^2 C_{\alpha f} + 2b^2 C_{\alpha r}}{V}$$

$$N_\phi = (2a\varepsilon_{Sf} C_{\alpha f} - 2b\varepsilon_{Sr} C_{\alpha r}) - (2a\,\varepsilon_{Cf} c\, C_{\alpha f} - 2b\,\varepsilon_{Cr} c\, C_{\alpha r}) - \mu_R k_S$$

where k_S is the suspension roll stiffness.

$$L_p = \frac{dL}{dp}$$

which depends primarily on the damper properties, and

$$L_\phi = k_S + m_S g d$$

where $m_S g d$ is the effect of roll in moving G from directly above the roll axis.

In addition, for steering effects:

$$Y_\delta = 2C_{\alpha f}$$

$$N_\delta = 2aC_{\alpha f} - \frac{dM_{Zf}}{d\alpha_f}$$

The equations given above are essentially the equations given by Segel, expressed in slightly different notation.

Knowing these actual values for the stability derivatives, the external forces can be inserted into the equations of motion to yield three simultaneous differential equations, for the three independent variables (e.g. lateral speed, yaw angle, roll angle). These are therefore solvable in principle, although analytic solutions are tedious to perform; the Laplace transform method was used by Segel, with results discussed later for various steer inputs.

In comparing this model with the simpler two-degrees-of-freedom model as expressed in Section 7.9 it should be appreciated that it is more complex for two separate reasons. Firstly, it includes more detail, e.g. aligning torque, and secondly, it includes roll as a separate variable giving three differential equations. This more complex model has been found to give very good agreement with experiment in the linear region of handling.

The natural motion and stability depend upon the characteristic equation. For the three-degrees-of-freedom model this is a quartic:

$$A_4D^4 + A_3D^3 + A_2D^2 + A_1D + A_0 = 0$$

As for the two-degrees-of-freedom model, it is found in practice that for fixed control dynamic instability does not occur, i.e. that if there is instability it is static, and hence depends simply on the sign of A_0, a positive value being stable, where

$$A_0 = -Y_\beta (N_r L_\phi + m_S dVN_\phi) - N_\beta \; [L_\phi (mV - Y_r) - m_S dVY_\phi]$$

There are two principal modes of oscillation, having natural frequencies of typically 0.5 Hz and 1 Hz, the lower frequency mode being primarily yawing and sideslipping, the other being primarily roll. Provided that the two natural frequencies are not too close, an approximate factorisation of the characteristic equation is reasonably accurate. Ellis developed the following approximation:

$$\left[mVI_zD^2 - (I_zY_\beta + mVN_r)D + (Y_\beta N_r + mVN_\beta - Y_r N_\beta) - \frac{m_S dV}{L_\phi}(N_\beta Y_\phi - Y_\beta N_\phi)\right]$$
$$\times \left[\left(I_x - \frac{m_S^2 d^2}{m}\right)D^2 - L_p D - L_\phi\right] = 0$$

where the first major factor corresponds to the yaw motion, and the second to the roll motion. The first factor, compared with the two-degrees-of-freedom characteristic equation, shows approximately how roll affects the yaw natural frequency.

7.11 Step steer

In practice it is not possible to perform a true step change to the steer angle, a ramp-step being more realistic. However, a step input is particularly simple theoretically. The step steer response is perhaps the most fundamental transient, because it corresponds to simple corner entry or exit conditions. After a step steer input, there is a new equilibrium state achieved after a transient depending on the natural frequency and damping in yaw. The most general step steer input is a shift from one corner radius to another, but the case usually analysed is that starting from straight running.

Figure 7.11.1(a) shows the transient yaw speed response for cases of understeer, neutral and oversteer below critical speed. The attitude response is similar. The best response is that of the neutral vehicle. For understeer there is a response overshoot because of the sub-critical damping, getting worse with increasing speed. For oversteer the damping is good but the response takes a long time to reach equilibrium. Physically, this is because the oversteering vehicle requires a smaller steer input for a given steady state, and the smaller input gives a smaller initial response. The slowness of oversteer vehicles to reach their steady-state levels is one of the factors that makes them generally more difficult to drive. Alternatively, Figure 7.11.1(b) shows the different response to equal steer inputs.

Immediately after a step steer input δ, before any yaw develops there is a front sideforce $2C_{\alpha f}\delta$ and yaw moment $2aC_{\alpha f}\delta$. There is a corresponding initial lateral acceleration $2C_{\alpha f}\delta/m$ and yaw angular acceleration $2aC_{\alpha f}\delta/I$. The yaw angular acceleration creates a yaw speed that gives a slip angle at the rear; this contributes a rear sideforce that aids the lateral acceleration. On the other hand, the sideforce induces a side velocity with a corresponding contrary rear slip angle. After a short time t the sideslip velocity and yaw speed are

$$v = \frac{2C_{\alpha f}\,\delta}{m}\,t$$

$$r = \frac{2aC_{\alpha f}\,\delta}{mk_z^2}\,t$$

where k_z is the yaw radius of gyration. The rear slip angle is therefore

$$\alpha_r = \frac{rb}{V} - \frac{v}{V}$$

$$= \frac{2C_{\alpha f}\,\delta\,(ab - k_z^2)\,t}{Vmk_z^2}$$

Figure 7.11.1. Yaw speed versus time for steer input: (a) different steer inputs, equal final yaw speeds, (b) equal steer inputs, different final yaw speeds

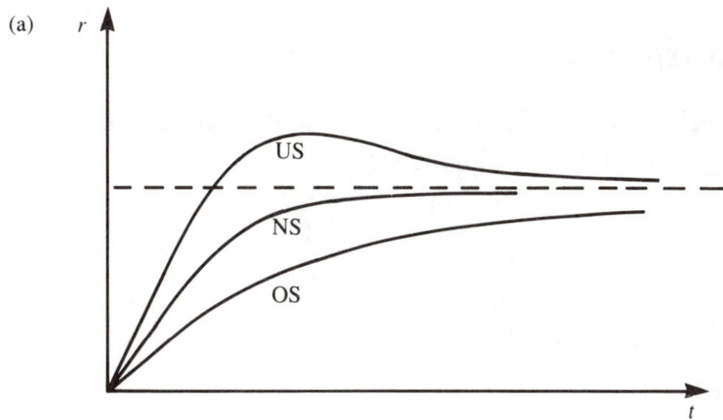

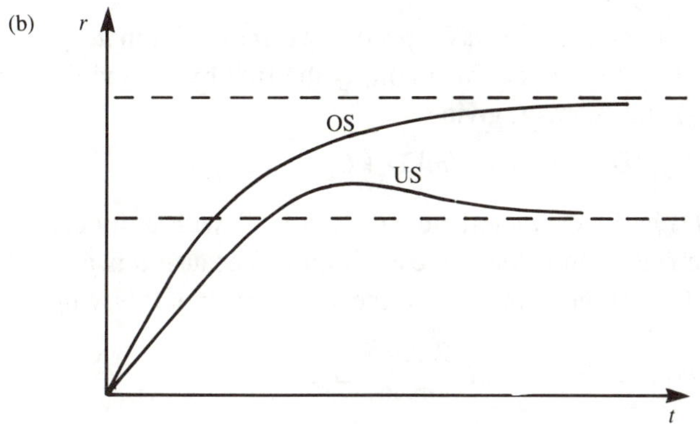

Therefore if $ab > k_z^2$ the rear slip angle will develop initially in the correct direction. If $ab < k_z^2$ it will initially develop in the wrong direction and will undergo a reversal before reaching steady state. Hence a large ab in relation to k_z^2 (a 'wheel on each corner') helps to produce an agile vehicle.

The full solution for yaw speed for the three-degrees-of-freedom model is too complex to reproduce here, but the two-degrees-of-freedom analysis is enlightening. After the steer angle step δ, in the vehicle-fixed axes the sideforce and yaw moment equations of motion are

$$(mVD + C_0)\,\beta \;+\; \left(mV + \frac{C_1}{V}\right) r \;=\; 2C_{\alpha f}\,\delta$$

$$(C_1)\,\beta \;+\; \left(ID + \frac{C_2}{V}\right) r \;=\; 2aC_{\alpha f}\,\delta$$

Remembering that the initial sideforce and moment are simply $2C_{\alpha f}\,\delta$ and $2aC_{\alpha f}\,\delta$, the initial conditions are

$$\beta_0 \;=\; \frac{v_0}{V} = 0$$

$$\dot{\beta}_0 \;=\; \frac{\dot{v}_0}{V} \;=\; \frac{2C_{\alpha f}\,\delta}{mV}$$

$$\psi_0 \;=\; 0$$

$$r_0 \;=\; 0$$

$$\dot{r}_0 \;=\; \frac{2aC_{\alpha f}\,\delta}{I}$$

To find, for example, the yaw speed r we must eliminate β from the equations of motion: to do this multiply the first by C_1 and the second by $(mVD + C_0)$ and subtract, giving

$$mIV f_c(D)\, r \;=\; \left[C_1 - (mVD + C_0)a\;\;\right] 2C_{\alpha f}\,\delta$$

where $f_c(D)$ is the characteristic equation in D with unity coefficient in D^2. On the right-hand side, we can simplify because δ is a constant and so $D\delta = 0$ (hence the simplicity of the step steer input), giving

$$f_c(D)\, r \;=\; (C_1 - aC_0)\,\frac{2C_{\alpha f}\,\delta}{mIV} \;=\; C$$

where C is constant for the motion. The solution to this differential equation comes in two parts – the particular integral corresponding to the final steady state, and the complementary function corresponding to the transient natural response. For a final steady-state yaw speed r_1, which is a constant, $Dr_1 = 0$ and $D^2 r_1 = 0$, so in terms of the undamped natural frequency the particular integral is given by

$$f_c(D)\, r_1 = \omega^2\, r_1$$

$$G_r \;=\; \frac{r_1}{\delta} \;=\; \frac{(C_1 - aC_0)\, 2C_{\alpha f}}{V\left(\dfrac{C_0 C_2}{V^2} \;-\; mC_1 \;-\; \dfrac{C_1^2}{V^2}\right)}$$

This expression for the yaw speed gain can be compared with that found in Chapter 6.

The solution for the transient is more complex, and depends on whether the vehicle is understeer, neutral or oversteer. In the case of understeer, assume a solution of damped oscillatory form:

$$r = e^{-\zeta\omega t}(A \sin \omega_d t + B \cos \omega_d t) + r_1$$

The initial condition at $t = 0$ is

$$r_0 = 0 = B + r_1$$

Differentiating to find $\dot{r}$, at $t = 0$

$$\dot{r}_0 = \frac{2aC_{\alpha f}\delta}{I} = -\zeta\omega B + A\omega_d$$

Hence $B = -r_1$

$$A = \frac{2aC_{\alpha f}\delta}{I\omega_d} - \frac{\zeta\omega r_1}{\omega_d}$$

This establishes the arbitrary constants, and hence the complete transient solution is known.

For an oversteer vehicle, the assumed form of solution is

$$r = A\, e^{D_1 t} + B\, e^{D_2 t} + r_1$$

where D_1 and D_2 will be negative. At $t = 0$

$$r_0 = 0 = A + B + r_1$$

Differentiating to find $\dot{r}$,

$$\dot{r} = D_1 A\, e^{D_1 t} + D_2 B\, e^{D_2 t}$$

At $t = 0$

$$\dot{r}_0 = \frac{2aC_{\alpha f}\delta}{I} = D_1 A + D_2 B$$

This suffices to find A and B, as required.

For the case of a neutral-steer vehicle, the standard solution is

$$r = A\, e^{D_1 t} + Bt\, e^{D_1 t} + r_1$$

At $t = 0$

$$r_0 = 0 = A + r_1$$

Differentiating:

$$\dot{r} = AD_1\, e^{D_1 t} + B\, e^{D_1 t} + BD_1 t\, e^{D_1 t}$$

At $t = 0$

$$\dot{r}_0 = \frac{2aC_{\alpha f}\delta}{I} = AD_1 + B$$

This suffices to find A and B for the neutral-steer case.

If it is required to find the attitude angle rather than the yaw speed, then r must be eliminated from the simultaneous differential equations of motion rather than β. To study the path of the vehicle, for example to find the path of G, the Earth-fixed axes formulation must be used. In all cases the method of solution is as above.

For the three-degrees-of-freedom model, the analysis is more complex but gives very similar results.

7.12 Oscillatory steer

The 'steady-state' response to an oscillatory (sinusoidal) steer input is not of great practical significance, since this is unlikely to arise in normal operation on the road. However a sinusoidal steer input is of some theoretical interest for a frequency domain analysis, and can be investigated experimentally.

The three-degrees-of-freedom analysis for this input was first performed by Segel (1955a,b). Figures 7.12.1 to 7.12.3 show how the yaw speed amplitude, roll speed amplitude and lateral acceleration amplitude varied with input frequency. Modern cars and tyres tend to give flatter responses up to rather higher frequencies. Modern coaches can produce results much as in these figures. For frequencies below 3 Hz, provided that the speed is not very low, then the tyre dynamic characteristic (lag between force and angle) can be neglected. As would be expected, the yaw velocity amplitude, Figure 7.12.1, diminishes with frequency, beginning with the steady-state response value. At high frequency it is inertia-limited, so there is only a small yaw amplitude, and the yaw moment tends to $2aC_{\alpha f}\delta$, where δ is the steer angle amplitude. The yaw acceleration amplitude tends to $2aC_{\alpha f}\delta/I$ and the yaw velocity amplitude to $2aC_{\alpha f}\delta/I\omega_f$, where ω_f is the forcing radian frequency of the steering motion. The phase of the response varies from in-phase at low frequency to 90° lag at high frequency.

The roll velocity response, Figure 7.12.2, begins at zero at low frequency, increases to a peak at around the roll natural frequency, and then declines. The phase lag goes from 90°, through 180° around resonant frequency, to 270° lag at high frequency.

The shape of the lateral acceleration response, Figure 7.12.3, is somewhat unexpected, being rather like an inverted resonance curve. At low frequency the value equals the steady-state response. At high frequency there is little yaw response, so for the 2-d.o.f. model the total sideforce tends to approximately $2C_{\alpha f}\delta$ and the lateral acceleration amplitude to $2C_{\alpha f}\delta/m$. In between, there is a minimum. The reason for this shape of curve is that it is the sum of three terms, two decreasing with frequency, and one, the linear lateral acceleration, increasing with frequency. The phase of the response begins at zero, goes to about 45° lead through the resonant frequency, and then declines to zero again.

If the steering forcing frequency is increased beyond about 3 Hz then the tyre dynamics will become significant, and there will be a further reduction in response and an extra phase lag.

Instead of using a sinusoidal steering motion of various given frequencies, an alternative frequency domain test is to use a random steer input. The vehicle motion is then the product of the random input spectrum and the vehicle transfer function, so by correlating the output motion with the input steering, the transfer function may be deduced. This is particularly easy if the input is 'white noise', i.e. having a uniform spectral distribution.

Figure 7.12.1. Yaw speed amplitude and phase: against steer frequency.

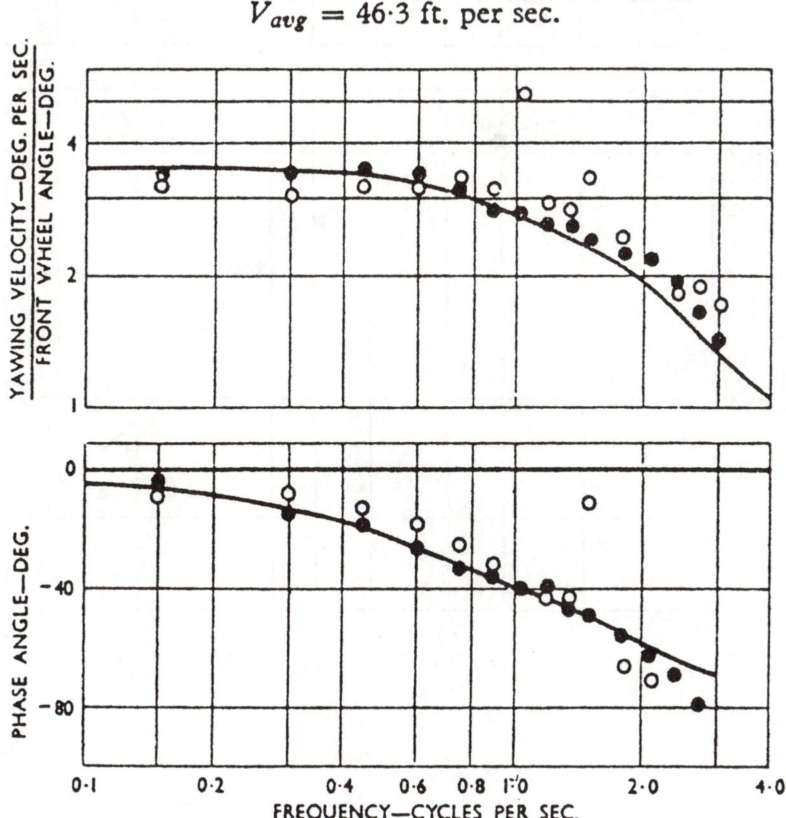

Figure 7.12.2. Roll speed amplitude and phase: against steer frequency.

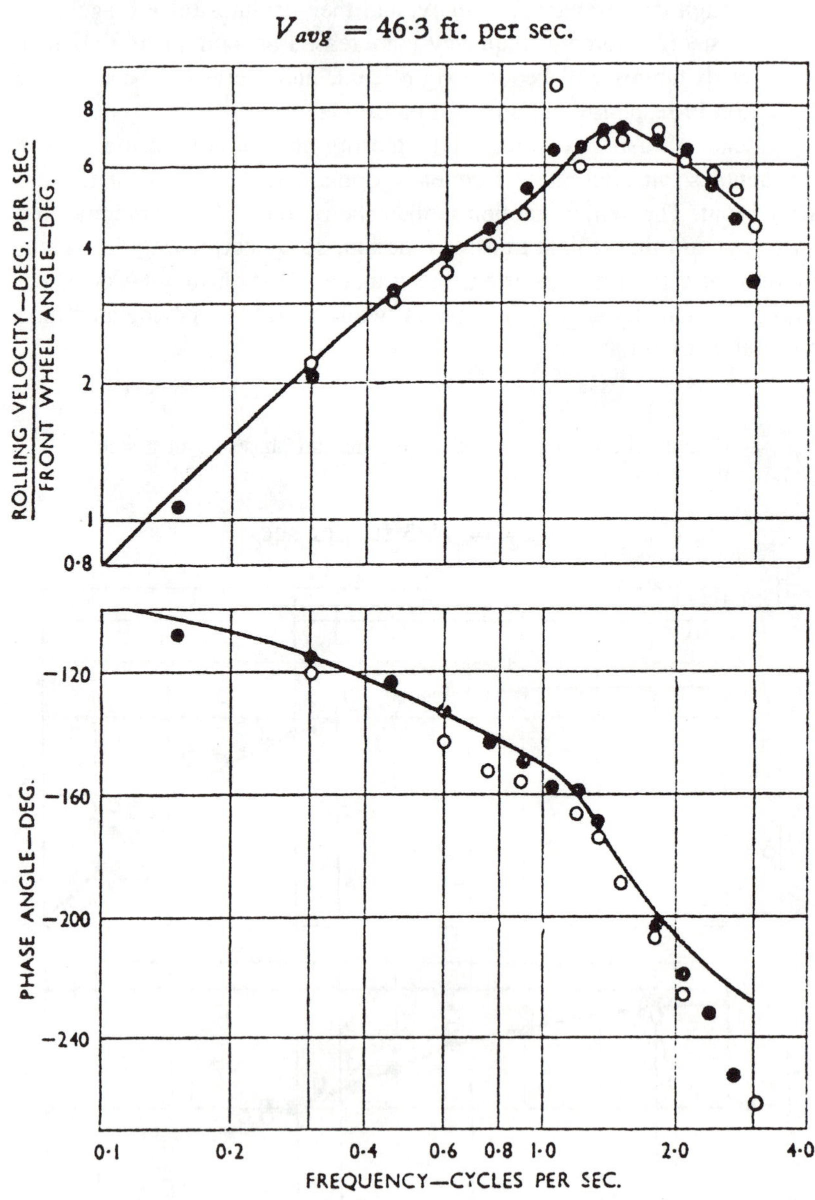

Figure 7.12.3. Lateral acceleration amplitude and phase: against steer frequency.

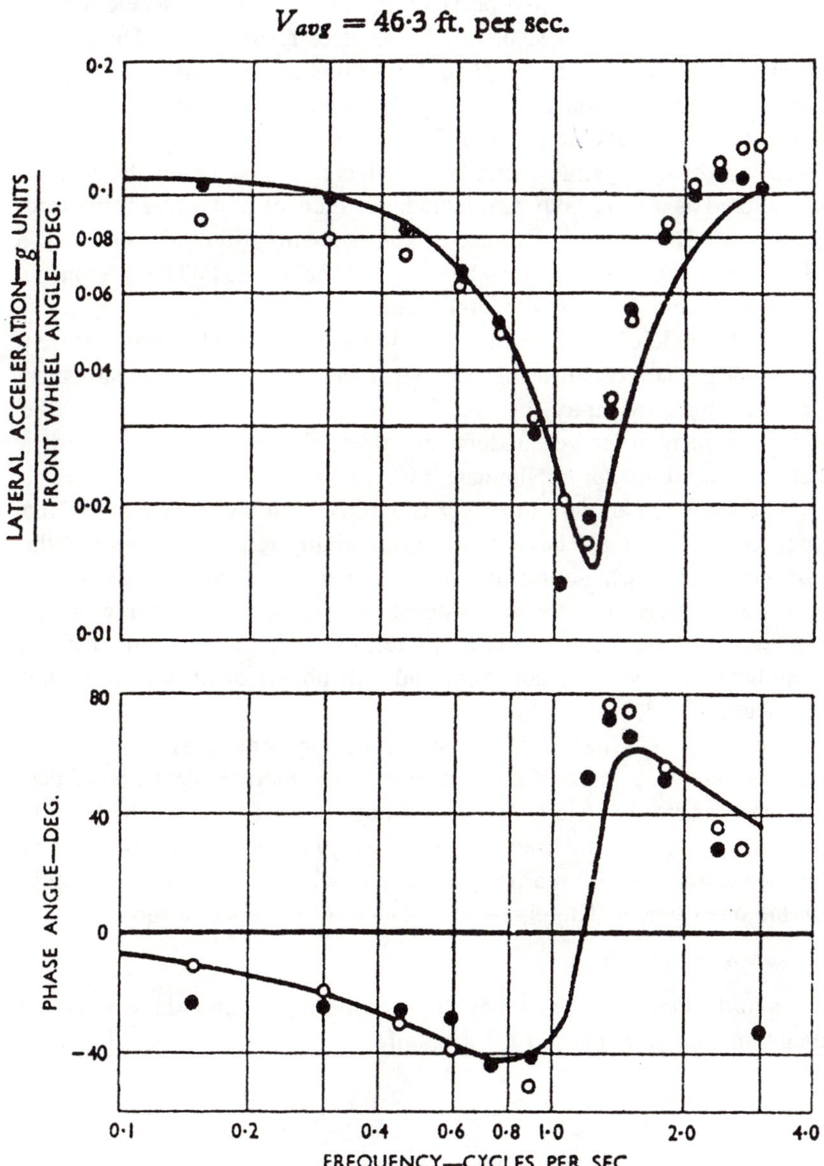

7.13 Power steer

Previous sections have dealt with the response to steering control, but there may also be steering response to the accelerator or brake, known as power steer and brake steer respectively. The former must obviously be distinguished from 'power steering', i.e. power-assisted steering. Such effects are generally undesirable and are designed out if possible (Section 5.13). The effects may be considered to arise in two ways. The first is by effects that occur on the vehicle considered as a rigid body and include differences in tractive force from side to side, for example because of a limited-slip differential or because of tractive force effects on the tyre characteristics. The second is because of vehicle internal compliances.

The dynamic effects are similar to those in the steady state, but may be much greater because the tractive forces are greater. In particular, 'lift-off tuck-in' may be very noticeable – really this is the disappearance of power understeer, although nowadays it is known how to control this, or to eliminate it if required.

Where a large amount of power is applied then the driven end of the vehicle will inevitably have its cornering ability reduced because of the tyre characteristics, giving oversteer for rear drive and understeer for front drive. For more moderate power applications then many details matter and one cannot really generalise. The effect of internal compliance is also very complex, and will not be dealt with here, but see Section 5.13.

The effect of a non-free differential on the vehicle as a rigid body may be analysed in the following way. The tractive force above that required for steady state is $F = mA_x$. Consider a rear-drive vehicle with a tractive force lateral transfer factor e_T; with $e_T = 0$ the traction is equally distributed as on a free differential, with $e_T = 1$ all the traction is on the outer wheel. The plan-view moment of the tractive force is

$$M = \tfrac{1}{2} F e_T T$$

To avoid changing the steady-state cornering radius there must be counteracting sideforces at the axles of

$$\frac{M}{l} = \frac{F e_T T}{2l}$$

The changes of slip angle are

$$\Delta\alpha_f = -\frac{F e_T T}{4 l C_{\alpha f}} \qquad \Delta\alpha_r = \frac{F e_T T}{4 l C_{\alpha r}}$$

The change of steer angle required is

$$\Delta\delta = \Delta\alpha_f - \Delta\alpha_r$$
$$= -\frac{Fe_T T}{4l}\left(\frac{1}{C_{\alpha f}} + \frac{1}{C_{\alpha r}}\right)$$

which is an oversteer effect for positive e_T. Using $F = mA_x$, for a car

$$\Delta\delta \approx 0.5 A_x e_T \text{ (deg)}$$

Now since e_T may be in the range 1 to –1 according to the type of differential and the power applied (Section 6.13), then $\Delta\delta$ may be as much as 0.5 deg/m s^{-2} (5 deg/g), a substantial change, potentially much greater than the differential effects in steady state. For good handling, the steer effect should not exceed 1 deg/g of longitudinal acceleration, and therefore non-free differentials may be problematic in this respect.

Power-steer effects are particularly problematic on racing cars, for several reasons. The lap time is very sensitive to car acceleration at exit from the corners, with combined lateral and longitudinal accelerations. The high power-to-weight ratios give strong longitudinal acceleration, which quickly reveals any handling shortcomings, and of course limited-slip differentials are used. Finally, many racing cars have aerodynamic forces that are very sensitive to pitch angle, and particularly to the height of the front wings from the ground.

7.14 Disturbance response

Transient disturbances include road roughness, changes of surface friction, and wind effects.

Theoretical investigations of road roughness effects have been very limited because of the shortage of data on tyre characteristics in conditions of fluctuating normal force. For small fluctuations one might expect the sideforces to adjust according to the response distance concept, as they do for varying slip angle, giving little change to the mean force. However, for large fluctuations, if the normal force goes to zero then there will be an immediate loss of sideforce, not compensated by the slow growing sideforce when the normal force becomes large, giving a reduced mean cornering force. Hence the dependence of tyre sideforce on normal force fluctuation amplitude and frequency seems to be an interesting area for investigation and a necessary prerequisite for detailed study of vehicle response in such conditions. Practical results will also depend upon how the vertical force actually fluctuates, depending on the suspension characteristics, including the dampers.

The effect of changes of surface friction, for example due to ice patches, is perhaps of wider importance. Again, this is not easily amenable to a useful general analytical approach, but is usually studied by time domain computer analysis.

Transient wind effects are of practical significance – much more so than steady wind effects. They are at their worst at high speed, because although for a given lateral wind speed the aerodynamic attitude angle is less, the dynamic pressure is greater. On motorways, when overtaking or being overtaken, one vehicle may temporarily shield another from a strong sidewind, or a large vehicle may subject a small adjacent vehicle to a strong flow even in the absence of wind. The effect of a lateral air speed component may be analysed in various ways. The consequent lateral force and moment can be inserted into the two-degrees-of-freedom dynamic model; if the vehicle deviation effect on the wind force is neglected, i.e. the force and moment are assumed constant, then this gives a manageable analytical solution. If a computer simulation of the vehicle is available then of course the wind force can be made a function of vehicle lateral speed and yaw angle.

The conclusions of such analysis are fairly simple. The steady-state response is not critical because it simply requires a fixed steering deflection (Section 6.20). The driver response time is about 0.25 s, although experimental results show that the path deviation takes typically 2 s to reach its maximum; it is this maximum that is the important result. This deflection is partly the result of aerodynamic side force but in most cases mainly of the yaw moment; hence a forward centre of mass eases the problem. The deflection also depends on the vehicle's handling characteristics, but since these are largely dictated by the wind-free handling requirements, the gust response can only really be controlled by tailoring the aerodynamic characteristics. Essentially this means minimising the yaw moment, which means keeping the lateral force centre of pressure as far back as possible. There is a conflict here, because low-drag shapes tend to have a relatively forward centre of pressure and it may be necessary to compromise on the detail design.

7.15 Testing

Unsteady-state testing can conveniently be divided into three types:

(1) Vehicle characteristics
(2) Task performance
(3) Subjective assessment

The first type, vehicle characteristic testing, is open-loop, whereas the last two types, by including the driver performance, are closed-loop tests.

In the open-loop tests, the vehicle motion is studied in response to specified control inputs. The most fundamental test is a step steer input; in practice a ramp-step is used because the step cannot be achieved instantaneously. This enables the rapidity of response and overshoot to be investigated, giving an experimental measure of the natural frequency and damping ratio. Alternatively a small amplitude single-cycle steering oscillation (open-loop version of a lane change) or a random steer input (white noise) can be used. The I.S.O. has published draft procedures for these tests. In practice the random steer test seems most practical because it does not require mechanically controlled steering, and can be performed on a two-lane wide road. A variation of the random input test is the steering sweep test, in which a basically sinusoidal steering motion is used, but with smoothly increasing frequency.

Another standard comparison of theory and experiment is the steer pulse, i.e. a sudden deflection of steering and prompt restoration to straight ahead. In Laplace transform analysis the steer deflection can be represented by a succession of three ramps with time delays, and gradients in the ratio 1:-2:1, or by a half period of a sine wave.

The slalom test is distinct from the oscillatory steer test in that the latter uses small steer amplitude and is linear. In the slalom test the vehicle is driven as quickly as possible on a sinuous course on alternating sides of a series of obstacles, and large lateral accelerations are required. Although in principle this is an open-loop test, in practice driver skill plays a significant part. One serious criticism of the slalom test is that comparative ranking of vehicles may vary with the spacing of the obstacles, because of the different vehicle natural yaw and roll frequencies. Hence it has not found wide application. In a variation of the test, the obstacle spacing is progressively reduced, and the vehicle is driven at constant speed; this does not overcome the previously mentioned objection.

Little has been published on results of testing on controlled roughness. The obvious test would be a complete circular test pad with added roughness elements, e.g. radial slats, that can be changed in severity and spacing. A more economical alternative that has been tried is to have a relatively short segment of the circular path slatted, and to drive into the rough region at or near to the maximum speed for the smooth section; in such tests the roughness seems to have surprisingly little effect. Tests using a single bump seem to suggest that this is an adequate test of roughness response.

A typical task performance test is the lane-change test. The vehicle is driven at fixed speed towards an obstruction requiring a lane change that could be either right or left; the driver is told as late as possible which way to go. The shortness of time or distance to the obstruction is a measure of the closed-loop lane change performance of the vehicle. This may be quite good for comparative testing of complete vehicles, or for testing detail design changes to the suspension or tyres.

Vehicle behaviour in sidewinds can be predicted quite well from wind tunnel yaw data. However, full-scale sidewind testing is still sometimes performed, for example by driving the vehicle into a sidewind generated by gas turbines exhausting into diffusers to widen the jet, which further expands by mixing with air (at M.I.R.A. one engine produces a 40 m wide 20 m/s jet at the road). To investigate the actual path deviation that will be experienced in practice, the vehicle should be in the sidewind for 2 s, the time to maximum deviation. At 20 m/s this requires an exposed path of about 40 m of relatively uniform wind. However, subjective testing is probably best performed in natural wind conditions.

The final test of dynamic handling is of course the subjective general driving test undergone by any new vehicle over a distance of many miles. Although theory is good at predicting the path of a vehicle in response to given control inputs, it is much less good at predicting whether the driver will feel confident in the vehicle. This is because the driver is sensitive to small dynamic effects, e.g. the rigidity of seat mountings. Therefore subjective testing remains of great importance.

7.16 Desirable results

Various research papers have addressed the question of how the driver's subjective opinion relates to the vehicle's dynamic characteristics, and what the characteristics should be for the best subjective and objective performance. Opinions vary. This may be for several reasons: some tests are performed with average drivers, others with experienced test drivers; investigations of a parameter over a limited range may wrongly suggest that it is not important; the test may not be long enough, so potentially superior features are rejected because drivers favour the familiar over the unfamiliar; finally, drivers are inherently highly adaptive, which tends to mask the effect of vehicle characteristics, whilst at the same time making them generally less critical. A good set of definitive objective tests that predict subjective quality has not yet emerged. On the other hand, there is often agreement as to what constitutes bad dynamic handling, and some conclusions can be drawn. The following indicates some of the results that have been published.

The yaw velocity gain under oscillatory steer conditions is given approximately by the simplified Laplace transfer function

$$G_r = \frac{G_{r\,ss}}{1 + T_r s}$$

where T_r is the effective time constant of the yaw speed behaviour, i.e. the lag of the yaw from the steer input. This expression is a reasonable approximation provided that the yaw damping is not too small, i.e. the vehicle is not too understeer. For both expert and average drivers, for a vehicle to be rated as satisfactory requires that $T_r < 0.5$ s, and a good rating requires $T_r < 0.3$ s.

In tests at 23 m/s (50 mile/hr), yaw velocity gains, referred to the handwheel, of 0.2 to 0.4 deg s^{-1}/deg are found to be desirable. This corresponds to understeer gradients of 3.4 deg/g down to zero. The undamped yaw natural frequency should be at least 1 Hz, and the yaw damping ratio at least 0.5 and preferably 1.0. For good yaw damping at high speed the understeer gradient needs to be small, under 2.0 deg/g. Vehicles notable for good handling have been observed to be close to neutral; some would even argue that a small amount of primary oversteer is justified to obtain high yaw damping. Rear roll understeer has, in particular, been implicated in detrimental reduction of yaw damping.

A good correlation has been found between subjective rating, objective performance and the 'TB value'. This is the product of $T_{\dot{\psi}}$ and β_{ss} where $T_{\dot{\psi}}$ is the time from step steer input to peak overshoot of yaw velocity, and β_{ss} is the steady-state attitude angle, these being measured at a speed of 31.8 m/s (70 mile/hr) and for a steer input sized to give 0.4g steady-state lateral acceleration. In practice, because the steer input is a ramp rather than a true step, time is measured from the moment when the steering is at half of its steady-state value. When a subjective handling poorness rating (1 being very easy handling, 5 being very difficult handling) is used, it is found that the poorness value is proportional to the TB value, reaching 5 at TB = 4.2 deg s.

Thus a TB value of 2 deg s or less is a desirable design target. However it is not clear that the TB value is really more significant than separately specifying values for $T_{\dot{\psi}}$ and β_{ss}, both of which should be as small as possible. Also $T_{\dot{\psi}}$ will be measurable only if there is a distinct yaw velocity overshoot, and this will not be the case if there is good yaw damping – itself desirable.

Good straight running definitely depends on the dynamic characteristics rather than just the steady-state understeer gradient. It is desirable to avoid large amounts of roll steer which can cause weaving. Two vehicles with the same overall understeer gradient may achieve the final value with very different component contributions. Under dynamic conditions, e.g. small wind disturbances or small steer corrections, these differences may manifest themselves, when it is generally found that a vehicle with large compensating steady-state effects will not behave well. Good straight running in windy conditions does however depend on the steady-state understeer gradient, amongst other things. Put another way, the relationship between the neutral steer point (the position of which is governed by the centre-of-mass position and the understeer gradient) and the aerodynamic centre of pressure for sideforces is very influential on straight running in gusty conditions.

Another problem with roll steer or bump steer is that it leads to increased tyre wear. Nevertheless, many vehicle manufacturers do use roll steer to obtain their desired handling characteristics. Provided that the effects are symmetrical from side to side (a matter of production tolerances) and not excessive, the effect on straight running may be acceptable.

Even if the vehicle itself has good straight running characteristics, it is possible for the driver to be unhappy, for example in the case of the 'oversteering seat' (Grylls 1972). This illustrates the great sensitivity of the driver to relatively small oversteering effects and how a small attitude gradient is important subjectively as well as objectively.

Obtaining good behaviour in gusty crosswinds is partly a matter of tuning the aerodynamic coefficients, basically getting the centre of pressure as far back as possible, and partly a matter of avoiding suspension arrangements that make the vehicle unduly responsive to wind stimulation.

In the case of racing cars, it is not just a case of obtaining maximum braking, cornering and traction, but also of making the handling predictable so that the driver can extract the full performance; to this end it is usual to eliminate compliance as much as possible, for example by replacing rubber bushes by metal ball-joints. This option is simply not available on passenger cars - the deliberate inclusion of compliance is essential to control noise, vibration and harshness. Until relatively recently the handling engineer saw this mostly as a handling problem with the compliance tolerated; in recent years with growing understanding it has become possible to use compliance advantageously, in the control of power steering and 'lift-off tuck-in' effects, which can now be held to low levels.

As already discussed, there is a correlation between subjective rating and the steering correction required because of braking longitudinal acceleration in a turn. For a good rating, the steer correction should not exceed 1 deg/g at the road wheels, say 20 deg/g at the handwheel, and preferably be less. Speculatively, a similar quantitative criterion could be applied to power steer on acceleration; in this case, non-free differentials with a steer effect of as much as 5 deg/g at the road wheels are clearly problematic.

7.17 Problems

Q 7.3.1 Establish the equation of motion for a 1-d.o.f. sideslip-only model, and show that there is damping but no oscillation.

Q 7.4.1 Establish the equation of motion for a 1-d.o.f yaw-only model, and describe the motion for understeer and oversteer cases.

Q 7.5.1 Establish the differential equations of motion for a 2-d.o.f. model in vehicle-fixed axes in v and r.

Q 7.5.2 For a 2-d.o.f. model in vehicle-fixed axes, derive the characteristic equation of motion.

Q 7.6.1 Establish the differential equations of motion for a 2-d.o.f. model in Earth-fixed axes, in u and ψ.

Q 7.6.2 Derive expressions for the tyre slip angles perceived in Earth-fixed axes and vehicle-fixed axes, and discuss the difference.

Q 7.7.1 Deduce the conditions for stable motion from the characteristic equation.

Q 7.7.2 Deduce equations for the damping ratio and damped natural frequency from the characteristic equation. Discuss how these vary with speed.

Q 7.7.3 'Dynamic yaw damping requirements place an upper limit on acceptable understeer gradient.' Discuss.

Q 7.7.4 'A neutral steer vehicle has a yaw damping ratio of unity.' Discuss.

Q 7.7.5 'Yaw damping is equally due to yaw and sideslip.' Discuss.

Q 7.7.6 When a free yaw oscillation is damped out, the energy of oscillation is dissipated. Where does it go?

Q 7.7.7 For a vehicle with $l = 3$ m, $C_{\alpha f} = 1000$ N/deg, $C_{\alpha r} = 1000$ N/deg, $m = 400$ kg, $I = 2744$ kg m^2, $a = 1.3$ m, at a speed of 22 m/s, find the undamped yaw frequency, damping ratio and damped yaw frequency. Also find the understeer gradient k, and V_{ch}.

Q 7.8.1 Explain how factors such as pneumatic trail can be incorporated into an improved 2-d.o.f. model.

Q 7.8.2 The vehicle of Q 7.7.7 is found experimentally to have a steady state understeer gradient of 1.74 deg/*g*. Suggest an effective value for C_1.

Q 7.9.1 Explain the stability derivative notation, and give expressions for the four stability derivatives for the simple 2-d.o.f. model and the improved version.

Q 7.9.2 For the vehicle of Q 7.7.7. evaluate the 2-d.o.f. stability derivatives at 15 m/s. State the units in each case.

Q 7.10.1 For the 3-d.o.f. model, list all the various possible stability derivatives, and discuss their relative importance. Give representative values.

Q 7.10.2 'The 3-d.o.f. model is more complex, more accurate and superior to the 2-d.o.f. model.' Discuss.

Q 7.10.3 Under what conditions is the 3-d.o.f. model stable?

Q 7.10.4 Explain quantitatively, giving equations and numerical values, how roll affects the yaw natural frequency. Describe the physical features that could be used to achieve this effect. What would be the disadvantages?

Q 7.11.1 Sketch the step steer response of various understeer, oversteer and neutral-steer vehicles, and discuss their relative merits.

Q 7.11.2 For a step steer input, explain how the rear slip angle develops for various vehicles.

Q 7.11.3 'Having a 'wheel on each corner' results in an agile vehicle.' Discuss.

Q 7.11.4 For an understeer vehicle, applying suitable initial conditions for a step steer input, derive an expression for the transient solution for 2-d.o.f.

Q 7.12.1 Describe how, for a real vehicle, yaw speed, roll speed and lateral acceleration depend on the frequency of a sinusoidal steer input of fairly small amplitude.

Q 7.13.1 Explain how traction affects transient handling.

Q 7.13.2 Derive equations describing how a limited-slip differential affects transient handling.

Q 7.14.1 Explain the effect of a step change of aerodynamic sideforce, and how this depends on the aerodynamic static margin.

Q 7.15.1 Give an overview of transient testing.

Q 7.15.2 Explain the difference between slalom testing and an oscillatory steer test.

Q 7.16.1 'It is not possible to have a definitive set of objective transient handling tests because of driver variability.' Discuss.

Q 7.16.2 'For the best overall handling, a vehicle should have a small amount of primary oversteer.' Discuss.

Q 7.16.3 Describe some of the transient handling criteria that have been suggested.

Q 7.16.4 Explain the 'TB' value, critically.

Q 7.16.5 'Compensating steer effects may not compensate under transient conditions, giving poor handling.' Discuss.

7.18 Bibliography

As far as books are concerned, some dynamic two-degrees-of-freedom analysis is included in Wong (1978). There is a fairly extensive dynamic analysis of two- and three-degrees-of-freedom models in Ellis (1969) and Ellis (1989). A useful introduction to the three-degrees-of-freedom model is given by Steeds (1960). The transient handling problems that have prevented four-wheel-drive success in racing are discussed by Henry (1975).

Of some historical interest, the first paper to study dynamic behaviour was by Stonex (1941). An early theoretical analysis was that by Rocard (1946). The real advance in dynamic analysis came in 1957. An extensive list of references up to that time is given in Milliken & Whitcomb (1956). An extensive analysis of the two-dimensional model is given in Whitcomb & Milliken (1957). The three-dimensional model is considered in detail in Segel (1957a and 1957b).

Some of the more interesting papers since that time include: Bergman (1965), Milner (1968), McHenry (1969), Dugoff *et al.* (1971), Carr *et al.* (1973), Bergman (1973), Pacejka (1975), Fancher *et al.* (1976), Bundorf & Leffert (1976), Yoshimori (1976), Good (1977), Weir & Zellner (1977), Chiang & Starr (1978), Weir & DiMarco (1978), Verma & Shepard (1981). The I.S.O. has (unpublished) draft procedures for steer response testing (I.S.O. 1979a,b,c)

The lane-change test has not achieved acceptance as an I.S.O. standard, but there is a technical report I.S.O. TR 3888: Lane Change Test. For braking-in-a-turn testing, see I.S.O. (1985).

The Peugeot 405 (1991) with 'Integrated Chassis Design' is broadly representative of modern car suspension designs. It uses struts at the front, but with longitudinal location by a wide-based wishbone rather than by the anti-roll bar which acts through drop links (hence strictly not MacPherson struts). The steering is rack-and-pinion, and is mounted on the same sub-frame as the suspension, reducing some adverse steering compliance effects. The rear uses trailing arms, also with anti-roll bar. With many modern cars using a similar layout, the difference between good and bad handling lies in careful optimisation of details of the motion geometry of the links (bump steer and bump camber), and of the compliance of the bushes (compliance steer), with a rigid body shell to integrate the whole system.

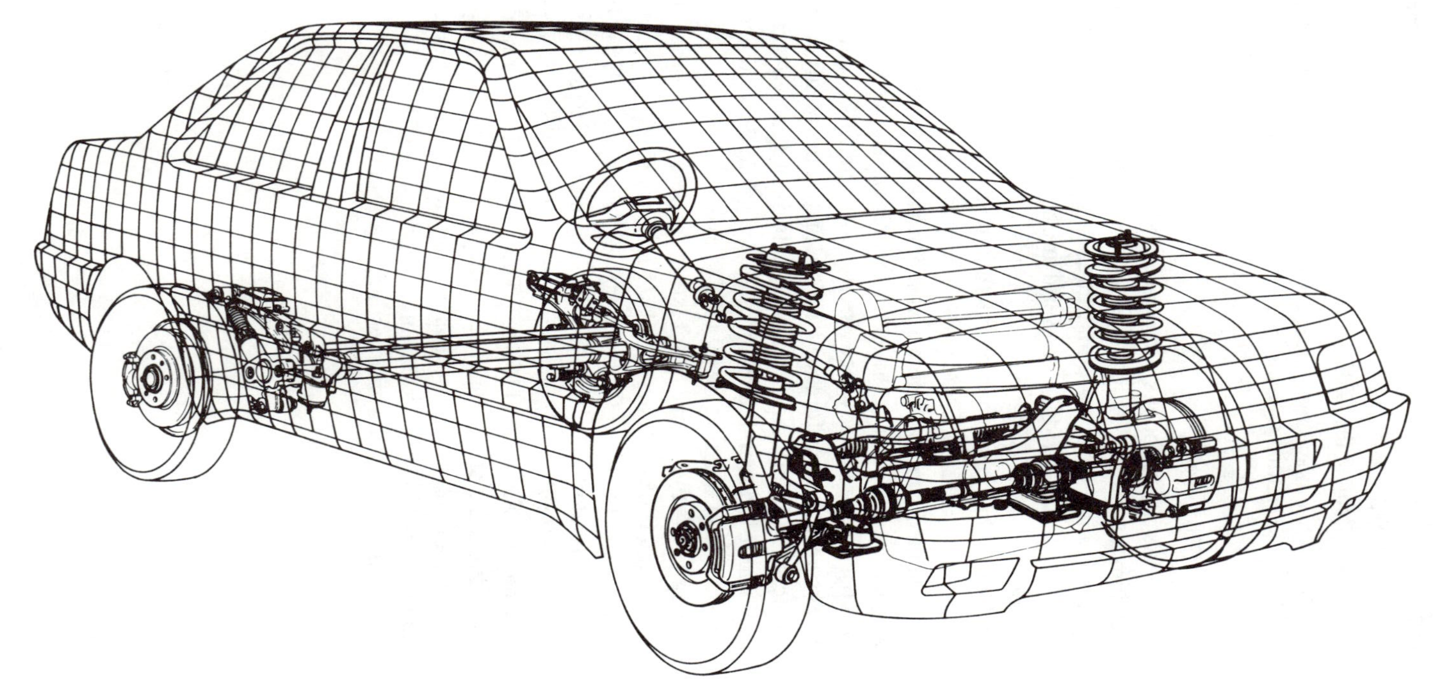

Appendix A

Nomenclature

Quantities such as acceleration and force are vector in nature. When the vector nature is implied, such variables are shown in bold type (e.g. $\mathbf{A}$, $\mathbf{F}_S$). Usually, however, the equations use the magnitudes or the scalar components of such vector quantities. These are therefore not vectors, and are printed in italic (sloping) type (e.g. A, A_X, F_S). Points are labelled in normal (upright) type. Hence: centre-of-mass point G, gear ratio G, linear momentum vector $\mathbf{G}$.

Chapter 1 Introduction

a	front axle to G	$\mathbf{H}$	angular momentum vector
A	acceleration	I	second moment of mass
A	area	k	radius of gyration
b	rear axle to G	K	stiffness
c	dimension	l	wheelbase
d	dimension	L	pendulum length
D	drag	m_k	mass of kth component
$\mathbf{F}$	force vector	m	vehicle mass
F	force magnitude	m_S	sprung mass
F_C	centripetal force	m_U	unsprung mass
F_T	tangential force	$\mathbf{M}$	moment vector
g	gravitational acceleration	M	moment magnitude
$\mathbf{G}$	linear momentum vector	N	wheel or axle vertical force
G	centre of mass	p	roll angular speed
H	height of G	P	power

q pitch angular speed
r yaw angular speed
R path radius, dimension
R_1 wheel loaded radius
T track, tension, tractive force
T_{sw} steering wheel torque
V speed
W weight force
$u\ v\ w$ velocity components
xyz vehicle-fixed axes
X_G x coordinate of G position
XYZ Earth-fixed axes

α tyre slip angle,
α angular acceleration
β attitude angle
δ steer angle
δ_{ref} reference steer angle
δ_{sw} steering wheel angle
θ pitch angle, angle
λ wavelength
μ_R rolling resistance coefficient
ν course angle
ρ air density
ϕ roll angle
ψ heading angle
ω angular speed
Ω wheel spin angular speed

Subscripts:
A aerodynamic
C centripetal, centrifugal, cradle
D drag
f front
k kth component
L left, lift
M maximum
m mean
O fixed axis
p principal
P pitch
r rear
R right
S sprung
T tyre, tangential
U unsprung
V vehicle
Y yaw

Chapter 2 The tyre

A area
c foundation stiffness
C a constant
C_C tyre camber stiffness coefficient
C_{FV} variation of F_Y with F_V
C_{FT} load transfer sensitivity of axle
C_M moment coefficient
C_S tyre cornering stiffness coefficient
C_T normalised load transfer sensitivity of axle
C_V normalised variation of F_Y with F_V
C_W wave speed
C_Y lateral force coefficient
C_α tyre cornering stiffness
C_γ tyre camber stiffness
C_ρ tyre curvature stiffness
d tread displacement
e_V vertical force transfer factor
F force magnitude
F_D tyre drag force
F_R rolling resistance force
F_S tyre central force
F_V tyre vertical force ($-F_Z$)
F_{VT} transferred vertical force
F_X, F_Y, F_Z tyre forces on $X'Y'Z'$ axes
F_X longitudinal force
F_Y lateral force

F_α	cornering force
F_γ	camber force
I	second moment of mass
k_1	load sensitivity of C_S
k_v	speed sensitivity of C_S
l	tyre footprint length
L	relaxation distance
M_X, M_Y, M_Z	tyre moments on $X'Y'Z'$ axes
M_X	overturning moment
M_Y	rolling resistance moment
M_Z	aligning moment (torque)
N_D	dynamic hydroplaning number
N_T	Turner number
N_V	viscous hydroplaning number
P	contact pressure,
p	load sensitivity of μ_C
p_i	inflation pressure
R_e	effective wheel radius (V/Ω)
R_l	loaded wheel radius
R_u	unloaded wheel radius
s	non-dimensional slip (S/S^*)
S	slip
S^*	characteristic slip
t	pneumatic trail
t_C	crown tension,
T	temperature, wheel drive torque
V	speed
x	position in tyre footprint
X	distance rolled
$X'Y'Z'$	wheel-aligned axes
$X''Y''Z''$	motion-aligned axes
α	slip angle
α^*	characteristic slip angle
γ	camber angle
θ	slope angle
λ	wavelength
μ	friction coefficient
μ_B	maximum braking force coefficient
μ_C	maximum cornering force coefficient
μ_D	dynamic friction coefficient
μ_R	rolling friction coefficient
μ_S	static friction coefficient
ρ	path curvature ($1/R$)
ρ_C	crown density per unit area
ρ_W	density of water
τ	water clearance time constant
Ω	wheel spin angular speed

Subscripts:

d	drum
M	maximum
m	mean
t	tyre
2	two wheels
α	related to slip angle
γ	related to camber angle

Chapter 3 Aerodynamics

A	reference or frontal area
A_D	drag area
C_D	drag coefficient
C_L	lift coefficient
C_P	pitch coefficient
C_R	roll coefficient
C_S	sideforce coefficient
C_Y	yaw coefficient
D	drag force
F	force magnitude
h	ground clearance
h_A	aerodynamic static margin from wheelbase mid-point
I	turbulent intensity

l	wheelbase
L	vertical lift force
M	molecular weight
Ma	Mach number
n	wind speed-height exponent
p	atmosphere pressure
P	pitch moment, probability
q	dynamic pressure
R	roll moment
R_G	universal gas constant
Re	Reynolds number
S	aerodynamic side force
T	temperature
T_K	absolute (kelvin) temperature
U	wind speed
v	velocity
v_a	ambient wind speed
v_r	relative wind speed
v_v	vehicle speed
V	speed
V_{ch}	characteristic wind speed
V_s	speed of sound
V_S	sustained wind speed
W	turbulent speed deviation
x	length dimension
Y	yaw moment
z	altitude from sea level
z_0	altitude constant
xyz	vehicle-fixed axes
XYZ	Earth-fixed axes
α	body pitch angle
β	Weibull wind shape factor, vehicle sideslip angle
β_A	aerodynamic sideslip angle
γ	ratio of specific heats
δ	steer angle
μ	dynamic viscosity
ν	kinematic viscosity
ν_a	ambient wind angle
ν_r	relative wind angle
ν_v	vehicle velocity angle
ρ	air density

Subscripts:

a	air
D	dry
f	front
M	maximum
m	mean
r	rear, relative
v	vehicle
W	water
0	reference

Chapter 4 Suspension components

c	damping coefficient
d	spring wire diameter
D	spring coil diameter
F	force
$F_{A/B}$	degrees of freedom A from B
G	shear modulus
h	suspension travel
I	second moment of mass
J	polar second moment of mass
l	link length
L	torsion bar length
m	component mass
m_e	vehicle end mass (m_f or m_r)
m_S	sprung mass
m_U	unsprung mass
N	number of working coils
r	torsion bar radius
R	motion ratio
T	torque
V	speed

$\mathbf{V}$	velocity vector	θ	angle, torsional deflection
x	spring deflection	ω	angular speed, wheel precession speed
xyz	coordinates	Ω	wheel spin angular speed
ζ	damping ratio		

Chapter 5 Suspension characteristics

a	front axle to G
A	lateral acceleration
A_B	braking deceleration
A_F	frontal area
A_M	maximum lateral acceleration
A_R	roll-over lateral acceleration
b	rear axle to G, kingpin offset
b_a	kingpin offset at wheel axis height
c	caster offset
C_A	aerodynamic roll-effect factor
C_L	vertical lift coefficient
C_P	pitch coefficient
C_R	roll coefficient
C_{Rra}	roll coefficient about roll axis
C_S	side lift coefficient
$\mathrm{C_T}$	turning centre point
$\mathrm{C_F}$	force centre point
C_Y	yaw coefficient
d	height of $\mathrm{G_S}$ above roll axis
D_1, D_2	diagonal sums of F_V
e	Panhard rod offset
e_B	tyre vertical force braking transfer factor
e_S	tyre sideforce lateral transfer factor
e_V	tyre vertical force lateral transfer factor
e_X	tyre vertical force longitudinal transfer factor
f	load transfer factor (h/T)
F	force
F_J	jacking force
F_{LAe}	aerodynamic lift force
F_S	sprung mass sideforce
F_{SAe}	aerodynamic side force
F_s	spring force
F_T	lateral load transfer
F_{TD}	diagonal load transfer (diagonal bias)
F_{TL}	lateral load transfer through suspension links
F_{TM}	lateral load transfer through sprung mass moment
F_{TS}	lateral load transfer through sprung mass
F_{TU}	lateral load transfer through unsprung mass
F_{TX}	longitudinal load transfer
F_{TXAe}	longitudinal aerodynamic load transfer
F_U	unsprung mass sideforce
F_V	wheel vertical force
F_Y	wheel lateral force
g	gravitational acceleration
G	centre of mass
G	overall steering gear ratio
G_m	mean steering gear ratio
h	roll axis height at $\mathrm{G_S}$, roll-centre height
h_A	roll axis height at aerodynamic sideforce position
H	height of G
J_{AR}	anti-roll coefficient
J_{ad}	front anti-dive coefficient
J_{al}	front anti-lift coefficient
J_{ar}	rear anti-rise coefficient
J_{as}	rear anti-squat coefficient

k	roll stiffness
k_A	axle roll stiffness
$k_{A\phi}$	axle roll gradient
k_R	total ride roll stiffness
k_S	suspension roll stiffness
$k_{S\phi}$	suspension roll gradient
k_β	attitude angle gradient $d\beta/dA$
$k_{\varepsilon S}$	roll-steer gradient
$k_{\varepsilon C}$	roll-camber gradient
k_η	compliance steer gradient
k_ϕ	roll-angle gradient $d\phi/dA$
K	stiffness
K_A	steering Ackermann factor
K_r	ride rate
K_s	spring stiffness
K_t	tyre vertical stiffness
K_w	wheel rate
l	wheelbase
L	aerodynamic lift
m	mass
M	steering moment about kingpin axis
M_A	aligning torque steer moment
M_L	lateral force steer moment
M_P	propshaft torque
M_{Pr}	propshaft moment
M_{PAe}	aerodynamic pitch moment
M_{RAe}	aerodynamic roll moment
M_S	sprung mass moment
M_{SAe}	aerodynamic roll moment about the roll axis
M_T	tractive force steer moment
M_V	vertical force steer moment
M_{YAe}	aerodynamic yaw moment
M_ϕ	roll moment from springs and anti-roll bar
N_f, N_r	axle vertical force from ground
p	front proportion of brake force
p_i	inflation pressure
q	dynamic pressure
R	cornering path radius
s	cross beam height
S	spring spacing, link spacing
S_R	safety factor against roll-over
t	front proportion of tractive force
T	track
T_R	torque to steer road wheels
T_s	handwheel torque
w	cross beam length
W	weight force
xyz	vehicle-fixed coordinates
XYZ	Earth-fixed coordinates
$\alpha_{\eta A}$	aligning moment compliance steer angle
$\alpha_{\eta S}$	sideforce compliance steer angle
β	attitude angle
β_{Ae}	aerodynamic attitude angle
β_C	bump camber coefficient (deg/m)
β_S	bump steer coefficient (deg/m)
$\gamma_{\eta C}$	sideforce compliance camber angle
$\gamma_{\eta O}$	overturning moment compliance camber angle
δ_{ref}	reference steer angle
δ_s	steering handwheel angle
ε_S	roll understeer coefficient (deg/deg)
ε_C	roll camber coefficient (deg/deg)
η_A	aligning moment compliance steer coefficient (deg/N m)
η_C	sideforce compliance camber (deg/N)
η_O	overturning moment compliance camber (deg/Nm)
η_S	sideforce compliance steer coefficient (deg/N)
θ_c	caster angle
θ_k	kingpin inclination angle
θ_{cf}	front anti-dive CE line inclination
θ_{cfi}	CE-line inclination for full front anti-dive effect
λ	Langensperger angle of axle

μ	friction coefficient
ρ_A	axle axis inclination
ρ_{ra}	roll axis inclination
ϕ	roll angle
ϕ_S	suspension roll angle
ϕ_A	axle roll angle
ϕ_{Ae}	aerodynamic roll angle
ϕ_{Pr}	propshaft roll angle

Subscripts:

A	axle
Ae	aerodynamic
c	caster
D	diagonal
f	front
i	incremental, inclination
J	jacking
L	left, links
M	maximum
m	mean
P	pitch
Pr	propshaft
r	rear
ra	roll axis
R	right, roll
S	sprung
T	transferred
U	unsprung
Y	yaw

Chapter 6 Steady-state handling

---

a	front axle to G
a'	front axle effective moment arm when steered
A	acceleration
A_{ax}	absolute acceleration in x direction
A_D	drag area
A_M	tyre grip maximum lateral acceleration
A_f	maximum front lateral acceleration
A_r	maximum rear lateral acceleration
A_R	roll-over maximum lateral acceleration
b	rear axle to G
C	coefficient
C_D	drag coefficient
C_F	force centre
C_L	lift coefficient
C_R	roll coefficient
C_P	pitch coefficient
C_S	side lift coefficient
C_S'	$dC_S/d\beta$
C_T	turning centre
C_x	tyre longitudinal stiffness
C_Y	yaw coefficient
C_Y'	$dC_Y/d\beta$
C_0, C_1, C_2	vehicle cornering stiffnesses
C_α	tyre cornering stiffness
C_γ	tyre camber stiffness
d_f, d_r	fractions of lateral load transfer
D	vehicle drag (aero + tyre)
D_A	aerodynamic drag
D_f, D_r	axle cornering compliance
D_R	tyre rolling resistance
D_T	tyre cornering drag
e_S	sideforce transfer factor
e_V	vertical force transfer factor
f	tyre C_α sensitivity to F_V
f_R	response factor
f_T	tractive force coefficient
F	force
F_S	steering wheel rim force

F_T	lateral load transfer
F_{TX}	longitudinal load transfer
F_X	tyre longitudinal (drive) force
g	gravity
G	overall steering ratio
G	centre of mass point
G_A	lateral acceleration gain
G_r	yaw velocity gain
G_ρ	path curvature gain
H	height of centre of mass
j	load transfer factor
k	understeer gradient/coefficient
k_C	steering column torsional compliance
k_t	tyre vertical stiffness
k_ϕ	roll gradient
k_β	attitude gradient
k_ε	roll steer gradient (with camber effect)
k_η	compliance steer gradient
l	wheelbase
L	aerodynamic lift
m	vehicle mass
m_f, m_r	front and rear endmasses
M	moment
N	total yaw moment
N_f, N_r	axle vertical reactions
N_U	understeer number
p	tyre maximum cornering force sensitivity to F_V
P	a power (index)
q	dynamic air pressure
r	yaw speed
R	path radius
R_V	vertical path radius
R^*	critical path radius
S	frontal or reference area
t	pneumatic trail
T	track
T_s	steering wheel torque
U	understeer factor ($1/f_R$)
V	speed
V^*	critical aerodynamic speed
V_c	V_{ch} or V_{cr} as appropriate
V_{ch}	characteristic speed
V_{cr}	critical speed
V_0	maximum cornering speed with no aerodynamic lift
w	transverse wind speed
x	forward movement of G
x_s	static margin
Y	sideforce

α	tyre slip angle
α_R	running toe angle
α_S	suspension steer angle
β	vehicle attitude angle
β_{Ae}	aerodynamic attitude angle
β_C	bump camber coef. (deg/m)
β_S	bump steer coef. (deg/m)
χ	tyre camber angle
δ	steer angle of road wheels
δ_b	steering backlash
δ_K	kinematic steer angle
δ_{ref}	reference steer angle
δ_s	steer angle of handwheel
δ_U	understeer angle
ε_S	roll understeer coefficient
ε_C	roll camber coefficient
Δ	change
θ	road slope
μ_A	tyre maximum cornering force coefficient
μ_L	tyre maximum longitudinal force coefficient
μ_R	rolling resistance coefficient
ρ	path curvature, air density
σ	compliance steer coefficient
ϕ	roll angle
ϕ_A	axle roll angle

Subscripts:

a	absolute (magnitude in XYZ)
D	driven
f	front
i	inner
K	kinematic
M	maximum
m	mean

o outer
r rear
R running, rolling
T traction
U understeer
V vertical
x, y, z components
0 reference state

Chapter 7 Unsteady-state handling

a front axle to G
A acceleration
A_f maximum front lateral acceleration
A_M maximum lateral acceleration
A_r maximum rear lateral acceleration
b rear axle to G
c tyre stiffness ratio C_γ/C_α
C damping coefficient (N s/m)
C_0, C_1, C_2 vehicle cornering stiffnesses
C_α tyre cornering stiffness
C_γ tyre camber stiffness
d height of G above roll axis
D time derivative operator
e_V tyre vertical force lateral transfer factor
f_R response factor
F force magnitude
F_{X2} tractive force of two wheels
G centre of mass
G_r yaw speed gain
h height of roll axis
I second moment of mass
k understeer gradient/coefficient, radius of gyration
k_s suspension roll stiffness
k_Z yaw radius of gyration
K spring stiffness
l wheelbase
L roll moment
L_p dL/dp
L_ϕ $dL/d\phi$
m vehicle mass
m_S sprung mass
m_U unsprung mass
M mass, moment
M_Z tyre self-aligning moment
N plan view moment
N_β $dN/d\beta$
N_r dN/dr
p roll speed
q dynamic air pressure
r yaw velocity ($\equiv \dot{\psi}$)
t time
T track
T_r directional time constant
u Y component of $\mathbf{V}$ ($\equiv V_Y$)
v lateral velocity component ($\equiv \dot{y} \equiv V_y$)
V vehicle speed
$\mathbf{V}$ vehicle velocity vector
V_c V_{ch} or V_{cr} as appropriate
V_{ch} characteristic speed
V_{cr} critical speed
V_t transition speed
x_s static margin
xyz vehicle-fixed axes
X amplitude of x
XYZ Earth-fixed axes
$\dot{y}$ lateral velocity component ($\equiv V_{ay} \equiv v$)
Y amplitude of y
Y sideforce (F_y)
Y_β $dY/d\beta$
Y_r dY/dr

α	tyre slip angle, damping factor (s^{-1})
β	attitude angle
β_{Ae}	aerodynamic attitude angle
β_C	bump camber coefficient (deg/m)
β_S	bump steer coefficient (deg/m)
δ	steer angle
ε	traction transfer coefficient
ε_S	roll understeer coefficient
ε_C	roll camber coefficient
ζ	damping ratio
μ_R	rolling resistance coefficient
ν	path angle
τ	time constant
ϕ	roll angle, phase angle
ψ	heading angle
ω	undamped natural frequency
ω_d	damped (actual) natural frequency
ω_f	forcing frequency

Subscripts:

a	absolute (magnitude in XYZ)
f	front
M	maximum
m	mean
r	rear
s	sprung
u	unsprung
x,y,z	components in vehicle-fixed axes
X,Y,Z	components in Earth-fixed axes
0	reference or specific condition
1	final or specific condition

Appendix B

Calculation of atmospheric properties

Standard sea-level properties for dry air are given in Section 3.2. In the range of ambient conditions, sufficiently accurate properties of air for engineering purposes may be calculated by the following procedure. Temperature, pressure and relative humidity are measured experimentally. The absolute (kelvin) temperature is

$$T_K = 273.15 + T$$

To find the saturated vapour pressure of water at this temperature, first:

$$f = 3.231 + 0.02807(T - 15) - 1.067 \times 10^{-4}(T - 15)^2$$

and then the saturated water vapour pressure is:

$$p_S = 10^f$$

Expressing the relative humidity as a simple decimal value r, not as a percentage, the actual water vapour pressure is

$$p_W = r\,p_S$$

The universal gas constant is

$$R_G = 8314.3 \text{ J/mole K}$$

Using the basic gas equation for the water vapour, the absolute humidity, i.e. the actual water vapour density in the air in kg/m^3, is

$$\rho_W = \frac{p_W M_W}{R T_K}$$

where $M_W = 18.015$ is the relative molecular mass (also known as the molecular weight or formula weight) of water. The pressure of the dry fraction of air is

$$p_D = p - p_W$$

The density of the dry air fraction is

$$\rho_D = \frac{p_D M_D}{R T_A}$$

where $M_D = 28.965$ is the relative molecular mass (molecular weight) of standard dry air. The total density is

$$\rho = \rho_D + \rho_W$$

The dynamic viscosity is

$$\mu = 17.75 \times 10^{-6} \left(\frac{T_K}{285}\right)^{0.76}$$

The kinematic viscosity is

$$\nu = \frac{\mu}{\rho}$$

The effective mean relative molecular mass is

$$M = \frac{\rho}{\rho_D / M_D + \rho_W / M_W}$$

The ratio of specific heats is

$$\gamma = 1.402$$

The speed of sound is

$$V_s = \sqrt{\frac{\gamma R T_K}{M}}$$

Appendix C

Example car specifications

The following tables give some sample data for a variety of cars. The vehicles as a whole do not represent particular real cars, although in many cases some of the data corresponds to real cases. In some cases a value has been omitted, either because it is inappropriate or because it is unknown to the author. Of the numerous parameters representing suspension and tyres, many have inevitably been omitted (for example, rate of roll centre vertical movement in heave, and camber change in bump). Omission is not intended to imply lack of importance.

Overall, the data are intended to indicate realistic values, and to provide a source of information for the setting of problems. The passenger car specifications are likely to remain realistic. For racing cars (vehicles G and H) the specifications are sensitive to rule changes, and are based on 1990 rules. The C1 sports prototype rules change significantly with a minimum mass reduction and engine capacity limitation in 1991.

Table C.1 *Example car specifications*

	Parameter		*Symbol*	*Units*	*Vehicle*							
1	Vehicle and type				A Saloon	B Saloon	C Coupé	D Saloon	E Saloon	F Sports	G Racing Formula One	H Racing C1 Sports Prototype
2	Wheelbase		l	m	2.040	2.398	2.580	2.770	3.075	2.500	2.718	2.910
3	Front track	f	T_f	m	1.206	1.362	1.440	1.480	1.560	1.540	1.804	1.600
4	Rear track	r	T_r	m	1.164	1.324	1.440	1.440	1.540	1.540	1.626	1.524
5	Total mass (part load)		m	kg	727	1045	1175	1435	1945	1600	625	1020
6	G behind front axle		a	m	0.775	1.031	1.161	1.302	1.568	1.250	1.631	1.717
7	G height		H	m	0.550	0.610	0.620	0.635	0.672	0.500	0.280	0.350
Suspension												
1	Wheel rate (no ARB)	f	k_{wf}	kN/m	16.00	12.40	12.80	12.95	13.40	20.00	300	230
	(per wheel)	r	k_{Wr}	kN/m	10.00	10.60	13.80	14.72	17.00	20.00	200	400
2	Roll stiffness of ARB	f	k_{Bf}	Nm/deg	0	146	248	210	205	1000	10000	4000
		r	k_{Br}	Nm/deg	0	0	0	73	0	250	2000	1000
3	Suspension roll stiffness	f	k_{Sf}	Nm/deg	203	347	480	458	490	1414	18520	9140
	(springs plus ARB)	r	k_{Sr}	Nm/deg	118	162	250	339	352	664	6611	9107
4	Unsprung mass	f	m_{Uf}	kg	50	80	92	105	115	90	46	70
	(per axle)	r	m_{Ur}	kg	40	72	120	110	130	100	60	90

					A Saloon	B Saloon	C Coupé	D Saloon	E Saloon	F Sports	G Formula 1	H C1
5	Unsprung G height	f	H_{Uf}	m	0.237	0.260	0.290	0.295	0.320	0.300	0.312	0.308
	(at loaded radius)	r	H_{Ur}	m	0.237	0.260	0.290	0.295	0.320	0.300	0.320	0.338
6	Roll-centre height	f	h_f	m	0.100	0.190	–0.020	0.040	0.062	0.070	–0.005	–0.012
		r	h_r	m	0.0	0.190	0.410	0.080	0.405	0.110	0.050	0.040
7	Roll understeer	f	ε_{Sf}	deg/deg	0.100	–0.08	0.150	0.070	0.000	0.040	0.0	0.0
		r	ε_{Sr}	deg/deg	0.0	0.120	0.040	0.100	0.140	0.020	0.0	0.0
8	Roll camber	f	ε_{Cf}	deg/deg	0.800	0.600	1.000	0.560	0.620	0.820	0.63	0.81
		r	ε_{Cr}	deg/deg	1.000	0.900	0.0	0.840	0.0	0.820	0.51	0.86
9	Lateral force	f	η_{Sf}	deg/kN	0.100	–0.080	0.060	0.180	–0.120	–0.040	0.0	0.0
	compliance understeer	r	η_{Sr}	deg/kN	–0.360	–0.240	0.120	0.200	0.180	–0.100	0.0	0.0
10	Aligning torque	f	η_{Af}	deg/kNm	0.800	1.120	0.650	1.040	0.640	0.200	0.0	0.0
	compliance understeer	r	η_{Ar}	deg/kNm	–0.120	–0.210	–0.080	–0.100	–0.140	–0.020	0.0	0.0
Aerodynamics												
1	Frontal area		A_F	m²	1.500	1.750	1.850	2.030	2.150	1.700	1.600	2.00
2	Vertical lift		C_L	–	0.610	0.310	0.280	0.120	0.210	–0.040	–2.640	–1.820
3	Pitch		C_P	–	–0.120	–0.060	0.040	0.020	–0.100	0.010	0.268	0.184
4	Drag		C_D	–	0.480	0.410	0.360	0.290	0.400	0.330	1.070	0.420
5	Side lift		C'_S	/deg	0.048	0.039	0.055	0.050	0.042	0.048	*	0.035
6	Yaw		C'_Y	/deg	0.0047	0.0041	0.0061	0.0080	0.0048	0.0065	*	0.003
7	Roll		C'_R	/deg	0.0062	0.0042	0.059	0.0060	0.0082	0.0055	*	0.002

Table C.1 *Example car specifications* (continued)

					A Saloon	B Saloon	C Coupé	D Saloon	E Saloon	F Sports	G Formula 1	H C1
Tyres												
1	Cornering stiffness	f	C_{Sf}	/deg	0.120	0.168	0.124	0.164	0.171	0.170	0.700	0.500
	coefficient (static F_V)	r	C_{Sr}	/deg	0.140	0.180	0.140	0.170	0.181	0.170	0.700	0.500
2	Camber stiffness	f	C_{Cf}	/deg	0.0175	0.0074	0.0172	0.0072	0.0075	0.0074	*	*
	coefficient	r	C_{Cr}	/deg	0.0175	0.0074	0.0170	0.0070	0.0071	0.0074	*	*
3	Max corn. force coef.	f	μ_{Cf}		0.900	0.940	0.900	0.880	0.850	1.100	1.60	1.60
	(static F_V)	r	μ_{Cr}	–	0.950	0.980	0.930	0.910	0.860	1.100	1.60	1.60
4	C_S sensitivity to F_V	f	k_{lf}	–	0.600	0.580	0.700	0.620	0.520	0.400	*	*
		r	k_{lr}	–	0.500	0.510	0.600	0.600	0.500	0.400	*	*
5	μ_C sensitivity to F_V	f	p_f	–	–0.14	–0.15	–0.16	–0.18	–0.14	–0.18	–0.24	–0.19
		r	p_r	–	–0.14	–0.15	–0.16	–0.18	–0.14	–0.18	–0.27	–0.24
6	Vertical stiffness	f	K_{tf}	kN/m	120	165	170	180	200	250	228	375
		r	K_{tr}	kN/m	100	150	150	170	200	250	258	398
7	Pneumatic trail	f	t_f	m	0.030	0.035	0.040	0.045	0.044	0.040	*	*
		r	t_r	m	0.025	0.032	0.035	0.043	0.044	0.040	*	*
Dampers												
1	Bump damping	f	C_{DBf}	kNs/m	1.1	1.0	1.0	1.2	1.1	4.1	6	8
		r	C_{DBr}	kNs/m	0.7	0.9	0.8	1.0	1.2	4.1	11	12
2	Rebound damping	f	C_{DRf}	kNs/m	1.5	1.2	2.0	1.8	2.0	5.8	10	12
		r	C_{DRr}	kNs/m	1.1	1.1	1.4	1.8	1.9	5.8	15	16

Table C.2 *Indicative range of passenger car parameter values*

Parameter		Symbol	Units	Lower	Upper
Mass (no load)		m	kg	570	1800
Wheelbase		l	m	2.00	3.00
G longitudinal position		a/l	—	0.33	0.60
G height		H	m	0.43	0.70
Product of inertia		I_{yz}	kg m^2	–200	200
Unsprung end-mass/total mass	f	m_{Uf}/m	—	0.06	0.08
Unsprung end-mass/total mass	r	m_{Ur}/m	—	0.06	0.12
Spring stiffness (wheel rate)	f/r	K_w	kN/m	10	30
Roll stiffness	f	k_f	Nm/deg	200	1400
Roll stiffness	r	k_r	Nm/deg	180	800
Roll stiffness front fraction		k_f/k	—	0.40	0.80
Roll centre height	f	h_f	m	–0.050	0.200
	r	h_r	m	–0.050	0.400
Bump steer (toe in positive)	f/r	β_S	deg/m	–5	40
Bump camber gradient	f/r	β_C	deg/m	0	500
Roll understeer gradient	f/r	ε_S	deg/deg	–0.10	0.20
Roll camber gradient	f/r	ε_C	deg/deg	0.00	1.00
Lateral force compliance US	f/r	η_S	deg/kN	–0.30	0.60
Aligning torque compliance US	f/r	η_A	deg/kNm	0.0	25.0
Lateral force compliance camber	f/r	η_C	deg/kN	0.20	0.80
Control gains:					
Steering angle gradient		$d\delta/dA$	deg/g	50	200
Understeer angle gradient		$d\delta_U/dA$	deg/g	0	10
Roll angle gradient		$d\phi/dA$	deg/g	3	12

Appendix D

Selected problem solutions

Chapter 1

Q 1.6.1

$a = 1.545$ m behind front axle

$y = 0.040$ m right of CL

Q 1.6.2

42.18 N

26.36 kg

Q 1.6.3

$a = 1.606$ m

$h = 0.836$ m

Q 1.6.4

$I_v = 7453$ kg m^2

Q 1.6.5

$I = 7397$ kg m^2

Q 1.6.6

$I = 7395$ kg m^2

$k = 2.050$ m

$k^2/ab = 1.87$

Q 1.6.7

$I_{VG} = 2580$ kg m^2

$k_P = 1.393$ m

Q 1.6.11

$m_f = 899$ kg

$m_r = 851$ kg

$m_S = 1350$ kg

$m_{Sf} = 769$ kg

$m_{Sr} = 581$ kg

$a_S = 1.257$ m

$H_S = 0.833$ m

Q 1.7.1

$a = 1.5794$ m

$h = 0.7053$ m

$I = 3904$ kg m^2

Chapter 2

Q 2.5.5

(1) $F_i = \frac{1}{2\pi} m_p R_u \Omega^2 \sin\theta$

(2) $(l/2)^2 = z(2R_u - x)$

(3) $F_i = \frac{\sqrt{2}}{2\pi} \frac{m_p V^2}{R_u^{3/2}} z^{1/2}$

Q 2.5.6

509 N

Q 2.6.2

$t = R\, p_i + \rho V^2$

$N_T = V/C = 1$

Q 2.6.3

87 kPa

Q 2.7.3

$\theta = 1.20°$ (in practice a little more, because of other losses, e.g. axle bearings).

Q 2.9.3

$S_1 = 0.0514$

$\alpha = 2.94°$

C_α = 977 N/deg

0.204/deg

4067 N

5760 N

Q 2.11.1

42 m/s

Q 2.14.3

Assuming a load sensitivity power of 0.5 gives 0.168/deg.

Q 2.14.4

751 N/deg

Q 2.14.6

(1) 0.205/deg

(2) 0.430/deg

Q 2.14.7

(1) 5405 N

(2) 854 N/deg

Q 2.14.9

(1) 0.25/deg

(2) 0.60

(3) 0.373/deg

(4) 1070 N/deg

Q 2.14.10

(1) 5263 N

(2) 850 N/deg

Q 2.19.2

50 m/s

Q 2.19.4

1.91°

0.3g

Q 2.20.2

396 N

1178 N

Q 2.20.3

0.2%

Chapter 3

Q 3.2.1

1.168 kg/m^3

Q 3.2.2

1.171 kg/m^3

Q 3.3.1

5.95 m/s

Q 3.3.2

7.3 to 20.7 m/s

Q 3.3.3

8%

Q 3.3.4

0.140

0.064%

Q 3.4.4

36.0 m/s

200.5°

Q 3.5.1

qA = 1185.8 N

L = 489 N

S = 190 N

D = 465 N

P = 295 N m

Y = 71 N m

Q 3.5.2

$C_{Lf} = 0.289$

$C_{Lr} = 0.123$

Q 3.5.3

$C_f = 0.419$

$C_r = 0.441$

Q 3.7.1

At $C_L = -0.5$, downforce is 20 kN, which equals the weight.

D = 24.5 kN = 1.24 × weight

Chapter 4

Q 4.9.1

C_{Dm} = 1.8 kN s/m

K_w = 13.95 kN/m

Q 4.12.2

C_e = 1.5 kN / (m/s)

Assuming typical 30/70 characteristic: 0.64 kN

Q 4.13.2

134 kN

Q 4.15.1

40.7 N m

Chapter 5

Q 5.2.1

18.46 N/mm

Q 5.2.2

–0.3

2.022 m

Q 5.2.6

1.9°

Q 5.4.1

0.108

216 N

Q 5.4.2

0.145

209 mm

Q 5.4.5

0.0971

177 mm

0.0987

180 mm

Q 5.4.6

$h = 0.1735$ m

$J_{AR} = 0.270$ (27%)

Q 5.6.5

0.120

218 mm

Q 5.7.2

348 mm

0.242

774 N

Q 5.10.5

9.05°

$F_{Tf} = 3167 + 269 = 3436$ N

$F_{Tr} = 1358 + 1049 = 2407$ N

59/41

7674, 802, 6017, 1203

Q 5.10.6

7.215°

$F_{Tf} = 2525 + 175 + 220$
$= 2920$ N

$F_{Tr} = 1082 + 267 + 790$
$= 2139$ N

58/42

Q 5.10.7

0.8°

$F_{Tf} = 192$ N

$F_{Tr} = -192$ N

Q 5.10.14

$\Omega_f = mVH/I$

Opposite rotation to wheels

$G = mR_1H/I$

Q 5.10.16

11.04 m/s²

$S_R = 1.36$

$S_R \approx 0.8 \times 1.36 = 1.09$

Q 5.12.2

Lf, Rf, Lr, Rr are 26, 30, 24, 20, respectively

Q 5.12.3

8, 44, 18, 30

Q 5.12.4

13, 41, 17, 29

Q 5.13.6

$e_V = 20/50 = 0.40$

$e_{Vf} = 12/25 = 0.48$

$e_{Vr} = 8/25 = 0.32$

Q 5.15.5

13.5 deg/m (toe-in on bump)

Q 5.16.9

$k_{\eta S} = \eta\, m_S = -m_S \sin\theta / KS$
$= -1.11$ deg/g

Q 5.16.10

0.84 deg/g

Q 5.18.3

1.506, 1.172, 1.296

Q 5.18.4

$a/l = 0.567$

Q 5.18.5

$F_{Tf} = 1713$ N

$F_{Tr} = 1224$ N

$F_T = 2937$ N

58.32/41.68 f/r

$e_{Vf} = 0.507$, $e_{Vr} = 0.216$

$H = 0.216$ m (seems a little too low)

Q 5.18.6

$L_f = 4209$ N $L_r = 8025$ N

$L_r/L = 0.656$

$A_L = 2.88$ m^2

Q 5.18.7

$p_f = -0.294$ (possible)

$p_r = -0.664$ (not realistic, one rear tyre not limiting?)

Chapter 6

Q 6.10.11

$k = 0$

5.7 deg/g

Q 6.10.12

0.5 deg/g

7.6 deg/g

Q 6.10.13

2.5 deg/g

4.3 deg/g

Q 6.10.14

1.55 deg/g

7.75 deg/g

Q 6.10.15

3.3 deg/g

Q 6.10.16

0.45 deg/g

−0.23 deg/g

Q 6.10.17

0.37 deg/g

−0.19 deg/g

Q 6.12.7

$\Delta A_f = -0.0103g$

$\Delta A_r = +0.0103g$

$\Delta A_M = -0.0103g$

$\Delta N_U = 0.026$

Q 6.12.8

−0.0427

Q 6.12.10

0.0704

Q 6.14.2

$\Delta k = 0.382$ deg/g (lift)

–0.184 deg/g (drag)

–0.0646 deg/g (sideforce)

–0.422 deg/g (yaw)

Q 6.15.2

–0.099

Q 6.17.2

0.42°

0.84°

Q 6.17.3

$\mu = 0.868$

51°

Q 6.18.2

$N_U = -0.12$

Q 6.18.3

–60%

Q 6.19.2

Initially 6 deg/g, 4 deg/g

Finally 6.75 deg/g, 3.25 deg/g

Q 6.20.2

$F_f = 1867$ N

$F_r = 1037$ N

1.867° 1.037°

$\Delta\delta_S = 16.6°$

Chapter 7

Q 7.7.7

8.96 rad/s, 1.43 Hz

0.900

3.897 rad/s, 0.620 Hz

$k = 1.628$ mrad/(m/s^2)
$= 0.916$ deg/g

$V_{ch} = 42.9$ m/s

Q 7.8.2

–87.1 kNm/rad

Q 7.9.2

$Y_\beta = -229.2$ kN/rad

$Y_r = -3056$ N s/rad

$N_\beta = 45.84$ kN m^2/rad

$N_r = -35.0$ kN m s/rad

References

Abbott, I.H. & von Doenhoff, A.E. (1959) *Theory of Wing Sections*, Dover, ISBN 486-60586-8.

Artamonov, M.D., Ilarinov, V.A. and Morin, M.M. (1976) *Motor Vehicles – Fundamentals and Design* (Trans. A. Troitsky), M.I.R. Publishers, Moscow.

Bastow, D. (1980) *Car Suspension and Handling*, Pentech Press, ISBN 0-7273-03050-8, pp. 89–95.

Bastow, D. (1987) *Car Suspension and Handling,* 2nd edition, Pentech Press, ISBN 0-7273-0316-3.

Bekker, M.G. (1956) *Theory of Land Locomotion*, Univ. of Michigan Press.

Bekker, M.G. (1960) *Off-the-Road Locomotion*, Univ. of Michigan Press.

Bekker, M.G. (1969) *Introduction to Terrain-Vehicle Systems*, Univ. of Michigan Press.

Bergman, W. (1965) *The Basic Nature of Vehicle Understeer–Oversteer*, S.A.E. Paper 957B.

Bergman, W. (1973) *Measurement and Subjective Evaluation of Vehicle Handling*, S.A.E. 730492.

BS AU 50: Part 2 (1979).

Bundorf, R.T. & Leffert, R.L. (1976) *The Cornering Compliance Concept for Description of Vehicle Directional Control Properties*, S.A.E. 760713.

Bundorf, R.T. (1967) *The Influence of Vehicle Design Parameters on Characteristic Speed and Understeer*, S.A.E. 670078.

Campbell, C. (1981) *Automobile Suspensions*, Chapman and Hall, ISBN 0-412-15820-5.

Carr, G.W., Rose, M.J. & Smith, N.P. (1973) 'Some aerodynamic aspects of safety in road vehicles' (in three parts), *Proc. I.Mech.E.* (*A.D.*) Vol. 187, No. 30, pp. 333–360.

Chiang, S.L. & Starr, D.S. (1978) *Using Computer Simulation to Evaluate and Improve Vehicle Handling*, S.A.E. 780009.

Clark, S.K. (1981) (ed.) *Mechanics of Pneumatic Tires,* 2nd edition, U.S. Dept of Transportation

Cogotti, A. (1983) 'Aerodynamic characteristics of car wheels', in Dorgham & Businaro (1983), pp. 173–196

Curtis, A. (1990) 'Anatomy of the 180 mph corner', *Autocar*, 22 August 1990, pp. 56–59.

Dixon, J.C. (1987a) 'The roll-centre concept in vehicle handling dynamics, *Proc. I.Mech.E.*, Vol. 201, No. D1, 22/87, pp. 69–78.
Dixon, J.C. (1987b) 'Limit steady state vehicle handling', *Proc. I.Mech.E.*, Vol. 201, No. D4, pp. 281–291.
Dixon, J.C. (1988) 'Linear and non-linear steady state vehicle handling', *Proc. I.Mech.E.*, Vol. 202, No. D3, pp. 173–186.
Dominy, J. & Dominy, R.G. (1984) 'Aerodynamic influences on the performance of the Grand Prix racing car', *Proc. I.Mech.E.*, Vol. 198, Part D, pp. 87–93.
Dorgham, M.A. & Businaro, U.L. (1983) (eds) *Impact of Aerodynamics on Vehicle Design, Proc. of the Int. Assoc. for Vehicle Design*, SP3, ISBN 0-907776-01-9.
Dugoff, H., Segel, L. & Ervin, R.D. (1971) *Measurement of Vehicle Response in Severe Braking and Steering Manouvers*, S.A.E. 710080.
Ellis, J.R. (1963) 'Understeer and oversteer', *Automobile Engineer*, May.
Ellis, J.R. (1969) *Vehicle Dynamics*, Business Books Ltd, SBN 220-99202-9.
Ellis, J.R. (1989) *Road Vehicle Dynamics,* published by J.R. Ellis, 4815 Provens Drive, Akron, Ohio 44319, U.S.A.
Evans, R.D (1935) 'Properties of tyres affecting riding, steering and handling', *Trans. S.A.E.*, Vol. 36, No. 2, February, pp. 41–49.
Fackrell, J.E. & Harvey, J.K. (1974) 'The aerodynamics of an isolated road wheel', in *Proc. A.I.A.A. 2nd Symposium on Aerodynamics of Sports and Competition Automobiles, 1974.*
Fancher, P., Segel, L., Bernard, J. & Ervin, R. (1976) *Test Procedures for Studying Vehicle Dynamics in Lane-Change Manouvers*, S.A.E. 760351.
Frank, F. & Hofferberth, W. (1967) 'Mechanics of pneumatic tyre', *Rubber Chemistry and Technology*, Vol. 40, No. 1, February, pp. 271–322.
French, T. (1989) *Tyre Technology*, Adam Hilger, ISBN 0-85274-360-2.
Frere, P. (1973) *The Racing Porsches*, Arco, N.Y., ISBN 0-668-02972-2.
Gardner, E.R. & Worswick, T. (1951) *Trans. Inst. Rubber Industry*, Vol. 27, p. 127.
Garrett, K. (1987) 'Traction control differentials', *Automotive Engineer*, February, pp. 20–26.
Giles, J.G. (Ed.) (1968) *Steering, Suspension and Tyres*, Iliffe.
Good, M.C. (1977) 'Sensitivity of driver–vehicle performance to vehicle characteristics revealed in open-loop tests', *Vehicle System Dynamics*, Vol. 6, pp. 245–277 (79 refs).
Goran, M.B. & Hurlong, G.W. (1973) 'Determining vehicle inertial properties for simulation studies', *Bendix Technical Journal*, Vol. 6, Spring, pp. 53–57.
Grylls, S.H. (1972) 'Traction v. stability in passenger cars', *Journal of Automotive Engineering*, May, pp. 8–15, also *Proc. I.Mech.E.*, Vol. 186, No. 17, pp. 169–177.
Hales, F.D. (1965) *The Handling and Stability of Motor Vehicles, Part 1, Handling, Stability and Control Response Definitions*, M.I.R.A., 1965/1.
Hales, F.D. (1969-70) 'Vehicle handling qualities', *Proc. I.Mech.E.*, Vol. 184, Pt 2A, No. 12, pp. 233–248.
Hall, L.C. (1986) 'The influence of limited slip differentials on torque distribution and steady-state handling of four-wheel drive military vehicles', I.Mech.E. Paper CO5/86, *Proc. I.Mech.E. Conf. on Vehicle Dynamics*, pp. 59–66.

Hays, D.F. & Browne, A.L. (1974) (eds) *The Physics of Tire Traction*, Plenum, N.Y., ISBN 0-306-30806-1.
Henry, A. (1975) *The 4-Wheel Drives*, Macmillan, SBN 333-17289-2.
Hoerner, S.F. & Borst H.V. (1975) *Fluid Dynamic Lift*, Hoerner Fluid Dynamics, Brick Town, N.J., U.S.A., LCCC 75-17441.
Hoerner, S.F. (1965) *Fluid Dynamic Drag*, Hoerner Fluid Dynamics, Brick Town, N.J., U.S.A.
Houghton, E.L. & Brock, A.E. (1970) *Aerodynamics for Engineering Students*, Arnold, SBN 7131-3227-2.
Howard, G. (1986) *Automobile Aerodynamics*, Osprey, ISBN 0-85045-665-7
Hucho, W.H. (1987) (ed.) *Aerodynamics of Road Vehicles*, Butterworths, ISBN 0-408-01422-9.
Huntingdon, R. (1981) *Design and Development of the Indy Car,* HP Books, Tucson, ISBN 0-89586-103-8.
I.Mech.E. (1988) Conference Publication C367 *International Conference on Advanced Suspensions*.
I.S.O. (1979a) ISO/TC22/SC9/N185 (Step/Ramp Input) May.
I.S.O. (1979b) ISO/TC22/SC9/N194 (Random Input) May.
I.S.O. (1979c) ISO/TC22/SC9/N219 (Sinusoidal Input) June.
I.S.O. (1982) *Road Vehicles – Steady State Circular Test Procedure*, ISO 4138.
I.S.O. (1985) *Road Vehicles – Braking in a Turn – Open Loop Test Procedure*, ISO 7975-1985, also issued as BS AU 205: 1986.
I.S.O. *Lane Change Test*, ISO TR 3888 (date unknown).
Irving, J.S. (1930) 'The Golden Arrow and the world's land speed record', *Automobile Engineer*, May, pp. 186–196.
Korff, W.H. (1980) *Designing Tomorrow's Cars*, M-C Publications, California, ISBN 0-9603850-0-2.
Lewis, R.P. & O'Brien, L.J. (1959) 'Limited slip differentials', S.A.E. Paper 590028 (was paper 29B), or *S.A.E. Trans*., Vol. 67, pp. 203–212.
Ludvigsen, K.E. (1970) *The Time Tunnel – A Historical Survey of Automobile Aerodynamics,* S.A.E. Paper 700035.
M.I.R.A. (1965) *Definition of Handling Terms*, M.I.R.A. 1965/1.
Massey, B.S. (1983) *Mechanics of Fluids*, Van Nostrand Reinhold, ISBN 0-442-30552-4.
McHenry, R.R. (1969) 'An analysis of the dynamics of automobiles during simultaneous cornering and ride motions', I.Mech.E. Conference on Handling of Vehicles under Emergency Conditions, Loughborough, 1969, in *Computer Aided Design*, Spring 1969, Vol. 1, pp. 19–32.
McLean, W.G. & Nelson, E.W. (1962) *Engineering Mechanics; Statics and Dynamics*, Schaum Outline Series, Schaum.
Milliken, W.F. Jr, Dell'Amico, F. & Rice, R.S. (1976) *The Static Directional Stability and Control of the Automobile*, S.A.E. 760712.
Milliken, W.F. Jr. & Rice, R.S. (1983) 'Moment method', *I.Mech.E. Conference on Road Vehicle Handling,* Paper C113/83.
Milliken, W.F. & Whitcomb, D.W. (1956) 'General introduction to a programme of dynamic research', *Proc. I.Mech.E. (A.D.)* pp. 287–309.
Milner, P.J. (1967) 'Steady state vehicle handling', *Automobile Engineer*, October.
Milner, P.J. (1968) 'Vehicle performance in crosswinds', *Automobile Engineer*, August, pp. 352–355.

Moore, D.F. (1975) *The Friction of Pneumatic Tyres,* Elsevier, ISBN 0-444-41323-5.

Norbye, J.P. (1979) *The Complete Handbook of Front Wheel Drive Cars*, Tab Books, ISBN 0-8306-2052-4.

Norbye, J.P. (1980) *The Car and Its Wheels – A Guide to Modern Suspension Systems*, Tab Books, ISBN 0-8306-2058-3.

Norbye, J.P. (1982) *The Michelin Magic*, Tab Books, ISBN 0-8306-2090-7.

Nye, D. (1978) *Theme Lotus*, Motor Racing Publications, London, ISBN 0-900549-40-8.

Odier, J. (1972) 'A dynamometer on which the dynamic behaviour of a passenger car can be simulated', *Proc I.Mech.E. (A.D.)*, Vol. 186, No. 7, pp. 87–96.

Olley, M. (1934) 'Independent wheel suspension – its whys and wherefores', *Trans. S.A.E.*, Vol. 34, No. 2, March, pp. 73–81.

Olley, M. (1961a) Notes on Suspensions, unpublished G.M. report, August.

Olley, M. (1961b) Steady State Steering, unpublished G.M. report, September.

Olley, M. (1962a) Notes on Vehicle Handling, unpublished G.M. report, February.

Olley, M. (1962b) Suspension Notes II, unpublished G.M. report, May.

Pacejka, H.B. (1973) 'Simplified analysis of steady-state turning behaviour of automobiles', *Vehicle Systems Dynamics*, Vol. 2, pp. 161–172, 173–183 (two parts).

Pacejka, H.B. (1975) 'Principles of plane motion of automobiles', *Proc. I.U.T.A.M. Symposium*, pp. 33–59.

Pershing, B. (ed.) (1968) *Proc. A.I.A.A. Symposium on the Aerodynamics of Sports and Competition Automobiles.*

Pershing, B. (ed.) (1974) *Proc. A.I.A.A. Second Symposium on the Aerodynamics of Sports and Competition Automobiles.*

Posthumus, C. & Tremayne, D. (1985) *Land Speed Record*, Osprey, London, ISBN 0-85045-641-X.

Prevost, R. (1928) Letter in *Automobile Engineer*, September 1928, p. 330.

Puhn, F. (1976/1981) *How to Make Your Car Handle*, HP Books, ISBN 0-912656-46-8.

Radt, H.S. & Pacejka, H.B. (1965) 'Analysis of the steady-state turning behaviour of an automobile', *Proc. I.Mech.E. Symposium on Control of Vehicles, 1965.*

Radt, H.S. & Milliken, W.F. (1983) 'Non-dimensionalising tyre data for vehicle simulation', *Proc. Conf. on Road Vehicle Handling, May 1983*, I.Mech.E. Conf. Publ. No. C113/83.

Rae, W.H. & Pope, A. (1984) *Low-Speed Wind Tunnel Testing*, Wiley, ISBN 0-471-87402-7.

Reimpell, J. (1982) *Fahrwerktechnik*, Vogel, Wurzburg, 3 volumes, ISBN 3-8023-0709-7.

Rice, R.S. & Milliken, W.F. Jr (1980) *Static Stability and Control of the Automobile Utilizing the Moment Method*, S.A.E. 800847.

Riede, P.M., Leffert R.L. and Cobb W.A. (1984) *Typical Vehicle Parameters for Dynamics Studies,* S.A.E. 840561.

Rocard, Y. (1946) 'Difficulties arising from auto-oscillation and instability during travel of vehicles', *La Revue Scientifique*, Vol. 84, No. 45, p. 15.

S.A.E. (1978) *Vehicle Dynamics Terminology*, S.A.E. J670e, June.

Sabey, B.E. (1969) 'The road surface in relation to friction and wear of tyres', *Road Tar*, Vol. 23, March, No. 1.

Sakai, H. (1981) 'Theoretical and experimental studies on the dynamic properties of tyres, Parts 1–3', *Int. J. of Vehicle Design*, Vol. 2, Nos 1–3.

Satchel, T.L. (1981) *The Design of Trailing Twist Axles*, S.A.E. 810420.

Scibor-Rylski, A.J. (1984) *Road Vehicle Aerodynamics*, Pentech Press, London, 2nd edition, ISBN 0-7273-1805-5.

Segel, L. (1957a) 'Theoretical prediction and experimental substantiation of the response of the automobile to steering control', *Proc. I.Mech.E.* (*A.D.*) pp. 310–330.

Segel, L. (1957b) 'Research in fundamentals of automobile control and stability', *S.A.E. Trans.*, Vol. 65, pp. 527–540.

Segel, L., Ervin R. & Fancher, P. (1980) *The Mechanics of Heavy Duty Trucks and Truck Combinations*, Int. Assoc. for Vehicle Design, course notes.

Segel, L., Ervin, R. & Fancher, P. (1981) *The Mechanics of Heavy Duty Trucks and Truck Combinations*, Int. Assoc. for Vehicle Design.

Setright, L.J.K. (1972) *Automobile Tyres*, Chapman and Hall, ISBN 412-09850-4.

Shearer, G.R. (1977) 'The rolling wheel – the development of the pneumatic tyre', *Proc. I. Mech. E. (A.D.)*, Vol. 191, No. 11/77, pp. 75-87.

Smith, C. (1978) *Tune to Win*, Aero Publishers Inc., U.S.A., LCCC 78-73549.

Smith, J.G. & Smith, J.E. (1967) 'Lateral forces on vehicles during driving', *Automobile Engineer*, December, pp. 510–515.

Steeds, W. (1958) 'Roll axes', *Automobile Engineer*, February, pp. 55–58.

Steeds, W. (1960) *Mechanics of Road Vehicles*, Iliffe.

Stonex, K.A. (1941) 'Car control factors and their measurement', S.A.E. Paper 410092, or *S.A.E. Trans.*, Vol. 48, No. 3, p. 81–93.

Tompkins, E. (1981) *The History of the Pneumatic Tyre*, Dunlop Ltd (Eastland Press), ISBN 0-903214-14-8.

Topping, R.W. (1974) *A Primer on Nonlinear Steady-State Vehicle Turning Behaviour*, S.A.E. 741096.

Verma, M.K. & Shepard, W.L. (1981) *Comparison of Transient Response Test Procedures for Motor Vehicles*, S.A.E. 810807.

Weir, D.H. & DiMarco, R.J. (1978) *Correlation and Evaluation of Driver/Vehicle Directional Handling Data*, S.A.E. 780010.

Weir, D.H. & Zellner, J.W. (1977) *The Application of Handling Requirements to an RSV-type Vehicle*, S.A.E. 770178 (see also S.A.E. 780010).

Whitcomb, D.W. & Milliken, W.F. (1957) 'Design implications of a general theory of automobile stability and control', *Proc. I.Mech.E.* (*A.D.*) pp. 367–391.

Winkler, C.B. (1973) *Measurement of Inertial Properties and Suspension Parameters of Heavy Highway Vehicles*, S.A.E. 730182.

Wong, J. (1978) *Theory of Ground Vehicles*, Wiley, ISBN 0-471-03470-3.

Wong, J. (1990) *Terramechanics and Off-Road Vehicles*, Elsevier.

Wright, P.G. (1983) 'The influence of aerodynamics on the design of Formula One racing cars', in Dorgham & Businaro (1983), pp. 158–172.

Yoshimori, K. (1976) *Vehicle Controllability and Human Response Characterstics*, S.A.E. 760780.

Index

Page numbers shown in italic (e.g. *196*) indicate the definition of a term, or a more significant discussion. After a page number, the letter f indicates a figure, t a table and e a significant equation. The Index does not list every mention of a term. The Nomenclature, References and Problem sections are generally not indexed.

B

D

E

M

N

O

P

T

Y

Z